AF361726

Integrated Surface Water and Groundwater Analysis

Integrated Surface Water and Groundwater Analysis

Editors

Il-Moon Chung
Sun Woo Chang
Yeonsang Hwang
Yeonjoo Kim

MDPI • Basel • Beijing • Wuhan • Barcelona • Belgrade • Manchester • Tokyo • Cluj • Tianjin

Editors
Il-Moon Chung
Department of Water
Resources and River Research
Korea Institute of Civil
Engineering and Building
Technology (KICT)
Goyang
Korea, South

Sun Woo Chang
Department of Water
Resources and River Research
Korea Institute of Civil
Engineering and Building
Technology (KICT)
Goyang
Korea, South

Yeonsang Hwang
Department of Civil
Engineering
Arkansas State University
Jonesboro
United States

Yeonjoo Kim
Department of Civil and
Environmental Engineering
Yonsei University
Seoul
Korea, South

Editorial Office
MDPI
St. Alban-Anlage 66
4052 Basel, Switzerland

This is a reprint of articles from the Special Issue published online in the open access journal *Hydrology* (ISSN 2306-5338) (available at: www.mdpi.com/journal/hydrology/special_issues/GWSW).

For citation purposes, cite each article independently as indicated on the article page online and as indicated below:

LastName, A.A.; LastName, B.B.; LastName, C.C. Article Title. *Journal Name* **Year**, *Volume Number*, Page Range.

ISBN 978-3-0365-5000-8 (Hbk)
ISBN 978-3-0365-4999-6 (PDF)

Cover image courtesy of Woniok Chung

Contents

About the Editors vii

Preface to "Integrated Surface Water and Groundwater Analysis" ix

Il-Moon Chung, Sun Woo Chang, Yeonsang Hwang and Yeonjoo Kim
Editorial for Special Issue: "Integrated Surface Water and Groundwater Analysis"
Reprinted from: *Hydrology* **2022**, *9*, 70, doi:10.3390/hydrology9050070 . . . . **1**

Lily M. Conrad, Alexander G. Fernald, Steven J. Guldan and Carlos G. Ochoa
A Water Balancing Act: Water Balances Highlight the Benefits of Community-Based Adaptive
Management in Northern New Mexico, USA
Reprinted from: *Hydrology* **2022**, *9*, 64, doi:10.3390/hydrology9040064 . . . . **5**

Daniel G. Gómez, Carlos G. Ochoa, Derek Godwin, Abigail A. Tomasek and María I. Zamora Re
Soil Water Balance and Shallow Aquifer Recharge in an Irrigated Pasture Field with Clay Soils
in the Willamette Valley, Oregon, USA
Reprinted from: *Hydrology* **2022**, *9*, 60, doi:10.3390/hydrology9040060 . . . . **27**

Rodrigo Rojas, Juan Castilla-Rho, Gabriella Bennison, Robert Bridgart, Camilo Prats and Edmundo Claro
Participatory and Integrated Modelling under Contentious Water Use in Semiarid Basins
Reprinted from: *Hydrology* **2022**, *9*, 49, doi:10.3390/hydrology9030049 . . . . **39**

Federico Cervi and Alberto Tazioli
Inferring Hydrological Information at the Regional Scale by Means of ^{18}O–δ^2H Relationships:
Insights from the Northern Italian Apennines
Reprinted from: *Hydrology* **2022**, *9*, 41, doi:10.3390/hydrology9020041 . . . . **67**

Esteban Caligaris, Margherita Agostini and Rudy Rossetto
Using Heat as a Tracer to Detect the Development of the Recharge Bulb in Managed Aquifer
Recharge Schemes
Reprinted from: *Hydrology* **2022**, *9*, 14, doi:10.3390/hydrology9010014 . . . . **87**

Jaewon Joo and Yong Tian
Impact of Stream-Groundwater Interactions on Peak Streamflow in the Floods
Reprinted from: *Hydrology* **2021**, *8*, 141, doi:10.3390/hydrology8030141 . . . . **101**

Nikita Tananaev, Vladislav Isaev, Dmitry Sergeev, Pavel Kotov and Oleg Komarov
Hydrological Connectivity in a Permafrost Tundra Landscape near Vorkuta, North-European
Arctic Russia
Reprinted from: *Hydrology* **2021**, *8*, 106, doi:10.3390/hydrology8030106 . . . . **109**

Lucio Di Matteo, Alessandro Capoccioni, Massimiliano Porreca and Cristina Pauselli
Groundwater-Surface Water Interaction in the Nera River Basin (Central Italy): New Insights
after the 2016 Seismic Sequence
Reprinted from: *Hydrology* **2021**, *8*, 97, doi:10.3390/hydrology8030097 . . . . **125**

Sun Woo Chang and Il-Moon Chung
Water Budget Analysis Considering Surface Water–Groundwater Interactions in the
Exploitation of Seasonally Varying Agricultural Groundwater
Reprinted from: *Hydrology* **2021**, *8*, 60, doi:10.3390/hydrology8020060 . . . . **143**

Rudy Rossetto, Alessio Barbagli, Giovanna De Filippis, Chiara Marchina, Thomas Vienken and Giorgio Mazzanti
Importance of the Induced Recharge Term in Riverbank Filtration: Hydrodynamics, Hydrochemical, and Numerical Modelling Investigations
Reprinted from: *Hydrology* **2020**, *7*, 96, doi:10.3390/hydrology7040096 . . . . **161**

Ian White, Tony Falkland and Taaniela Kula
Meeting SDG6 in the Kingdom of Tonga: The Mismatch between National and Local Sustainable Development Planning for Water Supply
Reprinted from: *Hydrology* **2020**, *7*, 81, doi:10.3390/hydrology7040081 . . . . **181**

Ryan T. Bailey, Katrin Bieger, Jeffrey G. Arnold and David D. Bosch
A New Physically-Based Spatially-Distributed Groundwater Flow Module for SWAT+
Reprinted from: *Hydrology* **2020**, *7*, 75, doi:10.3390/hydrology7040075 . . . . **207**

Gopal Chandra Saha and Michael Quinn
Integrated Surface Water and Groundwater Analysis under the Effects of Climate Change, Hydraulic Fracturing and its Associated Activities: A Case Study from Northwestern Alberta, Canada
Reprinted from: *Hydrology* **2020**, *7*, 70, doi:10.3390/hydrology7040070 . . . . **231**

Mun-Ju Shin, Soo-Hyoung Moon, Kyung Goo Kang, Duk-Chul Moon and Hyuk-Joon Koh
Analysis of Groundwater Level Variations Caused by the Changes in Groundwater Withdrawals Using Long Short-Term Memory Network
Reprinted from: *Hydrology* **2020**, *7*, 64, doi:10.3390/hydrology7030064 . . . . **255**

Mustafa El-Rawy, Okke Batelaan, Kerst Buis, Christian Anibas, Getachew Mohammed and Wouter Zijl et al.
Analytical and Numerical Groundwater Flow Solutions for the FEMME-Modeling Environment
Reprinted from: *Hydrology* **2020**, *7*, 27, doi:10.3390/hydrology7020027 . . . . **271**

About the Editors

Il-Moon Chung

Dr. Il-Moon Chung is a Groundwater Hydrologist and a Senior Research Fellow at the Korea Institute of Civil Engineering and Building Technology (KICT), South Korea. He received his Ph.D. from the Department of Civil Engineering at Yonsei University in 1998 with "Hydrogeological modeling of underground storage cavern and analysis of solute transport". After working at the SK Engineering & Construction Research Center for 3 years, he has been conducting research on groundwater management in Korea at KICT since 2002. His special interest is "surface water-groundwater integrated hydrological analysis", and he has developed a fully integrated SWAT-MODFLOW model. From 2012 to 2013, he conducted "Groundwater management options in Korea" research with Dr. Marios Sophocleous as a visiting professor at the Kansas Geological Survey. He has over 140 peer reviewed journal papers so far, and is serving as a Senior Editor of KSCE J. of Civil and Environmental Engineering Research (2022) and the Chief Editor of J. of Korea Water Resources Association (2021–2022).

Sun Woo Chang

Dr. Sun Woo Chang is currently the Research Fellow of Korea Institute of Civil Engineering and Building Technology (KICT), South Korea. Dr. Chang received her B.S. and M.Sc. in Seoul National University in 2001 and 2003, respectively. She majored in saltwater intrusion processes in saturated porous media systems in Department of Civil Engineering of Auburn University for her Ph.D. from 2008 to 2012. Before she joined in KICT, she worked as a researcher at Korea Institute of S&T Evaluation and Planning (KISTEP) from 2003 to 2006, and a post-doctoral researcher at Korea Institute of Geoscience and Mineral Resources (KIGAM) from 2010 to 2013. She has served on various scientific society including Korea Federation of Women's Sciences & Technology Association (KOFWST), Korean Society of Soil and Groundwater Environment (KOSSGE), and the Korean Society of Civil Engineers (KSCE). Recent research focuses on the flow and transport of multiple contaminant species, the dynamics of seawater intrusion (SWI), assessment of the vulnerability of coastal aquifers, and surface water-groundwater interactions.

Yeonsang Hwang

Prof. Yeonsang Hwang, an Associate Professor at Arkansas State University and a Hydrologist, is most interested in the study of the impact of climate variability on water systems. His early research works have been done through the support of two NOAA RISA (Regional Integrated Sciences and Assessments) programs in Colorado (WWA, Western Water Assessment) and South Carolina (CISA, Integrated Sciences and Assessments) in the area of uncertainties in daily rainfall data, the impact of climate variability in watershed modeling, ensemble forecast of drought indices, and a residual re-sampling technique. RISA teams are organized to develop and deliver water/climate-related information and research results to various stakeholders through an interdisciplinary setup. In recent years, he works on a large-scale comprehensive watershed study on the White River with the U.S. Army Corps of Engineers. As a research member and a water engineer, he worked for LG Engineering and Construction (currently GS E&C) from 1993 to 2001. He was involved in various projects related to water resources, water quality issues, and other general civil engineering problems. With these engineering experiences, he has been registered as a professional engineer in Korea.

Yeonjoo Kim

Prof. Yeonjoo Kim is currently the associate professor at the Department of Civil and Environmental Engineering, Yonsei University, South Korea. She earned her B.Sc. at Yonsei University in 2001 and her S.M. at MIT in 2003. Prof. Kim completed her Ph.D. at the University of Connecticut in 2006 with a dissertation titled soil moisture-vegetation-precipitation feedback at the seasonal time scale over North America. She worked as a postdoctoral fellow at Harvard University from 2006 to 2009 and was awarded the NASA Postdoctoral Program fellowship in 2009 for working at NASA/GISS. She then moved back to Korea to work as a senior research scientist at the Korea Environment Institute from 2010 until 2015, when she joined Yonsei University as a faculty member. Prof. Kim seeks to understand the interactions between hydrological processes and other physical, biological, and societal processes within the earth system. She developed and utilized various numerical models, including land surface-terrestrial biosphere models and watershed-scale hydrologic models, together with data from ground-based and remotely sensed measurements.

Preface to "Integrated Surface Water and Groundwater Analysis"

Comprehensive understanding of groundwater-surface water (GW–SW) interaction is essential for effective water resources management. Groundwater (GW) and surface water (SW) are closely connected components that constantly interact with each other within the Earth's hydrologic cycle. Many studies utilized observations to explain the GW-SW interactions by carefully analyzing the behavior of surface water (SW) features (streams, lakes, reservoirs, wetlands, and estuaries) and the related aquifer environments. However, unlike visible surface water, groundwater, an invisible water resource, is not easy to measure or quantify directly. Nevertheless, demand for groundwater that is highly resilient to climate change is growing rapidly. Furthermore, groundwater is the prime source for drinking water supply and irrigation, and hence critical to global food security. Groundwater needs to be managed wisely, protected, and especially sustainably used. However, this task has become a challenge to many hydrologic systems in arid to even humid regions because of added stress caused by changing environment, climate, land use, population growth, etc. In this issue, the editors present contributions on various research areas such as the integrated GW-SW analysis, sustainable management of groundwater, and the interaction between GW and SW. Methodologies, strategies, case studies as well as quantitative techniques for dealing with combined surface water and groundwater management are of interest for this issue.

Il-Moon Chung, Sun Woo Chang, Yeonsang Hwang, and Yeonjoo Kim

Editors

hydrology

Editorial

Editorial for Special Issue: "Integrated Surface Water and Groundwater Analysis"

Il-Moon Chung [1,*], Sun Woo Chang [1,*], Yeonsang Hwang [2] and Yeonjoo Kim [3]

1   Department of Water Resources and River Research, Korea Institute of Civil Engineering and Building Technology, Goyang-si 10223, Korea

2   College of Engineering and Computer Science, Arkansas State University, P.O. Box 1740, Jonesboro, AR 72467, USA; yhwang@astate.edu

3   Department of Civil and Environmental Engineering, Yonsei University, Seoul 03722, Korea; yeonjoo.kim@yonsei.ac.kr

*   Correspondence: imchung@kict.re.kr (I.-M.C.); chang@kict.re.kr (S.W.C.)

Citation: Chung, I.-M.; Chang, S.W.; Hwang, Y.; Kim, Y. Editorial for Special Issue: "Integrated Surface Water and Groundwater Analysis". *Hydrology* 2022, 9, 70. https://doi.org/10.3390/hydrology9050070

Received: 22 April 2022
Accepted: 25 April 2022
Published: 27 April 2022

Publisher's Note: MDPI stays neutral with regard to jurisdictional claims in published maps and institutional affiliations.

Comprehensive understanding of groundwater—surface water (GW–SW) interaction is essential for effective water resources management. Groundwater (GW) and surface water (SW) are closely connected components that constantly interact with each other within the earth's hydrologic cycle. Many studies utilized observations to explain the GW–SW interactions by carefully analyzing the behavior of surface water features (streams, lakes, reservoirs, wetlands, and estuaries) and the related aquifer environments. Surface water bodies gain water and solutes from groundwater systems, and in other cases surface water bodies recharge groundwater, which causes changes in groundwater quality. The interfaces between GW and SW environments, such as hyporheic—benthic zones and riparian corridors, often function as biogeochemical hotspots and can have significant influences on the entire stream ecology. Furthermore, groundwater is a major source of drinking water supply and irrigation, and hence critical to global food security. Groundwater needs to be wisely managed, protected, and especially sustainably used. However, the aforementioned tasks have become challenging to many hydrologic systems in various areas from arid to even humid regions because of added stress caused by changing environment, climate, land use, and population. The aim of the Special Issue "Integrated Surface Water and Groundwater Analysis" was to elevate integrated understanding of the science in GW–SW systems through healthy discussions in the relevant research communities.

In this Special Issue, researchers have contributed to the study of groundwater–surface water interactions on a variety of subjects and methods, such as analytical and explicit numerical approaches [1], groundwater level prediction via a long short-term memory (LSTM) network [2], the impact of hydraulic fracturing and climate change [3], modification of the SWAT+ watershed model [4], water management in small islands [5], fluctuation of induced aquifer recharge [6,7], response of river to the 2016 seismic sequence [8], hydrological connectivity in permafrost regions [9], groundwater and streamflow interactions during floods [10], heat transport in managed aquifer recharge (MAR) [11], isotope analysis for distinguishing different types of water [12], digital platform to support decision-making [13], and deep percolation in irrigated fields [14,15].

When evaluating SW–GW interactions, the accuracy of calibration or prediction has been demonstrated by new techniques or multidisciplinary techniques applied in site-specific regional studies. The hydrodynamic surface water module of the STRIVE package (stream river ecosystem) of FEMME (flexible environment for mathematically modelling the environment), combined with analytical/explicit numerical solutions for groundwater flows, successfully investigated the hydraulic GW–SW interaction [1]. Machine learning techniques predicted the groundwater level, revealing that the LSTM (long short-term memory) network approach can be very useful for one-day forecasting of groundwater fluctuations in Jeju Island, Korea [2]. Bailey [4] developed a new module called 'gwflow' for the SWAT+ modeling code and applied this module to simulate both land surface and subsurface hydrological processes of Little River Experimental Watershed (LREW) (327 km^2) in southern Georgia, USA. There was also a valuable case study that simultaneously employed water isotopes, dissolved organic carbon, and electrical resistivity tomography to analyze the hydrological connectivity in a permafrost region [9]. Oxygen and hydrogen isotope (δ^{18}O–δ^2H) relationships were characterized by means of various statistical approaches on the Northern Italian Apennines [12].

Important investigations were presented regarding the effects of natural and anthropogenic stress on GW–SW interactions. An integrated hydrologic model (MIKE-SHE and MIKE-11 models) and a cumulative effects landscape simulator (ALCES) were used to assess the impact of hydraulic fracturing on GW–SW interactions in a shale gas and oil play area (23,984.9 km^2) of northwestern Alberta, Canada during 2021–2036 under future climate change scenarios [3]. The impact of a 2016 seismic sequence was analyzed with stream discharge data and recession curves in Nera River Basin, Italy [8]. The hydrological–ecological integrated watershed-scale flow model (HEIFLOW) was tested to verify interactions between the groundwater and streamflow during flood events in 2013 in the Miho catchment, Korea [10].

The interaction of GW–SW was also understood by observing or assessing quantitative/qualitative changes in major hydrologic components. First, in the process of managed aquifer recharging (MAR), GW–SW interactions occur as a mechanism of induced recharge. Hydrodynamics, hydrochemical, and numerical modeling methods were used to analyze an induced aquifer recharge in riverbank filtration (RBF) at Serchio River in Italy [6]. Integrated MODFLOW and SWAT modeling quantitatively assessed induced aquifer recharge due to nearby rivers during the seasonal exploitation of groundwater water curtain cultivation sites in Korea, and it predicted that the aquifers were being depleted every year [7]. Groundwater heat and temperature were monitored in shallow aquifers in the alluvial plain of the Cornia River, Italy to detect the mechanism development of recharge in MAR operations [11]. Second, in addition to recharge, the SW–GW interaction can be explained by another component such as deep percolation (DP) from water balance analysis. In addition to recharge as a direct indicator, the SW–GW interaction can be explained by deep percolation. A two-year study on Willamette Valley in western Oregon, USA assessed DP and recharge into the aquifer [14]. Estimation of DP into shallow aquifers characterized the practice of water management of two flood-irrigated fields in northern New Mexico [15].

Development of tools for the decision-making process was also presented. White [5] found large water supply differences between small islands vulnerable to various natural disasters and climate change. The author compared the national Tonga Strategic Development Framework, 2015–2025 (TSDFII) and local community development plans (CDPs) with census and limited hydrological data in the study. Rojas et al. [13] focused on early involvement of stakeholders, and therefore developed a digital platform (SimCopiapo) that combined integrated modelling and participatory modelling to support decision making for water management in the Copiapó River Basin, northern Chile.

We believe that the insights from the latest research outcomes in the areas of SW–GW interaction observations, modeling calibration/analyses, and decision-making support systems presented in the articles published in this Special Issue can serve as a foundation for an integrated water resource management (IWRM) approach in the future.

Funding: This research received no external funding.

Acknowledgments: We want to thank the authors who contributed to this Special Issue on "Integrated Surface Water and Groundwater Analysis" and their anonymous reviewers who provided the authors with insightful and constructive comments.

Conflicts of Interest: The authors declare no conflict of interest.

References

1. El-Rawy, M.; Batelaan, O.; Buis, K.; Anibas, C.; Mohammed, G.; Zijl, W.; Salem, A. Analytical and Numerical Groundwater Flow Solutions for the FEMME-Modeling Environment. *Hydrology* **2020**, *7*, 27. [CrossRef]
2. Shin, M.; Moon, S.; Kang, K.; Moon, D.; Koh, H. Analysis of Groundwater Level Variations Caused by the Changes in Groundwater Withdrawals Using Long Short-Term Memory Network. *Hydrology* **2020**, *7*, 64. [CrossRef]
3. Saha, G.; Quinn, M. Integrated Surface Water and Groundwater Analysis under the Effects of Climate Change, Hydraulic Fracturing and its Associated Activities: A Case Study from Northwestern Alberta, Canada. *Hydrology* **2020**, *7*, 70. [CrossRef]
4. Bailey, R.; Bieger, K.; Arnold, J.; Bosch, D. A New Physically-Based Spatially-Distributed Groundwater Flow Module for SWAT+. *Hydrology* **2020**, *7*, 75. [CrossRef]
5. White, I.; Falkland, T.; Kula, T. Meeting SDG6 in the Kingdom of Tonga: The Mismatch between National and Local Sustainable Development Planning for Water Supply. *Hydrology* **2020**, *7*, 81. [CrossRef]
6. Rossetto, R.; Barbagli, A.; De Filippis, G.; Marchina, C.; Vienken, T.; Mazzanti, G. Importance of the Induced Recharge Term in Riverbank Filtration: Hydrodynamics, Hydrochemical, and Numerical Modelling Investigations. *Hydrology* **2020**, *7*, 96. [CrossRef]
7. Chang, S.; Chung, I. Water Budget Analysis Considering Surface Water–Groundwater Interactions in the Exploitation of Seasonally Varying Agricultural Groundwater. *Hydrology* **2021**, *8*, 60. [CrossRef]
8. Di Matteo, L.; Capoccioni, A.; Porreca, M.; Pauselli, C. Groundwater-Surface Water Interaction in the Nera River Basin (Central Italy): New Insights after the 2016 Seismic Sequence. *Hydrology* **2021**, *8*, 97. [CrossRef]
9. Tananaev, N.; Isaev, V.; Sergeev, D.; Kotov, P.; Komarov, O. Hydrological Connectivity in a Permafrost Tundra Landscape near Vorkuta, North-European Arctic Russia. *Hydrology* **2021**, *8*, 106. [CrossRef]
10. Joo, J.; Tian, Y. Impact of Stream-Groundwater Interactions on Peak Streamflow in the Floods. *Hydrology* **2021**, *8*, 141. [CrossRef]
11. Caligaris, E.; Agostini, M.; Rossetto, R. Using Heat as a Tracer to Detect the Development of the Recharge Bulb in Managed Aquifer Recharge Schemes. *Hydrology* **2022**, *9*, 14. [CrossRef]
12. Cervi, F.; Tazioli, A. Inferring Hydrological Information at the Regional Scale by Means of $\delta18O–\delta2H$ Relationships: Insights from the Northern Italian Apennines. *Hydrology* **2022**, *9*, 41. [CrossRef]
13. Rojas, R.; Castilla-Rho, J.; Bennison, G.; Bridgart, R.; Prats, C.; Claro, E. Participatory and Integrated Modelling under Contentious Water Use in Semiarid Basins. *Hydrology* **2022**, *9*, 49. [CrossRef]
14. Gómez, D.; Ochoa, C.; Godwin, D.; Tomasek, A.; Zamora Re, M. Soil Water Balance and Shallow Aquifer Recharge in an Irrigated Pasture Field with Clay Soils in the Willamette Valley, Oregon, USA. *Hydrology* **2022**, *9*, 60. [CrossRef]
15. Conrad, L.; Fernald, A.; Guldan, S.; Ochoa, C. A Water Balancing Act: Water Balances Highlight the Benefits of Community-Based Adaptive Management in Northern New Mexico, USA. *Hydrology* **2022**, *9*, 64. [CrossRef]

hydrology

Article

A Water Balancing Act: Water Balances Highlight the Benefits of Community-Based Adaptive Management in Northern New Mexico, USA

Lily M. Conrad [1], Alexander G. Fernald [1,*], Steven J. Guldan [2] and Carlos G. Ochoa [3]

1 Water Resources Research Institute, New Mexico State University, Las Cruces, NM 88003, USA; conradl@nmsu.edu
2 Sustainable Agriculture Science Center, New Mexico State University, Alcalde, NM 87511, USA; sguldan@nmsu.edu
3 Ecohydrology Laboratory, College of Agricultural Sciences, Oregon State University, Corvallis, OR 97331, USA; carlos.ochoa@oregonstate.edu
* Correspondence: afernald@nmsu.edu; Tel.: +1-575-646-4337

Abstract: Quantifying groundwater recharge from irrigation in water-scarce regions is critical for sustainable water management in an era of decreasing surface water deliveries and increasing reliance on groundwater pumping. Through a water balance approach, our study estimated deep percolation (*DP*) and characterized surface water and groundwater interactions of two flood-irrigated fields in northern New Mexico to evaluate the regional importance of irrigation-related recharge in the context of climate change. *DP* was estimated for each irrigation event from precipitation, irrigation input, runoff, change in soil water storage, and evapotranspiration data for both fields. Both fields exhibited positive, statistically significant relationships between *DP* and total water applied (TWA), where one field exhibited positive, statistically significant relationships between *DP* and groundwater level fluctuation (GWLF) and between GWLF and total water applied. In 2021, total *DP* on Field 1 was 739 mm, where 68% of irrigation water applied contributed to *DP*. Field 2's total *DP* was 1249 mm, where 81% of irrigation water applied contributed to *DP*. Results from this study combined with long-term research indicate that the groundwater recharge and flexible management associated with traditional, community-based irrigation systems are the exact benefits needed for appropriate climate change adaptation.

Keywords: flood irrigation; water management; deep percolation; surface water; groundwater; water balance

Citation: Conrad, L.M.; Fernald, A.G.; Guldan, S.J.; Ochoa, C.G. A Water Balancing Act: Water Balances Highlight the Benefits of Community-Based Adaptive Management in Northern New Mexico, USA. *Hydrology* **2022**, *9*, 64. https://doi.org/10.3390/hydrology9040064

Academic Editors: Il-Moon Chung, Sun Woo Chang, Yeonsang Hwang and Yeonjoo Kim

Received: 8 March 2022
Accepted: 12 April 2022
Published: 14 April 2022

Publisher's Note: MDPI stays neutral with regard to jurisdictional claims in published maps and institutional affiliations.

1. Introduction

Over 50% of the world's freshwater resources for human use and consumption rely on river discharge that can be greatly impacted by long-term changes in precipitation and temperature such as those caused by climate change, particularly in snow-dominated regions [1]. Much of the western United States depends on precipitation falling in the winter in mountainous regions as snow and subsequently released slowly as snowmelt throughout the following spring and summer seasons. However, long-term changes in temperature and precipitation are already affecting these crucial water resource systems by decreasing the maximum snowpack accumulation, shifting the timing of runoff to arrive earlier, and impacting the volume of river discharge [2,3] with changes amplified by a lack of reservoir storage [4]. More specifically, snow-dominated basins in the mid-high latitudes are the most vulnerable to the impacts of warming climates where maximum runoff is expected to arrive one month earlier by 2050 in the western United States [1].

For example, the Rio Grande and its tributaries are increasingly becoming water stressed due to the warming climate and the increasing demand from users in Colorado, New Mexico, Texas, and Mexico [5]. Rio Grande streamflow is vulnerable as it largely depends on snowpack conditions which are projected to decrease and melt earlier in the future [6–9]. This surface water resource must serve industrial, tourist, residential, agricultural, ecologic, and economic needs in the USA (e.g., Colorado, New Mexico, Texas) and Mexico. Under current climate conditions, New Mexico does not have water to spare between all users [10].

Oftentimes, agricultural sectors are the largest users of water and face greater pressure to develop new water management strategies to help non-agricultural sectors cope with future water scarcity caused by warming temperatures and climate uncertainty [11,12]. In New Mexico in 2015, irrigated agriculture accounted for 76% of total water use, 53% from surface water and 47% from groundwater. Flood irrigation is used on 45% of all irrigated fields in New Mexico [13].

Common water delivery systems for flood irrigation in New Mexico are *acequia* networks which face many socio-environmental challenges. First introduced to northern New Mexico in the 16th century, acequias are gravity-driven water delivery networks and also serve as the basis of community-managed water governance systems [14,15]. While acequias have many beneficial hydrologic (e.g., aquifer recharge) and social attributes (e.g., water sharing) that foster resilience [16], these ancient water systems still face the challenge of long-term, regional drought and difficult water policy [17,18]. Questions are continually raised at acequia irrigator meetings and posed to researchers regarding what the "right" management strategies are: Should we line the canals? Should we switch to drip? Should we pump groundwater? Irrigators find themselves stuck between cultural norms of propagating generational knowledge of traditional irrigation methods and pressures from decreasing water availability and outside agencies to modernize water delivery systems and maximize irrigation efficiency.

Agricultural irrigation practices involving surface water can cause percolation and groundwater recharge that significantly impact groundwater resources on regional scales [12,19–22]. A study by Bouimouass et al. (2020) focused on the acequia counterpart in Morocco—*seguias*—and concluded that flood irrigation of diverted surface water resulted in the dominant recharge process in mountain front landscapes [23]. Other studies from large agricultural drainages in China found that approximately 70% of applied flood irrigation water in maize fields recharged the groundwater during the growing season [24], and seepage from both irrigation canals and deep percolation (*DP*) from irrigation contributed to more than 90% of total annual shallow groundwater recharge [12]. Additionally, in a large traditional agricultural basin in Italy, irrigation water delivered through a system of canals provided 55 to 88% of groundwater recharge [25]. *DP* is the amount of water that travels below the effective root zone (ERZ) that can potentially reach the shallow aquifer [26]. One of our previous studies conducted in northern New Mexico showed that peak groundwater level response fluctuated up to 380 mm 8 to 16 h after the onset of flood irrigation [27], where another estimated 16% of unlined irrigation canal flows seeped into the subsurface, causing the water table to rise 1 to 1.2 m [28]. Additionally, annual shallow aquifer recharge ranged from 1044 to 1350 mm on a valley scale [22]. In these cases, *DP* from flood irrigation was a significant source of recharge to shallow groundwater. *DP* below the vegetative root zone can provide very important hydrologic and ecosystem benefits in irrigated valleys of semiarid and arid regions.

Conversely, groundwater may display evidence of interactions with surface water. As irrigation water infiltrates into the shallow aquifer, this *DP* can contribute groundwater return flows to the river. In northern New Mexico, this interaction is of particular interest considering *DP* can serve as temporary subsurface storage which provides delayed return flow during low-flow periods [22,29,30]. This serves as an important possible buffer for changing peak runoff timing associated with climate variability [19].

Considering interacting surface water and groundwater as one resource is essential for optimal protection of watersheds, sustaining water resources, and furthering integrated groundwater management [20,31]. This is critical within irrigation districts that are increasingly relying on pumping groundwater for agricultural and municipal uses, which can lead to the disconnection of surface water and groundwater [32]. More recently, groundwater recharge via flooding fields is becoming a more common conservation practice [33,34].

It is necessary to properly quantify aquifer recharge and foster an accurate understanding of DP and surface water and groundwater interactions in water-limited regions [35]. The water balance method is a technique commonly used to quantify groundwater recharge and characterize surface water and groundwater interactions [19,22,26,36,37]. Components of the water balance are precipitation, irrigation water applied, runoff, change in soil water storage, and evapotranspiration, where DP is unknown and calculated by the difference of these inputs and outputs [26].

Our first objective was to characterize and compare surface water and groundwater interactions and shallow aquifer response to irrigation events in flood-irrigated forage grass fields located within the same irrigated valley in northern New Mexico by estimating DP below the root zone with a water balance approach. Our second objective was to justify community-based adaptive management in the context of climate change by relating field-scale findings to regional climate change literature. The innovative approach of identifying tightly coupled objectives reflected the unique, tightly coupled natural and human irrigation system our study focused on. While cultivating a better understanding of available surface water resources is extremely important, irrigators and policy makers must also understand the effects of irrigation techniques on groundwater and surface water availability for downstream users [31]. Previous studies have quantified and compared DP across several crop fields, soil types, and valleys in northern New Mexico, USA [22,26]; however, more field observations of DP are needed to expand these studies from field-scale to valley or regional scales. We hypothesized that: (1) DP and total water applied and DP and groundwater response would be positively related on both fields; and (2) DP and total water applied would be significantly different across both study fields.

2. Materials and Methods

2.1. Study Area

This study was conducted on two acequia-irrigated fields in the Rio Hondo agricultural valley in northern New Mexico, USA. The Rio Hondo watershed drains an area of 185 km^2 [38] and is located 2200 m above sea level [18]. Snowmelt from the Sangre de Cristo Mountains serves as the Rio Hondo's main source of water and drains to the Rio Grande. Located in a semiarid steppe climate, 50% of the precipitation in this region falls during the monsoon season from June to September [39] with an annual average of 300 mm·year^{-1} [40]. The primary settlements in the Rio Hondo Valley are Valdez and Arroyo Hondo. The agricultural activity in the Rio Hondo watershed is small-scale in nature. Eight canals divert water from the Rio Hondo and deliver irrigation water to approximately 1161 ha through a system of branching acequias [38]. Typical crops include grasses (Mostly *Phleum pretense, Poa pratensis*), alfalfa (*Medicago sativa)*, orchards (e.g., plums, apples, apricot), and vegetables (e.g., squash, beets, greens, onions, radishes, etc.) [41].

Located in the community of Valdez within the Rio Hondo watershed, the first study field (F1) was approximately 27 km north of Taos, New Mexico, USA and covered 2.51 ha (Figure 1a). The main crops grown on the field were grasses, alfalfa, and clovers. The second study field (F2), located in the community of Arroyo Hondo within the Rio Hondo watershed, was approximately 20 km north of Taos, New Mexico (Figure 1b) and covered 1.62 ha. The main crops growing were grasses, alfalfa, and clovers.

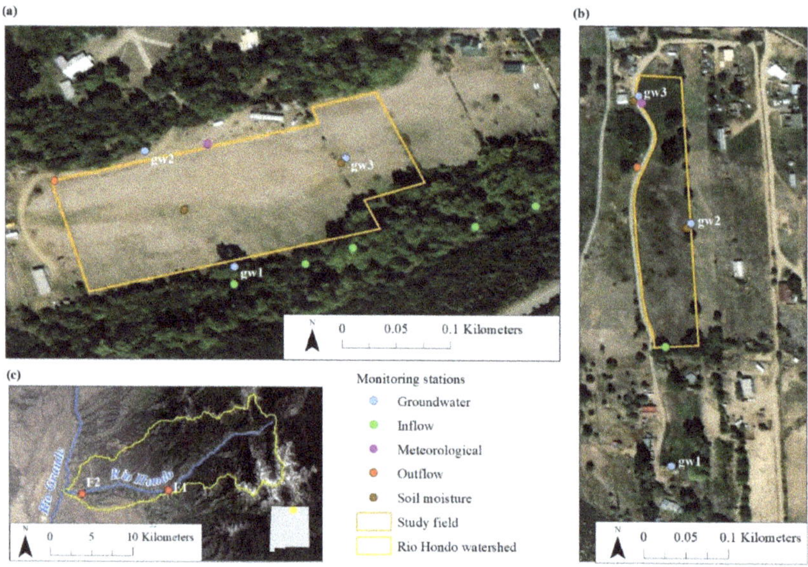

Figure 1. Water balance field study sites: (**a**) F1 (located at 36°32′05.3″ N, 105°34′04.5″ W); (**b**) F2 (located at 36°31′47.8″ N, 105°41′00.7″ W) and the corresponding monitoring stations. Both fields are located in the Rio Hondo watershed in Taos County, northern New Mexico ((**c**) inset). Monitoring station locations were selected to most accurately represent average field conditions of the irrigated area while also considering landowner needs for equipment maneuverability while cutting hay.

Soil Physical Properties

Of F1's total 2.51 ha, 2.35 ha were Manzano clay loam, and 0.16 ha were Loveland clay loam soil types. For the Manzano clay loam soil, slope values typically range from 3 to 5%, the soil is well-drained with medium runoff, average depth to the water table is more than 2 m, and a typical soil profile is clay loam for the top 1.5 m. For the Loveland clay loam soil, slope values typically range from 0 to 3%, the soil drains poorly and has a high runoff class, average depth to the water table is 0.15 to 0.46 m, and a typical soil profile is clay loam for the top 0 to 0.23 m, sandy clay loam for the middle 0.23 to 0.53 m, and very gravelly sand for the bottom 0.53 to 1.52 m [42].

Of F2's total 1.62 ha, 1.29 ha were Fernando silt loam, 0.24 ha were Fernando clay loam, and 0.08 ha were from the Sedillo–Silva association. The Fernando silt loam slope values typically range from 0 to 7% and are well drained with medium runoff. The average depth to the water table is greater than 2 m, and a typical soil profile is silt loam for the top 0 to 0.20 m, silty clay loam for the middle 0.20 to 0.91 m, and silt loam for the bottom 0.91 to 1.52 m. The Fernando clay loam generally has slope values from 3 to 5%, is well drained with medium runoff, depth to the water table is more than 2 m, and a typical soil profile is clay loam for the top 0 to 0.18 m, silty clay loam for the middle 0.18 to 0.64 m, and silty loam for the bottom 0.64 to 1.52 m. The Sedillo–Silva association soil typically has a slope of 10 to 25%, is well drained with high runoff, depth to the water table is greater than 2 m, and a typical soil profile is very gravelly loam for the top 0 to 0.08 m, gravelly clay loam for the middle 0.08 to 0.28 m, and very cobbly sandy loam for the bottom 0.28 to 1.52 m [43].

Soil bulk density varied between the two fields, whereas soil texture remained relatively consistent (Table 1). For F1, bulk density ranged from 1.46×10^9 Mg·m^{-3} in the topsoil to 1.23×10^9 Mg·m^{-3} toward the bottom of the soil profile. Within the field F2 soil profile, bulk density ranged from 1.19×10^9 Mg·m^{-3} to 1.27×10^9 Mg·m^{-3} from top to bottom. Soil texture was sandy clay loam for all soil depths except the top layer of the F1 soil profile which was sandy loam. Soil texture components exhibited the same trends through the soil profile for sand and silt but differed for clay. Sand content decreased, and silt content increased toward the bottom of the soil profile for both fields, while clay content increased in F1 and decreased in F2 (Table 1).

Table 1. Soil physical properties for the two field sites from manual soil sample collection (see Section 2.2.3). Laboratory analysis determined soil bulk density, soil particle distribution, and soil texture for each sensor depth in the soil profile. Values for each soil depth represent the averaged value between the two soil-monitoring stations on each field.

Field	Soil Depth (m)	Bulk Density (Mg·m^{-3})	Sand (%)	Silt (%)	Clay (%)	Soil Texture
	0.2	1.46×10^9	72.6	17.6	9.90	Sandy loam
F1	0.5	1.35×10^9	55.6	27.5	16.9	Sandy clay loam
	0.8	1.23×10^9	51.7	31.5	16.9	Sandy clay loam
	0.2	1.19×10^9	60.6	27.4	12.0	Sandy clay loam
F2	0.5	1.26×10^9	59.6	32.5	7.90	Sandy clay loam
	0.8	1.27×10^9	57.5	32.4	10.0	Sandy clay loam

2.2. Field Data Collection

We monitored various parameters at both study sites to calculate DP using a water balance approach for irrigation events over the 2020 and 2021 irrigation seasons. The water balance method was an appropriate approach for our study given our goals of estimating recharge for individual irrigation events within an irrigation season and subsequently relating our findings to community adaptive management and climate change. DP is the water that infiltrates into the subsurface, past the ERZ. ERZ varies depending on crop root development, effective soil depth, soil fertility or fertility management, and soil physical properties [44]. We recorded ERZ measurements of root systems at each site during soil volumetric water content (θ) sensor installation where F1 ERZ was 0.51 m and F2 ERZ was 0.53 m. Data collected throughout the 2020 and 2021 irrigation seasons returned a groundwater recharge estimate for each irrigation event through a field-scale water balance approach:

$$DP = PPT + IRR - RO - \Delta S - AET \tag{1}$$

where PPT is the amount of rainfall during the time interval (mm), IRR is irrigation water applied during the time interval (mm), RO is the amount of irrigation runoff during the time interval (mm), ΔS is the change of storage or change in θ during the time interval (mm), and AET is the actual evapotranspiration during the time interval (mm). The time interval for each irrigation event begins with the onset of irrigation and extends to 24 h after the end of the irrigation water delivery to achieve an assumed state of field capacity.

2.2.1. Precipitation

Precipitation falling on the study sites during the irrigation season was mainly rainfall measured by weather stations on each field. Both weather stations were equipped with a tipping bucket rain gauge (ClimaVUE50, Campbell Scientific, Inc.; Logan, UT, USA) programmed to record incremental precipitation every five minutes.

2.2.2. Irrigation Inflow and Outflow

Property owners and field managers for both fields used acequia-delivered surface water to flood irrigate throughout the growing season and decided to irrigate based on water allocations, environmental conditions, and crop needs. Surface water is diverted from the acequia onto fields through a series of wooden and metal headgates depending on the size and orientation of the field with respect to the acequia. F1 had five irrigation inflow monitoring stations and one irrigation outflow station. F2 had one irrigation inflow monitoring station and one irrigation outflow station. Rectangular Samani–Magallanez flumes [45] installed at inflow and outflow locations, each equipped with a CS451 pressure transducer and a CR300 datalogger (Campbell Scientific, Inc.; Logan, UT, USA) and programmed to record water level at five-minute increments measured *IRR* and *RO* on each field.

2.2.3. Soil Water Content and Physical Properties

Derived from soil volumetric water content data, the change in storage was determined as:

$$\Delta S = \sum_{i=1}^{n} (\theta_2 - \theta_1)_i \Delta d_i \tag{2}$$

where n is the number of layers represented by a soil sensor in the ERZ profile, θ_1 is the soil volumetric water content at the onset of irrigation ($m^3 \cdot m^{-3}$), θ_2 is the soil volumetric water content 24 h after irrigation ends or the average soil volumetric water content at field capacity ($m^3 \cdot m^{-3}$), and Δd_i is the soil layer thickness (mm). Equation (2) converted θ at each sensor location to the amount of water (mm) held in the ERZ.

Each field had two monitoring stations measuring θ. At each station, a CR300 datalogger and three horizontally placed CS655 (Campbell Scientific, Inc.; Logan, UT, USA) soil sensors were arranged vertically in the ERZ at depths of 0.2 m, 0.5 m, and 0.8 m and recorded changes in θ every minute and averaged data at 30-min increments.

Soil samples collected while installing the sensor network underwent laboratory analysis to determine soil texture and bulk density. Three soil cores were collected at each sensor depth on the opposite wall of the pit where sensors were installed with a split soil core sampler and analyzed with the Blake and Hartge bulk density method [46] and the Gee and Bauder hydrometer method to determine soil texture [47].

2.2.4. Evapotranspiration

We used the following equation to calculate the amount of actual evapotranspiration (*AET*):

$$AET = K_c ET_0 \tag{3}$$

where ET_0 is the total evapotranspiration (mm) calculated with the Penman–Monteith equation programmed into a CR1000 datalogger (Campbell Scientific, Inc.; Logan, UT, USA). The Penman–Monteith equation outperforms others by including more factors that influence crop water loss (e.g., absorbed radiant energy, wind, atmospheric vapor deficit) and is therefore expected to provide more accurate estimates [48]. Post-processing the ET_0 values with crop coefficient (K_c) values calculates *AET*. We used crop coefficient curves for grass at different stages of the growing season presented in a previous study that took place near our study area [49] (p. 151). ET_0 values were recorded, and *AET* values were calculated for hourly data.

2.2.5. Groundwater Level

Three monitoring wells equipped with water level loggers (HOBO Logger U20-001-01, Onset; Bourne, MA, USA) recorded water table fluctuations on each field. All monitoring wells on F1 were steel drive-point wells 2 to 3 m deep. Two of these wells were installed by previous researchers [26]. We installed two steel drive-point wells 2 to 3 m deep on F2 and used the landowner's residential drinking well that was 13 m deep for the third monitoring well. This residential well has been used for long-term groundwater monitoring, where the data clearly show groundwater level response to the irrigation season.

The groundwater level data helped characterize shallow aquifer response to *DP* from irrigation inputs. Calculated for each irrigation event, groundwater level fluctuation (GWLF) (mm) was the difference between groundwater level prior to the irrigation onset (averaged over the 6 h prior to the irrigation onset) and maximum water level rise until the following irrigation event. Negative GWLF values indicate declining groundwater levels.

2.3. Statistical Analyses

Specific parameters that characterize surface water and groundwater interactions underwent linear regression and ANOVA statistical analyses to delineate any significant relationships within and across fields. Linear regression models evaluated and compared interactions between total water applied ($TWA = IRR + PPT - RO$) and DP, DP and GWLF, TWA and GWLF. ANOVA analyses identified significant differences in means between the two study fields and different stations. Differences were considered significant at $\alpha = 0.05$. The 2020 irrigation season data collection only spanned mid-June through October (partial season), whereas the 2021 irrigation season data collection spanned April through October (complete season). Therefore, only 2021 data were included in the statistical analysis and presented in the Results section of this paper for optimal scientific consistency and comparability.

3. Results

3.1. Irrigation Events and Deep Percolation Estimates

The number of irrigation events and DP varied between both fields over the 2021 irrigation season (Tables 2 and 3). Eight irrigation events took place on F1 (Table 2). A total of 24 irrigation events took place on F2 (Table 3). The average IRR was 137 mm, and the DP was 92 mm per irrigation event on F1 (Table 2). The average IRR was 64 mm, and the DP was 52 mm per irrigation event on F2 (Table 3). The F1 DP estimates total was 739 mm, where 68% of the IRR contributed to DP (Table 4). For F2, the DP estimates total was 1249 mm, where 81% of the IRR contributed to DP (Table 4).

Table 2. DP results calculated with the water balance method for each irrigation event in the 2021 irrigation season for F1. This table shows the total time of irrigation, change in θ (ΔS), total irrigation water applied (IRR), tailwater runoff (RO), total precipitation (PPT), and total AET from the beginning of each irrigation event to 24 h after the end of irrigation. DP estimates that resulted in negative values likely due to large ΔS values were considered to be 0, where no recharge occurred.

Date	Irrigation Duration (h)	ΔS (mm)	IRR (mm)	RO (mm)	PPT (mm)	AET (mm)	DP (mm)
27 April 2021	49	55	158	0	0	8	94
4 May 2021	49	11	185	0	0	11	162
11 May 2021	48	105	197	2	0	14	76
18 May 2021	58	−3	138	1	7	13	134
24 May 2021	45	23	235	47	0	18	147
1 June 2021	83	−6	135	0	1	21	122
23 July 2021	70	18	12	0	22	11	5
31 July 2021	165	129	34	0	5	22	0
Average	71	42	137	6	4	15	92

Table 3. DP results calculated with the water balance method for each irrigation event in the 2021 irrigation season for F2. This table shows the total time of irrigation, change in θ (ΔS), total irrigation water applied (IRR), tailwater runoff (RO), total precipitation (PPT), and total AET from the beginning of each irrigation event to 24 h after the end of irrigation. DP estimates that resulted in negative values likely due to large ΔS values were considered to be 0, where no recharge occurred.

Date	Irrigation Duration (h)	ΔS (mm)	IRR (mm)	RO (mm)	PPT (mm)	AET (mm)	DP (mm)
16 April 2021	70	46	313	0	8	6	270
10 May 2021	5	−1	0	0	0	5	0
10 May 2021	8	−1	5	0	0	1	6
11 May 2021	11	0	2	0	0	6	0
11 May 2021	11	0	2	0	0	0	2
12 May 2021	36	3	51	0	0	11	37
13 May 2021	9	0	2	0	0	0	2
14 May 2021	98	164	341	0	22	24	175
18 May 2021	10	−55	1	0	4	2	68
19 May 2021	40	−23	18	0	2	15	27

Table 3. *Cont.*

Date	Irrigation Duration (h)	ΔS (mm)	*IRR* (mm)	*RO* (mm)	*PPT* (mm)	*AET* (mm)	*DP* (mm)
23 May 2021	2	−3	0	0	0	2	1
24 May 2021	1	−2	1	0	0	4	0
24 May 2021	18	−6	0	0	0	10	0
29 May 2021	33	−5	8	0	0	7	5
31 May 2021	23	−4	3	0	16	7	15
5 June 2021	78	−5	216	3	2	31	189
26 June 2021	112	82	397	8	11	10	308
1 August 2021	58	6	26	0	7	12	15
18 August 2021	7	−3	1	0	0	3	1
18 August 2021	39	−4	39	0	0	6	37
27 August 2021	8	−4	12	0	0	6	10
29 August 2021	18	−4	20	0	0	7	17
8 September 2021	50	−4	70	0	0	11	63
11 September 2021	8	−1	2	0	0	6	0
Average	31	7	64	0	3	8	52

Table 4. Summary table displaying total number of irrigation events, cumulative *IRR*, *DP*, and percent of *IRR* that contributed to *DP* for each field over the 2021 irrigation season.

Field	Year	Number of Irrigation Events	*IRR* (mm)	*DP* (mm)	Percent *DP* (%)
F1	2021	8	1093	739	67.7
F2	2021	24	1541	1249	81.1

While annual variability is common due to differing environmental conditions, surface water availability, and irrigation scheduling, monthly irrigation summaries and averages on both fields demonstrate similar ranges of water balance parameters between 2020 and 2021 (Table 5). On F1 in 2020, *DP* averaged 12 mm and 29 mm per irrigation event in July and August, respectively, with no irrigations in September. In 2021, the average *DP* for July was 2 mm with no irrigations in August or September. No irrigations took place on F2 in July in 2020 and 2021. In 2020, F2 *DP* averaged 9 mm in August and 1 mm in September. In 2021, *DP* averaged 16 mm in August and 32 mm in September.

Table 5. Comparison of monthly number of irrigation events, average ΔS, average *IRR*, average *RO*, average *PPT*, average *DP*, and total *DP* for F1 and F2 for three months in 2020 and 2021. The months chosen for comparison are July through September because these were the first complete monthly records after data collection began in 2020 (data collection began early-June 2020) to ensure optimal comparability between the two irrigation seasons on both fields.

Field	Month	Number of Irrigation Events	Avg ΔS (mm)	Avg *IRR* (mm)	Avg *RO* (mm)	Avg *PPT* (mm)	Avg *AET* (mm)	Avg *DP* (mm)	Sum *DP* (mm)
F1	July 2020	2	−3	13	0	1	4	12	24
	August 2020	1	−3	31	0	5	10	29	29
	September 2020	0			(no irrigation events)				
	July 2021	2	73	23	0	13	16	2	5
	August 2021	0			(no irrigation events)				
	September 2021	0			(no irrigation events)				

Table 5. *Cont.*

Field	Month	Number of Irrigation Events	Avg ΔS (mm)	Avg *IRR* (mm)	Avg *RO* (mm)	Avg *PPT* (mm)	Avg *AET* (mm)	Avg *DP* (mm)	Sum *DP* (mm)
F2	July 2020	0				(no irrigation events)			
	August 2020	3	2	11	0	5	7	9	27
	September 2020	2	−1	6	0	0	6	1	3
	July 2021	0				(no irrigation events)			
	August 2021	5	−2	19	0	1	7	16	80
	September 2021	2	−3	36	0	0	8	32	63

The linear regression analysis showed a positive, significant relationship between *DP* and *TWA* and *TWA-ΔS* for both F1 ($p = 8.37 \times 10^{-3}$ and $p = 3.48 \times 10^{-4}$, respectively) and F2 ($p = 7.88 \times 10^{-16}$ and $p < 2.00 \times 10^{-16}$, respectively) (Table 6). Previous research in the region found that prior θ significantly impacted *DP* [50], which is why we included *TWA-ΔS* in the linear regression. Additionally, F2 exhibited a significant positive relationship between *DP* and irrigation duration ($p = 5.42 \times 10^{-8}$). ANOVA showed statistically significant differences in the mean irrigation duration and mean number of irrigation events between F1 and F2 (Table 7).

Table 6. Statistics from linear regression models comparing *DP* and *TWA*, *DP* and *TWA-ΔS*, and *DP* and irrigation duration for irrigation events on each field in 2021. Significant relationships are highlighted by *p* values with an asterisk (*).

Field	t	R^2	p
	TWA (mm)		
F1	3.86	0.713	8.37×10^{-3} *
F2	16.4	0.925	7.88×10^{-16} *
	TWA-ΔS (mm)		
F1	7.26	0.898	3.48×10^{-4} *
F2	67.9	0.995	$<2.00 \times 10^{-16}$ *
	Irrigation duration (h)		
F1	−2.12	0.427	0.0787
F2	8.04	0.746	5.42×10^{-8} *

Table 7. ANOVA tests conducted with the field as the independent variable and different variables of interest as dependent variables to identify significant differences in means between surface water and groundwater interactions and irrigation management across both fields for 2021 irrigation events. Significant differences are highlighted by *p* values with an asterisk (*).

Dependent Variable	F	R^2	p
DP (mm)	1.40	0.0447	0.245
TWA (mm)	2.23	0.0691	0.146
TWA-ΔS (mm)	0.811	0.0263	0.375
Irrigation duration (h)	8.17	0.214	7.67×10^{-3} *
Number of irrigation events	9.66	0.244	4.09×10^{-3} *

3.2. Shallow Groundwater Response to Irrigation Inputs

GWLF and response to irrigation inputs were observed for both study fields over the 2020 and 2021 irrigation seasons (Figure 2). In 2021 on F1, gw1 GWLF averaged 533 mm, gw2 GWLF averaged 262 mm, and gw3 GWLF averaged 863 mm. The greatest observed GWLF of the F1 monitoring wells was 1699 mm on 11 May 2021 (Table 8).

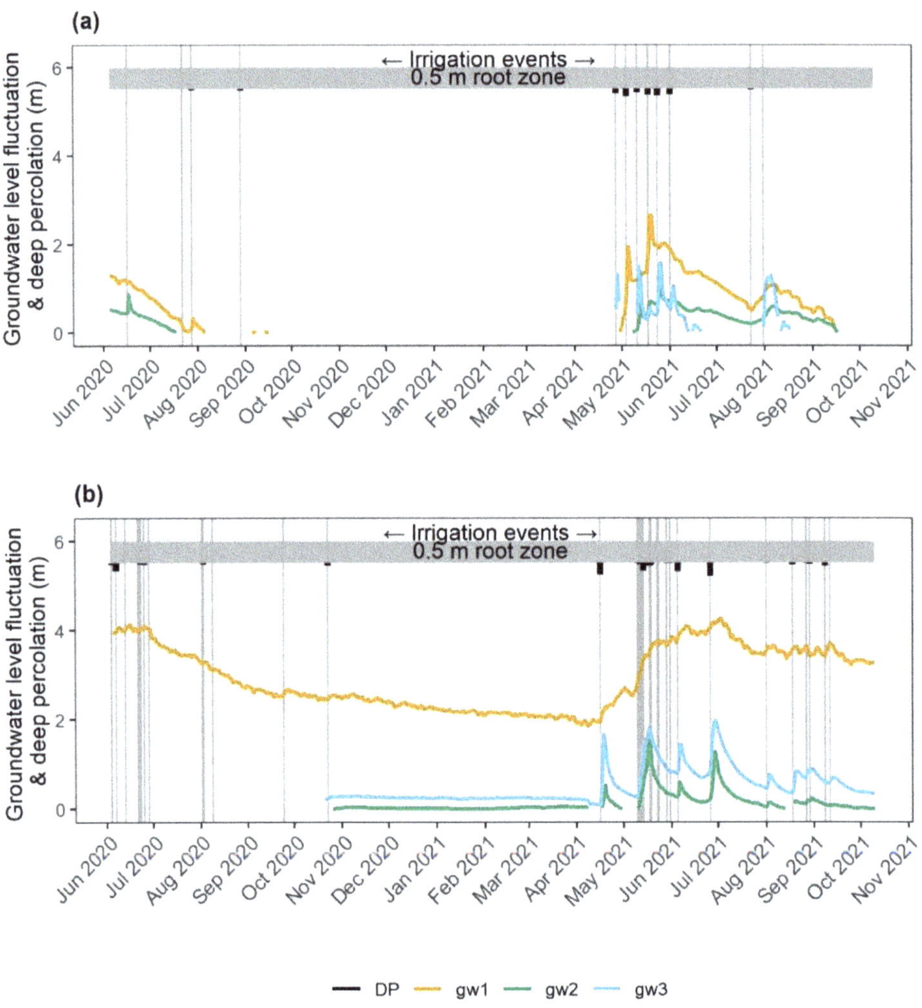

Figure 2. Shallow groundwater levels for all monitoring wells, irrigation events (vertical gray lines), and DP estimates for the study sites: (**a**) F1; (**b**) F2 from early June 2020 through October 2021.

Table 8. GWLF (mm) in response to irrigation events in 2021 for all wells on F1 calculated as the difference between groundwater level prior to the irrigation onset (averaged over the 6 h prior to the irrigation onset) and maximum water level rise until the following irrigation event.

Date	GWLF gw1 (mm)	GWLF gw2 (mm)	GWLF gw3 (mm)
27 April 2021	452	0	1549
4 May 2021	1601	55	0
11 May2021	197	781	1699
18 May 2021	1363	280	288
24 May 2021	55	428	1350
1 June 2021	−6	108	618
23 July 2021	320	150	114
31 July 2021	282	290	1282
Average	533	262	863

Both study fields exhibited sharp groundwater response to irrigation events and *DP* (Figure 2). F1 groundwater levels would generally show a moderate decline after the peak GWLF (Figure 2a). F1 gw1, located next to the irrigation canal, maintained more elevated groundwater levels for longer than the other two wells on this field. Gw3 displayed the "flashiest" response to irrigation events and *DP* both in the rise and fall around the peak GWLF.

F2 groundwater levels—specifically gw2 and gw3—would decline more rapidly following the peak GWLF (Figure 2b). F2 gw1 showed more short-term fluctuation due to pumping water for residences on the property and more gradual rise and fall to the beginning and end of the irrigation season due to its deeper reach, upgradient position, and closer tie to ditch seepage from nearby acequias and water delivery canals rather than irrigation events. On F2, GWLF in gw1 differed from the groundwater response in the other monitoring wells to irrigation events (Figure 2b). This well (gw1) was the landowner's drinking water well that was 13 m deep and located upgradient of the irrigated field (Figure 1). GWLF from F2 gw1 levels were likely related to acequia flow as opposed to irrigation events. The main acequia flowed along the south border of the property, and the intermediate ditch that delivered water from the acequia onto F2 flowed next to gw1, so ditch seepage from delivery canals likely supplied this well. F2 gw1 GWLF averaged 167 mm in 2021. For the other F2 monitoring wells gw2 and gw3, GWLF averaged 210 mm and 272 mm, respectively. The greatest observed GWLF of the F2 monitoring wells was 1697 mm on 14 May 2021 (Table 9).

Table 9. GWLF (mm) in response to irrigation events in 2021 for all wells on F2 calculated as the difference between groundwater level prior to the irrigation onset (averaged over the 6 h prior to the irrigation onset) and maximum water level rise until the following irrigation event.

Date	GWLF gw1 (mm)	GWLF gw2 (mm)	GWLF gw3 (mm)
16 April 2021	876	602	1677
10 May 2021	100	7	42
10 May 2021	87	21	87
11 May 2021	76	141	129
11 May 2021	30	132	103
12May 2021	278	275	288
13 May 2021	53	−3	20
14 May 2021	242	1697	969
18 May 2021	3	−45	−23
19 May 2021	254	−40	47
23 May 2021	40	−7	−9
24 May 2021	−35	−8	−15
24 May 2021	57	3	1
29 May 2021	198	−1	−4
31 May 2021	164	30	33
5 June 2021	362	418	795
26 June 2021	247	1179	1204
1 August 2021	93	140	345
18 August 2021	41	0	43
18 August 2021	168	220	430
27 August 2021	30	96	147
29 August 2021	250	145	76
8 September 2021	368	3	28
11 September 2021	33	33	106
Average	167	210	272

Linear regression statistical analysis identified any significant relationships between *DP* and GWLF of each monitoring well as well as *TWA* and GWLF of each monitoring well for both study fields (Table 10). No significant relationships were identified between GWLF and *DP* nor GWLF and *TWA* for any F1 monitoring wells. This is likely related to the land manager's use of several headgates spread out along the southern field border used at different times unevenly applying irrigation water. All wells on F2 exhibited positive significant relationships between GWLF and *DP*, and GWLF and *TWA* (Table 10). However, F2 gw1 GWLF was likely related to acequia flow and ditch seepage rather than irrigation events due to its upgradient position.

Table 10. Statistics from linear regression models comparing GWLF and *DP* as well as GWLF and *TWA* for all monitoring wells on each field from data collected over the 2021 irrigation season. Significant relationships are highlighted by *p* values with an asterisk (*).

	GWLF gw1 (mm)			GWLF gw2 (mm)			GWLF gw3 (mm)		
	t	R^2	p	t	R^2	p	t	R^2	p
DP (mm)									
F1	1.19	0.190	0.280	-0.252	0.0105	0.810	-0.426	0.0294	0.685
F2	4.60	0.490	1.40×10^{-4} *	5.78	0.603	8.09×10^{-6} *	11.5	0.858	8.77×10^{-11} *
TWA (mm)									
F1	0.540	0.0464	0.608	0.720	0.0795	0.499	0.639	0.0638	0.546
F2	3.93	0.413	7.08×10^{-4} *	10.3	0.828	7.37×10^{-10} *	11.4	0.856	9.91×10^{-11} *

Figure 3 provides a visualization of how the groundwater level data from the three monitoring wells compare across the two study fields. The greatest variation is apparent between gw1 on F1 and F2 because the well on F2 is a residential drinking well and is much deeper (see Section 2.2.5 for more detailed metrics regarding the groundwater monitoring wells included in this study).

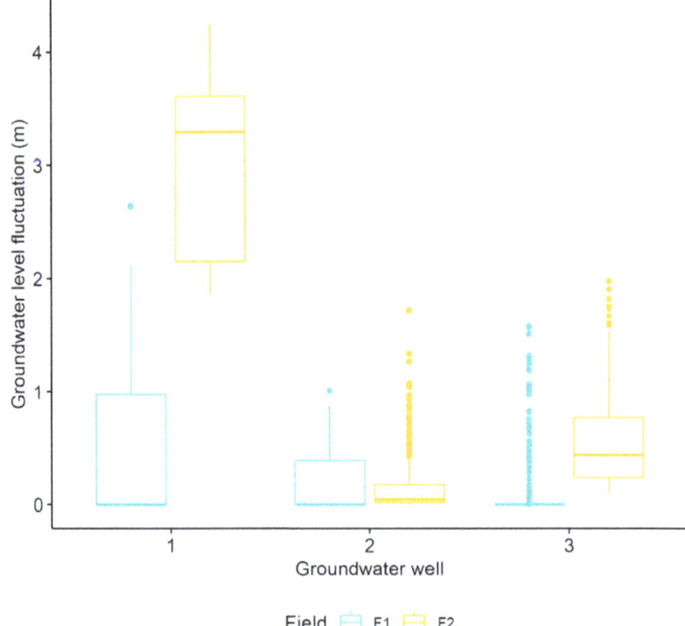

Figure 3. Boxplots created using daily averages of GWLF data from April 2021 through October 2021 visually comparing the medians, quartiles, and ranges of the three monitoring wells across both study fields.

4. Discussion

Our results showed both fields have significant relationships between *DP* and *TWA* and between *DP* and *TWA-ΔS* (Table 6). One field, F2, exhibited significant relationships between *DP* and irrigation duration (Table 6), GWLF and *DP*, and GWLF and *TWA* (Table 10). Antecedent soil moisture and soil conditions are particularly important factors to consider when discussing *DP*. More irrigation water is needed to saturate the ERZ when antecedent soil moisture is low or at times of greater plant water use which results in potentially less groundwater recharge from a given amount of irrigation water applied. *DP* was not significantly different when compared across both fields. The only significant differences when comparing irrigation events and *DP* estimates across both fields were irrigation duration and the number of irrigation events (Table 7). These results indicate that surface water and groundwater are tightly connected in this area, but variation in *DP* and groundwater response exists between land managers and fields due to differing irrigation practices.

Although we only report data and results from the 2021 irrigation season, our data collection began in June 2020. When comparing monthly averages and totals for months with complete records for both 2020 and 2021, several patterns emerge regarding irrigation scheduling, average ΔS, and average *DP* (Table 5). F1 irrigation frequency tapered off toward the end of the irrigation season for both 2020 and 2021, while F2 irrigation frequency increased toward the end of the season. Average ΔS was constant between the two fields over both years of data collection, ranging from −3 to 2 mm with a notably large value for F1 in July 2021 (73 mm) as an outlier perhaps related to frequent rainfall that occurred around that time of year and uneven irrigation water application. On F1, the average *DP* ranged from 2 to 29 mm over 2020 and 2021. Similarly, the F2 average *DP* ranged from 1 to 32 mm. These patterns help validate our water balance results (Tables 2 and 3) by demonstrating consistent and comparable water balance components and *DP* estimates across both fields over 2020 and 2021.

Previous acequia research in northern New Mexico forage fields that also used water balance methodology to estimate *DP* reflects similar results (Table 11), illustrating that we appropriately captured acequia surface water and groundwater interactions and irrigation practices. Our results reflected the greatest *DP* season totals (739 and 1249 mm) and the greatest percentage of *IRR* that contributed to *DP* (68 and 81%), critically filling in the range of possible seasonal *DP* values and characteristics by refining our understanding of acequia irrigation-related recharge in the context of long-term field data collection in northern New Mexico.

Table 11. A comparison of how our *DP* estimates compare to similar studies that used a water balance approach to estimate *DP* in forage grass fields in northern New Mexico.

Study	Location & Year	Average DP per Irrigation Event (mm)	Sum DP over Irrigation Season (mm)	Percent DP over Irrigation Season (%)
Ochoa et al. (2013)	Alcalde 2005	107	533	46
Ochoa et al. (2013)	Alcalde 2006	119	476	48
Gutiérrez-Jurado et al. (2017)	Rio Hondo (F1) 2013	53	531	51
Gutiérrez-Jurado et al. (2017)	Alcalde 2013	55	382	39
Gutiérrez-Jurado et al. (2017)	El Rito 2013	77	462	31
Conrad et al. (this paper)	Rio Hondo (F1) 2021	92	739	68
	Rio Hondo (F2) 2021	52	1249	81

Observations and projections of changing climate and snowmelt dynamics within the Rio Grande Basin—specifically the Upper Rio Grande headwaters region—are of particular interest to researchers and stakeholders due to the reliance of downstream users on snowmelt-dominated subbasins to meet water availability needs. For example, streamflow at Fort Quitman, Texas, USA has decreased 95% relative to the river's native streamflow [51]. In the Colorado River Basin, temperature-driven "hot droughts" have been connected to increased sublimation of snow which results in less runoff from a given snowpack [52]. Similarly, the interannual variability of streamflow related to snow water equivalent (SWE) has decreased by 40% in the Upper Rio Grande Basin, indicating that the connection between peak SWE and runoff volume is substantially weaker [7]. This drift between SWE and runoff is particularly critical because a large portion—50 to 75%—of the Rio Grande streamflow is sustained by seasonal snowpack accumulation [53]. Through paleoclimate reconstructions published in 2017, researchers identified a 30-year declining trend in runoff ratio since the 1980s which appeared unprecedented in the context of the last 440 years [54]. Observed, historical mean winter and spring temperatures have significantly increased in the Upper Rio Grande Basin [7], and temperatures rose at an alarming rate of 0.4 °C per decade from 1971 through 2011, informing temperature predictions of a 2 to 3 °C increase in average temperature by the end of the 21st century [8]. The SWE on April 1 has significantly decreased by 25% [7], where the mean melt season snow covered area is predicted to decrease 57 to 82%, and peak flow is predicted to arrive 14 to 24 days earlier than usual [6].

The combination of increasing temperature and more variable precipitation inputs are expected to create a decrease in summertime flows and increase the frequency, intensity, and duration of floods and droughts in the Upper Rio Grande Basin [8]. Elias et al. (2015) found that total annual runoff volume of Upper Rio Grande subbasins and tributaries could increase 7% in wetter scenarios but decrease 18% in drier scenarios. In the Rio Hondo watershed, annual daily mean streamflow has significantly decreased 0.85% per year since water year 1976 [55]. Another study found that the Rio Hondo baseflow, runoff, and streamflow have also significantly decreased since water year 1980 due to decreasing snowmelt rates [56].

Decreasing surface water flows in the Upper Rio Grande region will have negative effects on acequia water availability for acequia communities in this region. A previous study conducted in the Rio Hondo Valley found statistically significant relationships between river and acequia flows [57]. Similarly, spatial analysis of the normalized difference vegetation index (NDVI) found that the irrigated landscape within the Rio Hondo Valley expanded and contracted in response to wet or dry years, showing that irrigation intensity varied with available surface water [58,59]. Therefore, in the Rio Hondo Valley, acequia flow is directly related to river flow, and the variability of acequia irrigation intensity is apparent in wet and dry years. As a result, the irrigated landscape and acequia irrigation decrease as surface water resources decrease.

If surface water river flows continue to decrease, then acequia water availability and the acequia-irrigated landscape will decrease, as will regular DP and groundwater recharge [60]. As a mechanism that temporarily stores surface water in the subsurface which eventually returns to the river system as delayed return flow, DP can serve as a very important buffer against climate change; however, mean recharge in Taos County first significantly decreased in 1996 [61]. Baseflow is also an extremely critical element of the hydrologic regime in the Upper Rio Grande Basin where baseflow contributions account for 49% of total discharge upstream of Albuquerque, New Mexico [56]. Surface water and groundwater connectivity is critical for continued baseflows, and acequia-related DP and return flows play an important role in maintaining this connection. As climate change continues to negatively impact surface water availability and groundwater recharge in northern New Mexico and the Upper Rio Grande Basin, both acequia communities and the state of New Mexico will have to decide how to adapt to new climatic and hydrologic regimes.

When considering water use and management practices, either as a water manager or for modeling purposes, it is critical to determine the type and direction of adaptation (e.g., adaptation or maladaptation) occurring in response to climate change stressors [62]. Maladaptive actions are enacted to prevent or reduce vulnerability associated with climate change but ultimately have adverse impacts or increase vulnerabilities in the same or related systems. Examples of adverse impacts include: (1) an increase greenhouse gas emissions; (2) a disproportionate impact on vulnerable populations; (3) high environmental opportunity costs; (4) reduced incentives to adapt; and (5) dependencies that limit future generations [63]. Unfortunately, all too often, water management adaptation and governance strategies are maladaptive, such as water operations in Flint, Michigan [64], water deliveries in California's San Joaquin Valley [65], and development in Australian coastal cities [66].

Adaptive management practices are more prepared for climate change by incorporating flexibility and responsiveness into water management institutions and governance structures [67–72]. While some suggest doing this through intraregional contracts and mergers [67], acequias have been doing this for centuries through a concept known as *repartimiento*—the ability to employ flexible and dynamic water deliveries to distribute water as equitably as possible by sharing water shortages either within a single acequia or between different acequias throughout a given watershed. Cruz et al. (2019) documented this phenomenon by showing that the water available in acequias is directly correlated to the water available in the stream system [57]. When not enough surface water is available to irrigate crops, landowners will typically irrigate a smaller parcel of their total crop land as opposed to the entire area. This shows the inherent adaptability embedded within traditional acequia irrigation frameworks that is and will continue to be crucial in the context of a changing climate, growing seasons, and streamflow regimes.

The flood irrigation regime these two fields and the greater Rio Hondo Valley—as well as other acequia communities—follow experience groundwater recharge benefits inadvertently associated with managed aquifer recharge (MAR). Recently, many research articles [73] have featured different MAR techniques and pilot programs [34,74]. MAR is an approach used to replenish groundwater resources and is becoming more common in areas of heavy groundwater pumping and declining aquifer levels. There are many different techniques and objectives within this overarching mitigation approach. One promising approach that utilizes already existing infrastructure is applying MAR to irrigated agricultural lands, where surface water is applied over large areas as opposed to the more traditional MAR approach of facilitating high recharge at dedicated recharge sites [34]. This form of MAR reduces costs associated with infrastructure, piping, and energy given the gravity-driven water distribution [75]. This framework is naturally mirrored on a regional scale in northern New Mexico's acequia networks. Acequia networks divert surface water through a system of (typically earthen) canals to fields for flood irrigation, where seepage occurs throughout time in the canals and application in the fields. Acequia irrigators greatly value these contributions to groundwater for the many environmental and water storage benefits the recharge provides (Figure 4). Acequias are not without their challenges, but they can serve as a model for sustainable, integrated water management that implicitly employs MAR and welcomes groundwater recharge as a benefit rather than an inefficiency [76].

Characterized by regular shallow aquifer recharge and flexible and dynamic water management that reflects equity and current environmental conditions, acequias offer several reasons why we should consider maintaining traditional irrigation systems in the face of climate change (Figure 5). In times of reduced surface water availability, acequia irrigators only irrigate smaller parcels of their total irrigated land and typically invest in deep rooted, drought-tolerant crops that can persist through growing seasons without much irrigation water application. Acequia communities have followed this model traditional flood irrigation model and persisted through drought for hundreds of years in northern New Mexico. However, when thinking about the future, the question then becomes: How should acequia communities adapt to meet reduced surface water availability and changing streamflow regime challenges that the current prolonged and unprecedented drought presents if traditional acequia irrigation practices are no longer sufficient?

Traditional acequia operations are typically associated with resilience [77], but many acequia irrigators and managers are unsure of how sustainable certain adaptations are moving forward (e.g., lining earthen irrigation canals, switching from flood to drip irrigation, greater reliance on groundwater pumping) and the implications of any detrimental effects on groundwater levels (i.e., lowering the water table) which are ultimately connected to surface water availability (Figure 5). Lining irrigation ditches, switching from flood to drip irrigation, and supplemental groundwater pumping are commonly called into question by acequia community members which is why these adaptation strategies are highlighted in this paper. While these three strategies can be beneficial, lining ditches and drip irrigation reduce pathways for surface water to seep into the groundwater, and a growing reliance on groundwater pumping will negatively impact surface water and groundwater connectivity by lowering the water table (Figure 5). When used simultaneously in a region where baseflow is a crucial component of sustaining Rio Grande streamflow [56] and traditional acequia irrigation related recharge serves as delayed return flow [22], the reduction in groundwater recharge and the increase in groundwater pumping will negatively impact surface water availability for downstream users and begin propagating a cycle of maladaptation. It will be critical to prioritize traditional flood irrigation approaches and benefits such as groundwater recharge as much as possible when acequia communities or similar community-based irrigation systems are seeking solutions under conditions of reduced surface water availability.

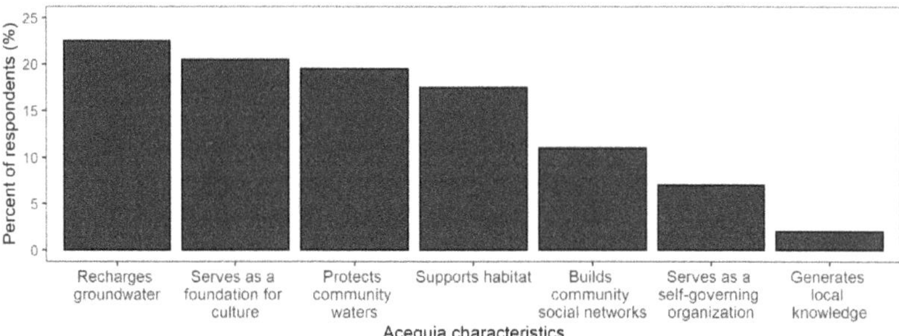

Figure 4. When posed the question: "In addition to providing irrigation water for local uses, which of the following characteristics of acequias are most important to you?", Rio Hondo Valley acequia community members (22.5%) reported groundwater recharge as the most valued attribute out of a variety of environmental, social, cultural, and governance options (n = 25). These data were collected as background information from adaptive capacity pre- and post-survey instruments distributed to the Rio Hondo acequia community. The final percent of respondents were averaged across the two surveys [55].

It is important to distinguish between modernization of irrigation *infrastructure* and modernization of irrigation *management*. While lining ditches, switching to drip, and supplemental groundwater pumping focus on using surface water more efficiently through engineering and infrastructure improvements, water managers and irrigators must be provided with tools, resources, and information that enable efficient and adaptive water management and allocation. One example of this is real-time monitoring accessible through a web interface which has been shown to increase adaptive capacity indicators within the Rio Hondo acequia community [55]. With water scarcity only becoming a more pressing issue in the Southwest within the context of climate change, it will be critical to continue evaluating the adaptability of water management and agricultural production approaches, reflect findings in new and transformative policy, and ask ourselves if we should be modernizing infrastructure or management to avoid falling into the irrigation efficiency paradox trap [78,79].

A key element for the success of acequia and other community-based irrigation systems is community water management system functionality (see Figure 5). To have a functioning community water management system, there must first be a community to manage and use the water, so individuals must see value in acequias or acequia irrigation. When researchers explored capital gained within acequia communities, they found that only about 30% of family income was generated from acequia agriculture and that external income helped sustain acequia-irrigated properties and agriculture [80]. Surveys and interviews revealed that connection to land, water, and community were the values that drove acequia community members to respond to adverse circumstances (e.g., economic hardship, population growth, drought, increased development), demonstrating that acequia communities are founded and fueled by values within the *moral* economy rather than the typical market economy [30,59,80]. Therefore, identifying appropriate irrigation modernization recommendations must consider irrigation community motivations or values and be tailored toward enabling water management system functionality. While acequias foster long-term resilience, short-term vulnerabilities that impact acequia irrigation are surface water shortages. More work is needed to assess the specific impacts of changing irrigation regimes and technologies in acequia regions and identify adaptations that optimize groundwater recharge while also taking declining surface water availability into consideration.

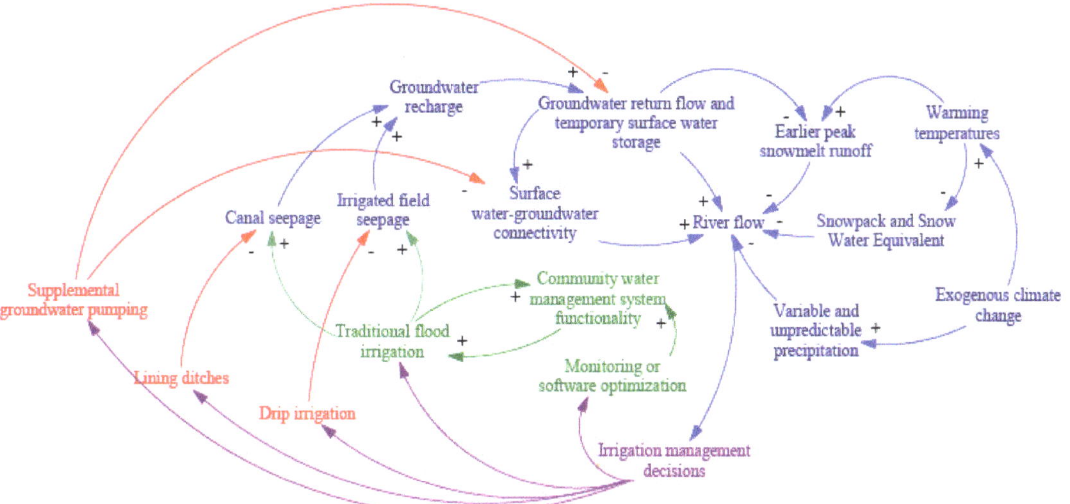

Figure 5. Causal loop diagram (CLD) showing the interactions and connections between environmental phenomena (blue), decision making (purple), adaptive management (green), and potentially maladaptive management (red) of acequia irrigation systems where traditional flood irrigation is assumed to only use surface water. Please note that the potentially maladaptive management options might be considered adaptive for other regions and irrigation regimes outside the scope of this paper.

5. Conclusions

In this study, we compared surface water and groundwater interactions and shallow aquifer response to irrigation events in two flood-irrigated forage grass fields located within the same acequia-irrigated valley in northern New Mexico, USA. Our results indicate that surface water and groundwater are tightly connected in this area, but variations in *DP* and groundwater response exists between land managers and fields due to differing flood irrigation scheduling and management. Additionally, while our results are consistent with previous water balance studies conducted in acequia-irrigated forage grass fields in northern New Mexico, this is the first paper to relate the findings from all the similar studies in the region since the first study was conducted in 2005. Because recharge acequia irrigation-related recharge eventually becomes delayed return flow to rivers [22], studies such as this are critical for determining how surface water and groundwater connectivity changes over time as it directly impacts surface water availability for downstream users [56]. We expect less *DP* will occur in acequia-irrigated fields if future climate change predictions and warming trends continue.

The *DP* and shallow groundwater recharge that occur as a byproduct of acequia flood irrigation are the exact management benefits needed for appropriate climate change adaptation. By maintaining recurring and consistent groundwater recharge, stream systems stay watered, which enables valley and regional cooperation between acequia-governing systems to continue. Alternatively, if acequia regions begin relying more heavily on groundwater pumping (for example), water tables would drop, making less water available in stream systems as surface water and groundwater become disconnected. These actions would propagate a cycle of maladaptation by undermining the hydrologic functions and community collaboration that make acequias so sustainable.

Author Contributions: Conceptualization, L.M.C., A.G.F. and C.G.O.; methodology, L.M.C., A.G.F. and C.G.O.; validation, L.M.C., A.G.F., S.J.G. and C.G.O.; formal analysis, L.M.C.; investigation, L.M.C., A.G.F. and C.G.O.; resources, L.M.C., A.G.F. and C.G.O.; data curation, L.M.C.; writing—original draft preparation, L.M.C.; writing—review and editing, L.M.C., A.G.F., S.J.G. and C.G.O.; visualization, L.M.C.; supervision, A.G.F. and C.G.O.; project administration, L.M.C. and A.G.F.; funding acquisition, A.G.F. All authors have read and agreed to the published version of the manuscript.

Funding: This research was funded by the state of New Mexico, through special and state appropriations made to the New Mexico Water Resources Research Institute NMWRRI2019-2020 and NMWRRI2021, and New Mexico State University's College of Agricultural, Consumer and Environmental Sciences.

Institutional Review Board Statement: Not applicable.

Informed Consent Statement: Not applicable.

Data Availability Statement: The data and statistical analysis presented in this study are openly available within the New Mexico Water Resources Research Institute's Data Set Repository at the following link: https://nmwrri.nmsu.edu/ds-002/ (accessed on 8 March 2022).

Acknowledgments: We would like to sincerely thank local stakeholders for their involvement, support, collaboration, and for their interest in understanding how irrigation practices relate to surface water and groundwater interactions.

Conflicts of Interest: The authors declare no conflict of interest. The funders had no role in the design of the study; in the collection, analyses, or interpretation of data; in the writing of the manuscript, or in the decision to publish the results.

References

1. Barnett, T.P.; Adam, J.C.; Lettenmaier, D.P. Potential impacts of a warming climate on water availability in snow-dominated regions. *Nature* **2005**, *438*, 303–309. [CrossRef] [PubMed]
2. Harpold, A.; Brooks, P.; Rajagopal, S.; Heidbuchel, I.; Jardine, A.; Stielstra, C. Changes in snowpack accumulation and ablation in the intermountain west. *Water Resour. Res.* **2012**, *48*, 1–11. [CrossRef]
3. Clow, D.W. Changes in the timing of snowmelt and streamflow in Colorado: A response to recent warming. *J. Clim.* **2010**, *23*, 2293–2306. [CrossRef]
4. Payne, J.T.; Wood, A.W.; Hamlet, A.F.; Palmer, R.N.; Lettenmaier, D.P. Mitigating the effects of climate change on the water resources of the Columbia River basin. *Clim. Change* **2004**, *62*, 233–256. [CrossRef]
5. Hurd, B. Climate vulnerability and adaptive strategies along the Rio Grande/Rio Bravo border of Mexico and the United States. *J. Contemp. Water Res. Educ.* **2012**, *149*, 56–63. [CrossRef]
6. Elias, E.H.; Rango, A.; Steele, C.M.; Mejia, J.F.; Smith, R. Assessing climate change impacts on water availability of snowmelt-dominated basins of the Upper Rio Grande basin. *J. Hydrol. Reg. Stud.* **2015**, *3*, 525–546. [CrossRef]
7. Chavarria, S.B.; Gutzler, D.S. Observed changes in climate and streamflow in the Upper Rio Grande basin. *J. Am. Water Resour. Assoc.* **2018**, *54*, 644–659. [CrossRef]
8. Llewellyn, D.; Vaddey, S. *West-Wide Climate Risk Assessment: Upper Rio Grande Impact Assessment Report*; U.S. Department of the Interior, Bureau of Reclamation, Upper Colorado Region, Albuquerque Area Office: Albuquerque, NM, USA, 2013.
9. Bai, Y.; Fernald, A.; Tidwell, V.; Gunda, T. Reduced and earlier snowmelt runoff impacts traditional irrigation systems. *J. Contemp. Water Res. Educ.* **2019**, *168*, 10–28. [CrossRef]
10. Hurd, B.H.; Coonrod, J. *Climate Change and Its Implications for New Mexico's Water Resources and Economic Opportunities*; Technical Report 45; New Mexico State University: Las Cruces, NM, USA, 2008.
11. Herrero, J.; Robinson, D.A.; Nogués, J. A regional soil survey approach for upgrading from flood to sprinkler irrigation in a semi-arid environment. *Agric. Water Manag.* **2007**, *93*, 145–152. [CrossRef]

12. Xu, X.; Huang, G.; Qu, Z.; Pereira, L.S. Assessing the groundwater dynamics and impacts of water saving in the Hetao Irrigation District, Yellow River basin. *Agric. Water Manag.* **2010**, *98*, 301–313. [CrossRef]
13. Magnuson, M.L.; Valdez, J.M.; Lawler, C.R.; Nelson, M.; Petronis, L. *New Mexico Water Use By Categories 2015*; New Mexico Office of the State Engineer: Santa Fe, NM, USA, 2019.
14. Fernald, A.; Tidwell, V.; Rivera, J.; Rodríguez, S.; Guldan, S.; Steele, C.; Ochoa, C.; Hurd, B.; Ortiz, M.; Boykin, K.; et al. Modeling sustainability of water, environment, livelihood, and culture in traditional irrigation communities and their linked watersheds. *Sustainability* **2012**, *4*, 2998–3022. [CrossRef]
15. Rivera, J.A.; Glick, T.F. The Iberian Origins of New Mexico's Community Acequias. In Proceedings of the XIII Economic History Congress of the International Economic History Association, Buenos Aires, Argentina, 22–26 July 2002.
16. Rosenberg, A.; Guldan, S.; Fernald, A.G.S.; Rivera, J. *Acequias of the Southwestern United States: Elements of Resilience in a Coupled Natural and Human System*; College of Agricultural, Consumer and Environmental Sciences, New Mexico State University: Las Cruces, NM, USA, 2020.
17. Guldan, S.J.; Fernald, A.G.; Ochoa, C.G.; Tidwell, V.C. Collaborative community hydrology research in northern New Mexico. *J. Contemp. Water Res. Educ.* **2013**, *152*, 49–54. [CrossRef]
18. Sabie, R.P.; Fernald, A.; Gay, M.R. Estimating Land Cover for Three Acequia-irrigated Valleys in New Mexico using Historical Aerial Imagery between 1935 and 2014. *Southwest. Geogr.* **2018**, *21*, 36–56.
19. Fernald, A.G.; Cevik, S.Y.; Ochoa, C.G.; Tidwell, V.C.; King, J.P.; Guldan, S.J. River hydrograph retransmission functions of irrigated valley surface water-groundwater interactions. *J. Irrig. Drain. Eng.* **2010**, *136*, 823–835. [CrossRef]
20. Hamilton, P. Groundwater and surface water: A single resource. *Water Environ. Technol.* **2005**, *17*, 37–41.
21. Kendy, E.; Bredehoeft, J.D. Transient effects of groundwater pumping and surface-water-irrigation returns on streamflow. *Water Resour. Res.* **2006**, *42*, 1–11. [CrossRef]
22. Ochoa, C.G.; Fernald, A.G.; Guldan, S.J.; Tidwell, V.C.; Shukla, M.K. Shallow aquifer recharge from irrigation in a semiarid agricultural valley in New Mexico. *J. Hydrol. Eng.* **2013**, *18*, 1219–1230. [CrossRef]
23. Bouimouass, H.; Fakir, Y.; Tweed, S.; Leblanc, M. Groundwater recharge sources in semiarid irrigated mountain fronts. *Hydrol. Process.* **2020**, *34*, 1598–1615. [CrossRef]
24. Li, D. Quantifying water use and groundwater recharge under flood irrigation in an arid oasis of northwestern China. *Agric. Water Manag.* **2020**, *240*, 106326. [CrossRef]
25. Rotiroti, M.; Bonomi, T.; Sacchi, E.; McArthur, J.M.; Stefania, G.A.; Zanotti, C.; Taviani, S.; Patelli, M.; Nava, V.; Soler, V.; et al. The effects of irrigation on groundwater quality and quantity in a human-modified hydro-system: The Oglio River basin, Po Plain, northern Italy. *Sci. Total Environ.* **2019**, *672*, 342–356. [CrossRef]
26. Gutiérrez-Jurado, K.Y.; Fernald, A.G.; Guldan, S.J.; Ochoa, C.G. Surface water and groundwater interactions in traditionally irrigated fields in Northern New Mexico, U.S.A. *Water* **2017**, *9*, 102. [CrossRef]
27. Ochoa, C.G.; Fernald, A.G.; Guldan, S.J.; Shukla, M.K. Deep percolation and its effects on shallow groundwater level rise following flood irrigation. *Am. Soc. Agric. Biol. Eng.* **2007**, *50*, 73–82.
28. Fernald, A.G.; Baker, T.T.; Guldan, S.J. Hydrologic, riparian, and agroecosystem functions of traditional acequia irrigation systems. *J. Sustain. Agric.* **2007**, *30*, 147–171. [CrossRef]
29. Fernald, A.G.; Guldan, S.J. Surface water-groundwater interactions between irrigation ditches, alluvial aquifers, and streams. *Rev. Fish. Sci.* **2006**, *14*, 79–89. [CrossRef]
30. Fernald, A.; Guldan, S.; Boykin, K.; Cibils, A.; Gonzales, M.; Hurd, B.; Lopez, S.; Ochoa, C.; Ortiz, M.; Rivera, J.; et al. Linked hydrologic and social systems that support resilience of traditional irrigation communities. *Hydrol. Earth Syst. Sci.* **2015**, *19*, 293–307. [CrossRef]
31. Jakeman, A.J.; Barreteau, O.; Hunt, R.J.; Rinaudo, J.; Ross, A. *Integrated Groundwater Management: Concepts, Approaches and Challenges*; Springer Open: Berlin/Heidelberg, Germany, 2016.
32. Fuchs, E.H.; King, J.P.; Carroll, K.C. Quantifying disconnection of groundwater from managed-ephemeral surface water during drought and conjunctive agricultural use. *Water Resour. Res.* **2019**, *55*, 5871–5890. [CrossRef]
33. Hashemi, H.; Berndtsson, R.; Persson, M. Artificial recharge by floodwater spreading estimated by water balances and groundwater modelling in arid Iran. *Hydrol. Sci. J.* **2015**, *60*, 336–350. [CrossRef]
34. Ghasemizade, M.; Asante, K.O.; Petersen, C.; Kocis, T.; Dahlke, H.E.; Harter, T. An integrated approach toward sustainability via groundwater banking in the southern Central Valley, California. *Water Resour. Res.* **2019**, *55*, 2742–2759. [CrossRef]
35. Winter, T.C.; Harvey, J.W.; Franke, O.L.; Alley, W.M. *Ground Water and Surface Water: A Single Resource*; USGS Circular 1139; U.S. Geological Survey: Reston, VA, USA, 1999.
36. Jaber, F.H.; Shukla, S.; Srivastava, S. Recharge, upflux and water table response for shallow water table conditions in southwest Florida. *Hydrol. Process.* **2006**, *20*, 1895–1907. [CrossRef]
37. Scanlon, B.R.; Healy, R.W.; Cook, P.G. Choosing appropriate techniques for quantifying groundwater recharge. *Hydrogeol. J.* **2002**, *10*, 18–39. [CrossRef]
38. Fleming, W.M.; Rivera, J.A.; Miller, A.; Piccarello, M. Ecosystem services of traditional irrigation systems in northern New Mexico, USA. *Int. J. Biodivers. Sci. Ecosyst. Serv. Manag.* **2014**, *10*, 343–350. [CrossRef]
39. WRCC Taos, New Mexico-Climate Summary. Available online: https://wrcc.dri.edu/cgi-bin/cliMAIN.pl?nm8668 (accessed on 11 June 2021).

40. Cox, M. Applying a social-ecological system framework to the study of the Taos Valley irrigation system. *Hum. Ecol.* **2014**, *42*, 311–324. [CrossRef]
41. Sabu, S. Modeling Acequia Water Use in the Rio Hondo Watershed. Master's Thesis, University of New Mexico, Albuquerque, NM, USA, March 2014.
42. United States Department of Agriculture, Natural Resources Conservation Service. *Custom Soil Resource Report for Valdez, New Mexico*; USDA-NRCS Web Soil Survey. Available online: http://websoilsurvey.nrcs.usda.gov/ (accessed on 28 July 2020).
43. United States Department of Agriculture, Natural Resources Conservation Service. *Custom Soil Resource Report for Arroyo Hondo, New Mexico*; USDA-NRCS Web Soil Survey. Available online: http://websoilsurvey.nrcs.usda.gov/ (accessed on 28 July 2020).
44. United States Department of Agriculture, Natural Resources Conservation Service. *New Jersey Irrigation Guide*; USDA-NRCS: Somerset, NJ, USA, 2005.
45. Samani, Z. Three simple flumes for flow measurement in open channels. *J. Irrig. Drain. Eng.* **2017**, *143*, 04017010-4. [CrossRef]
46. Blake, G.R.; Hartge, K.H. Bulk Density. In *Methods of Soil Analysis, Part 1. Physical and Mineralogical Methods*; Klute, A., Ed.; SSSA Book Ser. 5.1.; SSSA, ASA: Madison, WI, USA, 1986; pp. 377–382.
47. Gee, G.W.; Bauder, J.W. Particle-size Analysis. In *Methods of Soil Analysis, Part 1. Physical and Mineralogical Methods*; Klute, A., Ed.; SSSA Book Ser. 5.1.; SSSA, ASA: Madison, WI, USA, 1986; pp. 383–411.
48. Campbell Scientific, Inc. *On-Line Estimation of Grass Reference Evapotranspiration with the Campbell Scientific Automated Weather Station*; Application Note 4-D.; Campbell Scientific, Inc.: Logan, UT, USA, 1999.
49. Cevik, S.Y. A Long-Term Hydrological Model for the Northern Espanola Basin, New Mexico. Ph.D. Thesis, New Mexico State University, Las Cruces, NM, USA, December 2009.
50. Ochoa, C.G.; Fernald, A.G.; Guldan, S.J.; Shukla, M.K. Water movement through a shallow vadose zone: A field irrigation experiment. *Vadose Zo. J.* **2009**, *8*, 414–425. [CrossRef]
51. Moeser, C.D.; Chavarria, S.B.; Wootten, A.M. *Streamflow Response to Potential Changes in Climate in the Upper Rio Grande Basin*; USGS Scientific Investigations Report 2021-5138; U.S. Geological Survey: Reston, VA, USA, 2021.
52. Udall, B.; Overpeck, J. The twenty-first century Colorado River hot drought and implications for the future. *Water Resour. Res.* **2017**, *53*, 2404–2418. [CrossRef]
53. Rango, A. Snow: The real water supply for the Rio Grande basin. *New Mex. J. Sci.* **2006**, *44*, 99–118.
54. Lehner, F.; Wahl, E.R.; Wood, A.W.; Blatchford, D.B.; Llewellyn, D. Assessing recent declines in Upper Rio Grande runoff efficiency from a paleoclimate perspective. *Geophys. Res. Lett.* **2017**, *44*, 4124–4133. [CrossRef]
55. Conrad, L. Collaborative Community Hydrology: Integrating Stakeholder Engagement, Hydrology, and Social Indicators to Support Acequia Water Management in Northern New Mexico. Master's Thesis, New Mexico State University, Las Cruces, NM, USA, May 2022.
56. Rumsey, C.A.; Miller, M.P.; Sexstone, G.A. Relating hydroclimatic change to streamflow, baseflow, and hydrologic partitioning in the Upper Rio Grande Basin, 1980 to 2015. *J. Hydrol.* **2020**, *584*, 124715. [CrossRef]
57. Cruz, J.J.; Fernald, A.G.; VanLeeuwen, D.M.; Guldan, S.J.; Ochoa, C.G. River-ditch flow statistical relationships in a traditionally irrigated valley near Taos, New Mexico. *J. Contemp. Water Res. Educ.* **2019**, *168*, 49–65. [CrossRef]
58. Fernald, A.; Guldan, S.J.; Sabie, R. Traditional Hydro-Cultural Water Management Systems Increase Agricultural Capacity and Community Resilience in Drought Prone New Mexico, USA. In Proceedings of the American Geophysical Union Fall Meeting, Washington, DC, USA, 10–14 December 2018.
59. Fernald, A.; Ochoa, C.G.; Guldan, S.J. Coupled Natural and Human System Clues to Acequia Resilience. In *Acequias of the Southwestern United States: Elements of Resilience in a Coupled Natural and Human System*; Rosenberg, A., Guldan, S., Fernald, A.G., Rivera, J., Eds.; New Mexico State University, College of Agricultural, Consumer and Environmental Sciences: Las Cruces, NM, USA, 2020; pp. 72–78.
60. Tolley, D.G.; Frisbee, M.D.; Campbell, A.R. Determining the importance of seasonality on groundwater recharge and streamflow in the Sangre de Cristo Mountains using stable isotopes. In *Guidebook 66-Geology of the Las Vegas Area*; Lindline, J., Petronis, M., Zebrowski, J., Eds.; New Mexico Geological Society Guidebook, 66th Field Conference; Geology of the Meadowlands: Las Vegas, NM, USA, 2015; pp. 303–312.
61. Li, X.; Fernald, A.G.; Kang, S. Assessing long-term changes in regional groundwater recharge using a water balance model for New Mexico. *J. Am. Water Resour. Assoc.* **2021**, *57*, 807–827. [CrossRef]
62. Olmstead, S.M. Climate change adaptation and water resource management: A review of the literature. *Energy Econ.* **2014**, *46*, 500–509. [CrossRef]
63. Barnett, J.; O'Neill, S. Maladaptation. *Glob. Environ. Change* **2010**, *20*, 211–213. [CrossRef]
64. Biddle, J.C.; Baehler, K.J. Breaking bad: When does polycentricity lead to maladaptation rather than adaptation? *Environ. Policy Gov.* **2019**, *29*, 344–359. [CrossRef]
65. Christian-Smith, J.; Levy, M.C.; Gleick, P.H. Maladaptation to drought: A case report from California, USA. *Sustain. Sci.* **2015**, *10*, 491–501. [CrossRef]
66. Torabi, E.; Dedekorkut-Howes, A.; Howes, M. Adapting or maladapting: Building resilience to climate-related disasters in coastal cities. *Cities* **2018**, *72*, 295–309. [CrossRef]
67. Haddad, B.M.; Merritt, K. Evaluating regional adaptation to climate change: The case of California water. In *The Long-Term Economics of Climate Change*; Hall, D.C., Howarth, R.B., Eds.; Elsevier Science: Amsterdam, The Netherlands, 2001; pp. 65–93.

68. McCaffrey, S.C. The need for flexibility in freshwater treaty regimes. *Nat. Resour. Forum* **2003**, *27*, 156–162. [CrossRef]
69. Jacobs, K.L.; Snow, L. Adaptation in the water sector: Science & institutions. *Daedalus* **2015**, *144*, 59–71. [CrossRef]
70. Sultana, P.; Thompson, P.M. Adaptation or conflict? Responses to climate change in water management in Bangladesh. *Environ. Sci. Policy* **2017**, *78*, 149–156. [CrossRef]
71. Cody, K.C. Flexible water allocations and rotational delivery combined adapt irrigation systems to drought. *Ecol. Soc.* **2018**, *23*, 47. [CrossRef]
72. Carlsson, L.; Berkes, F. Co-management: Concepts and methodological implications. *J. Environ. Manag.* **2005**, *75*, 65–76. [CrossRef]
73. Dillon, P.; Escalante, E.F.; Megdal, S.B.; Massmann, G. Managed Aquifer Recharge for Water Resilience. *Water* **2020**, *12*, 1846. [CrossRef]
74. Kwoyiga, L.; Stefan, C. Institutional feasibility of managed aquifer recharge in northeast Ghana. *Sustainability* **2019**, *11*, 379. [CrossRef]
75. Escalante, E.F.; Sebastián Sauto, J.S.; Gil, R.C. Sites and indicators of MAR as a successful tool to mitigate climate change effects in spain. *Water* **2019**, *11*, 1943. [CrossRef]
76. Pérez-Blanco, C.D.; Loch, A.; Ward, F.; Perry, C.; Adamson, D. Agricultural water saving through technologies: A zombie idea. *Environ. Res. Lett.* **2021**, *16*, 114032. [CrossRef]
77. Gunda, T.; Turner, B.L.; Tidwell, V.C. The influential role of sociocultural feedbacks on community-managed irrigation system behaviors during times of water stress. *Water Resour. Res.* **2018**, *54*, 2697–2714. [CrossRef]
78. Grafton, R.Q.; Williams, J.; Perry, C.J.; Molle, F.; Ringler, C.; Steduto, P.; Udall, B.; Wheeler, S.A.; Wang, Y.; Garrick, D.; et al. The paradox of irrigation efficiency. *Science* **2018**, *361*, 748–750. [CrossRef] [PubMed]
79. Bai, Y.; Langarudi, S.P.; Fernald, A.G. System dynamics modeling for evaluating regional hydrologic and economic effects of irrigation efficiency policy. *Hydrology* **2021**, *8*, 61. [CrossRef]
80. Mayagoitia, L.; Hurd, B.; Rivera, J.; Guldan, S. Rural community perspectives on preparedness and adaptation to climate-change and demographic pressure. *J. Contemp. Water Res. Educ.* **2012**, *147*, 49–62. [CrossRef]

 *hydrology*

Article

Soil Water Balance and Shallow Aquifer Recharge in an Irrigated Pasture Field with Clay Soils in the Willamette Valley, Oregon, USA

Daniel G. Gómez [1,2], Carlos G. Ochoa [1,*], Derek Godwin [1,3], Abigail A. Tomasek [4] and María I. Zamora Re [3]

1 Ecohydrology Lab, College of Agricultural Sciences, Oregon State University, Corvallis, OR 97331, USA; gomezdan@oregonstate.edu (D.G.G.); derek.godwin@oregonstate.edu (D.G.)
2 Water Resources Graduate Program, Oregon State University, Corvallis, OR 97331, USA
3 Department of Biological and Ecological Engineering, Oregon State University, Corvallis, OR 97331, USA; maria.zamorare@oregonstate.edu
4 Department of Crop and Soil Science, Oregon State University, Corvallis, OR 97331, USA; abigail.tomasek@oregonstate.edu
* Correspondence: carlos.ochoa@oregonstate.edu

Abstract: Quantifying soil water budget components, and characterizing groundwater recharge from irrigation seepage, is important for effective water resources management. This is particularly true in agricultural fields overlying shallow aquifers, like those found in the Willamette Valley in western Oregon, USA. The objectives of this two-year study were to (1) determine deep percolation in an irrigated pasture field with clay soils, and (2) assess shallow aquifer recharge during the irrigation season. Soil water and groundwater levels were measured at four monitoring stations distributed across the experimental field. A water balance approach was used to quantify the portioning of different water budget components, including deep percolation. On average for the four monitoring stations, total irrigation applied was 249 mm in 2020 and 381 mm in 2021. Mean crop-evapotranspiration accounted for 18% of the total irrigation applied in 2020, and 26% in 2021. The fraction of deep percolation to irrigation was 28% in 2020 and 29% in 2021. The Water Table Fluctuation Method (WTFM) was used to calculate shallow aquifer recharge in response to deep percolation inputs. Mean aquifer recharge was 132 mm in 2020 and 290 mm in 2021. Antecedent soil water content was an important factor influencing deep percolation. Study results provided essential information to better understand the mechanisms of water transport through the vadose zone and into shallow aquifers in agricultural fields with fine-textured soils in the Pacific Northwest region in the USA.

Keywords: water balance; water table fluctuation method; irrigated pastures; deep percolation; aquifer recharge; clay soils

Citation: Gómez, D.G.; Ochoa, C.G.; Godwin, D.; Tomasek, A.A.; Zamora Re, M.I. Soil Water Balance and Shallow Aquifer Recharge in an Irrigated Pasture Field with Clay Soils in the Willamette Valley, Oregon, USA. *Hydrology* **2022**, *9*, 60. https://doi.org/10.3390/hydrology9040060

Academic Editors: Ilmoon Chung, Sunwoo Chang, Yeonsang Hwang and Yeonjoo Kim

Received: 9 March 2022
Accepted: 31 March 2022
Published: 4 April 2022

Publisher's Note: MDPI stays neutral with regard to jurisdictional claims in published maps and institutional affiliations.

1. Introduction

Numerous investigations have demonstrated that irrigation can lead to deep percolation, and recharge shallow aquifers while also providing return flows to nearby streams [1–4]. Deep percolation is highly dependent on soil physical characteristics, extraction patterns of the roots, ponding time at the surface, and depth to the water table [5–7]. Clay soils are especially important because their high field capacity allows for more water storage while their lower transmissivity rates slow water percolation through the soil profile, thereby potentially increasing water lost to evapotranspiration [8,9]. By contrast, clay soils are sensitive to drying and wetting cycles that can create cracks in the soil profile and cause macropore flow paths that rapidly deliver water, nutrients, and pollutants down to the water table [10].

As the western USA continues to experience exceptional drought, it is imperative to understand the relationship between water use and transport. Greater understanding of surface water–groundwater interactions (SW-GW) will be essential for farmers and water managers to better estimate field water budget components, recharge-to-irrigation ratios, and potential pollutant leaching (e.g., nitrogen), while improving overall water management decisions affecting irrigation water supply and return flows to surface water and groundwater reservoirs [11,12]. The water balance method (WBM) can be used to estimate deep percolation below the root zone when reliable field observations are available. In many studies, deep percolation has been associated with aquifer recharge estimates [5,13]. Groundwater recharge is commonly quantified using approaches such as the Water Table Fluctuation Method (WTFM) [13,14]. The WTFM is often used because water level data is relatively easy to measure, and the WTFM assumes that rises in the water table are caused by actual recharge [14,15]. The method relies on the specific yield of an aquifer, defined as "the volume of water released from a unit volume of saturated aquifer material drained by a falling water table," multiplied by changes in the water level [14]. Recharge estimates using the WTFM are based on the premise that observed groundwater-level rises are directly related to irrigation or precipitation recharge arriving to the water table [13,16]. The WFTM's limitations are, firstly, the difficulty in obtaining an accurate specific yield value for a particular aquifer and, secondly, that specific yield varies by depth [15,17].

The specific connections between SW-GW, as they relate to water transport through the vadose zone and into the shallow aquifer, have not been fully explored in pasturelands of the Willamette River Basin. In this investigation, soil physical properties (e.g., soil texture and bulk density) and soil water content were utilized to assess water movement through the vadose zone and into the shallow aquifer of an irrigated, livestock-grazed pastureland in the Willamette Valley in western Oregon, USA. Objectives of this two-year study were to (1) determine deep percolation in an irrigated pasture field with clay soils, and (2) assess shallow aquifer recharge during the irrigation season.

2. Materials and Methods

2.1. Site Description

This two-year (2020 and 2021) study was conducted in a 2.1 ha pasture field (44.568 Lat.; 123.301 Long.) at the Oregon State University (OSU) Dairy Center in Corvallis, Oregon, USA. The site is located in western Oregon, in the Willamette Valley. The pasture field drains south, and is bordered by a discharge channel to the west, gravel roads to the north and east, and Oak Creek to the south (Figure 1). The field is irrigated with water pumped from Oak Creek. Streamflow in the discharge channel is negligible during the summer. No groundwater pumping for agriculture exists in the area. Depending on streamflow and soil water conditions following winter precipitation, irrigation at the OSU Dairy Center typically starts in early summer. However, due to modifications conducted on the irrigation pipes at the stream pumping site, the onset of the 2020 irrigation was delayed several weeks. The irrigation seasons ran from 27 July to 12 September in 2020, and from 15 June to 9 September in 2021. Vegetation at the study site included a mixture of balansa clover (*Trifolium michelianum balansae*), subterranean clover (*Trifolium subterraneum*), white clover (*Trifolium repens*), perennial ryegrass (*Lolium perenne*), annual ryegrass (*Lolium multiflorum*), orchard grass (*Dactylis glomerata*), and common chicory (*Cichorium intybus*) that was used for dairy cattle grazing in late spring and summer. Two soils series, as described in the USDA official series description [18], were present at the study site: Bashaw clay covered 55.8% of the experimental field while Holcomb silt loam covered 44.2%. Both soil series show slope values of 0% to 3%. Average depth to water table ranges between 0 to 76 mm in the winter rainy season, while the drainage class falls within the 'somewhat-poorly-drained' category for the Bashaw Clay and 'poorly-drained' category for the Holcomb silt loam [18]. Depth to water table, measured at the lowest level at the onset of the 2020 irrigation season, ranged between 1.2 and 1.6 m. The region has a Mediterranean type of climate with a warm and dry season in the summer and a mild and wet winter season. Most precipitation occurs as rainfall between November and April. Mean annual precipitation in the basin ranges from 2500 mm at higher elevations to 1000 mm in the valley where our study site was located. The monthly-averaged lowest temperature happens in January (0.67 °C), while the highest occurs in August (27.4 °C). The lowest and highest total monthly precipitations happen in July (9.1 mm) and December (181.4 mm), respectively, [19].

Figure 1. Study site illustrating the location of the four monitoring stations used to measure soil water content and groundwater levels. Study site (44.568 Lat.; −123.301 Long.) is in Benton County, Oregon, USA.

2.2. Field Data Collection

Multiple field-based water, soil, and weather data were used to determine soil properties, calculate the field water budget, and estimate shallow aquifer recharge during the two years of the experiment.

2.2.1. Soil Water Content and Soil Physical Properties

Four soil water stations (North, South, West, and East) were installed on the pasture field (see Figure 1). Each station included a vertical network of three soil volumetric water content (θ) sensors (Model CS455, Campbell Scientific Inc., Logan, UT, USA) installed at 0.2, 0.5, and 0.8 m depths. At each station, the sensors were connected to a CR300 datalogger (Campbell Scientific Inc., Logan, UT, USA) programed to hourly record θ data. Three soil samples were collected at each sensor depth for characterizing soil physical properties (i.e., dry bulk density (ρ_b) and texture). The samples were obtained using a split soil core sampler (50 mm × 100 mm) (AMS Inc., American Falls, ID, USA). Soil samples at 0.2 and 0.5 m depths were collected from 43 additional locations spaced every 25 m to create a grid covering the entire pasture field. All soil samples were analyzed for ρ_b, using the core method [20], and for soil texture, using the hydrometer method [21].

Ordinary Kriging (OK), a geospatial interpolation method using ArcGIS Pro (version 2.8; Redlands, CA, USA), was used to project clay content distribution across the entire pasture field at selected dates during the irrigation season. Clay content was chosen as the soil texture variable of interest due to its influence on water-holding capacity, and therefore in θ. The assumption, in using the OK method, was that statistical and spatial relationships among measured points exist, and value predictions in neighboring spaces are possible due to the existence of spatial correlations.

2.2.2. Irrigation

The pasture field was irrigated using a pod sprinkler system (K-Line North America© 2016). The K-Line irrigation system consisted of two lines of sprinkler pods extending from two irrigation pipe risers in the center of the field. One line had 9 sprinkler pods while the other had 8, and each line was rotated approximately every 24 h to cover the entire field in four days. After a 3-day resting period, a new 4-day irrigation cycle would begin. When the soil conditions were drier, the sprinklers were kept running for about 48 h before being moved to the next location within the field. For example, due to modifications conducted on the irrigation pipes at the pumping site in the creek, the onset of the 2020 irrigation was delayed by several weeks. As a result, initial soil conditions were much drier, and each subsection was initially irrigated for 48 h to raise soil moisture conditions to an adequate level. Four transects, consisting of metal and plastic containers (108 mm diameter), were used as water-collectors to measure the amount of water applied during each 24 or 48 h irrigation application. The water-collectors were placed at 1.5, 3, 4.5, 6, 7.5, and 9 m from the center of two sprinklers on each line. The location of the water-collector transects was rotated to different sprinklers throughout the season to capture potential water-application variability. An additional plastic gauge was installed at each soil water station to measure the amount of irrigation water reaching the sensors' location. Regardless of the irrigation application duration (24 vs. 48 h), all water-collectors were measured approximately every 24 h. In addition to the daily-measured irrigation applications, an in-flow meter (UltraMag; McCrometer Inc., Hemet, CA, USA) was installed in each of the two irrigation lines to measure total water application during the 2021 irrigation season.

2.2.3. Groundwater Level

Data from shallow monitoring wells were used to characterize irrigation season aquifer recharge during the two years evaluated. One well was installed next to each of the θ stations at approximately 5 m deep and consisted of 50 mm diameter PVC pipes with a 1.5 m screen section in the bottom. These wells were equipped with CTD-10 (Decagon Devices Inc., Pullman, WA, USA) water level sensors. All water level sensors were connected to the CR300 dataloggers in each location and were programmed to hourly record water level data. All wells were surveyed to determine soil-surface and water-table elevations.

2.3. Soil Water Balance Method (SWBM)

A soil water balance approach was used to calculate *DP*, defined as the water passing below the 0.8 m sensor depth:

$$DP = IRR + P - RO - DS - AET \tag{1}$$

where DP = deep percolation (mm); IRR = irrigation depth (mm); P = precipitation (mm); DS = change in soil water storage (mm); RO = field runoff (mm); and AET = actual evapotranspiration (mm). DP was calculated for each soil-monitoring station, following individual irrigation applications to the corresponding subsection being irrigated. IRR was obtained from the water-collector-measured irrigation applications. P was obtained from the weather station records; no quantifiable RO occurred during either irrigation season. DS was calculated for each sensor depth (0.2, 0.5, and 0.8 m) based on the θ difference between the onset of irrigation and 48 h after the end of irrigation. DS was then averaged across all three sensor depths to represent the entire 0.8 m profile. AET was calculated using the reference ETo for short grass estimated by the weather station and multiplied by crop coefficient (Kc) values (0.25 to 0.68) developed by the USDA Agriculture Research Service for pastures in the Columbia-Pacific Northwest Region [22]. All the individual SWBM results obtained for each irrigation application were aggregated each year to obtain an overall seasonal SWBM estimate. Figure 2 illustrates the soil water- and groundwater-level instrumentation, and the water budget components evaluated.

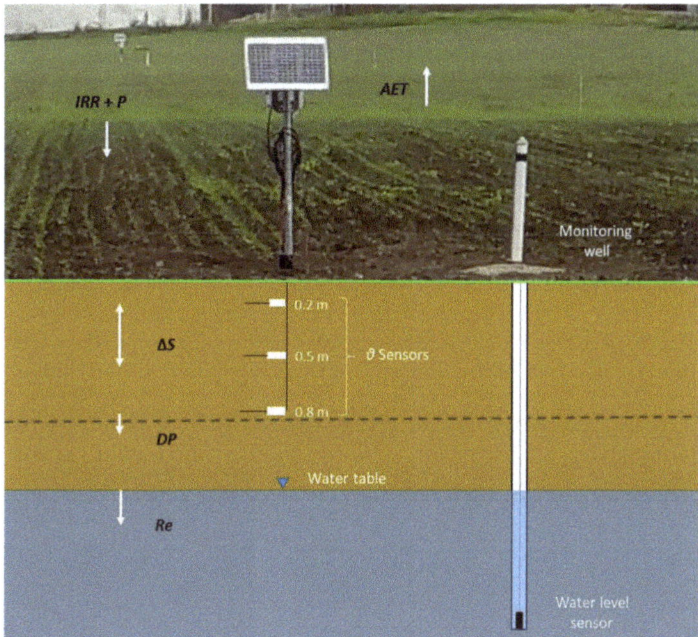

Figure 2. Schematic illustrating soil water (θ) and groundwater-levels measurement at each monitoring station. The main water budget components–irrigation (*IRR*), precipitation (*P*), change in soil water (ΔS), deep percolation (*DP*), actual evapotranspiration (*AET*), and aquifer recharge (*Re*)–are shown. Figure not to scale.

2.4. Shallow Aquifer Recharge

Groundwater-level fluctuations during the irrigation season were characterized using data from each monitoring well on the pasture field. The recharge (*Re*, in mm) of the shallow aquifer was calculated using the groundwater-level data from the wells in the pasture field, and the WTFM,

$$Re = \Delta h \times Sy \tag{2}$$

where *Re* = aquifer recharge (mm); Δh = change in water level (mm); and *Sy* = specific yield of the unconfined aquifer. Data recorded by the water level sensors installed in the wells were used to determine changes in water level. As described in Sophocleous [14], dividing the potential recharge values (i.e., *DP*) by the associated rises in the water table over several events can provide a "site-calibrated effective storativity value". A mean *Sy* value of 0.06, with most values ranging from 0.03 to 0.06, was estimated based on the *DP* events observed during the two irrigation seasons. The exception was a *Sy* value of 0.13 obtained for the South well location in 2020. Our field based *Sy* values were similar to the *Sy* mean value of 0.06, and ranging from 0.01 to 0.18, for clay materials reported in Dingman [23].

2.5. Statistical Analyses

An Analysis of Variance (ANOVA) was performed to explore the relationships of seasonal aquifer recharge, *Re*, observed at each monitoring well in 2020 and 2021. For each soil-monitoring station, we also conducted a linear-regression analysis to test the relationship between total water applied (*TWA*) and *DP*. Our previous research [2,5] had shown that antecedent soil water can have a significant effect on deep percolation. Therefore, we also ran the linear-regression analysis, subtracting ΔS from *TWA*. SigmaPlot® version 14.0 (Systat Software Inc., San Jose, CA, USA) was used for all the statistical analyses.

3. Results

3.1. Soil Properties

Soil texture and ρ_b varied across soil water stations and sensor depths (Table A1). Finer-texture soils were found in the East and West stations in the middle of the field. Clay loam soils were found for all sensor depths at the top (North station) and at 0.5 and 0.8 m depths at the bottom (South station) of the field. Coarser (54% sand) texture was observed at 0.2 m depth in the South station, which also had the lowest ρ_b values.

Field-scale clay content distribution analysis showed the highest clay content in the middle of the field (Figure 3). This was the case for the 0.5 m depth, with values ranging from 32% to 35% near the North station and from 35% to 38% near the South station (Figure 3b). In addition, higher clay content values (38% to 41%) were estimated for areas near the stations in the middle of the field, which were consistent with the 43% and 45% values obtained for the East and West stations. More discrepancy was observed at the 0.2 m depth. Although higher than the clay content values at the top and bottom of the field, the highest values of 32% to 35% observed in the middle of the field were below the 43% and 45% values obtained at the East and West stations, respectively, (Figure 3a; Table A1).

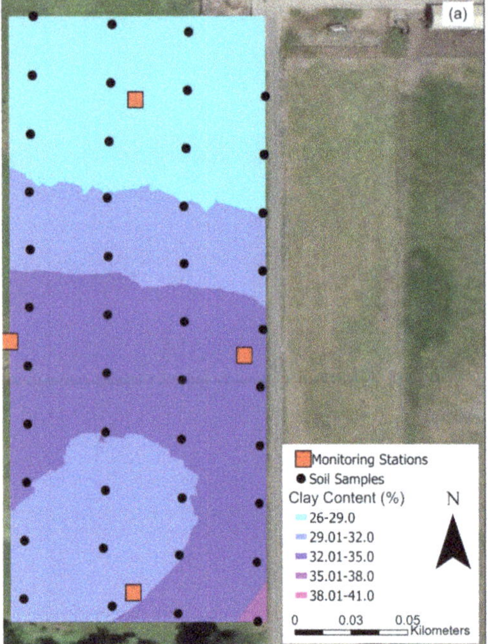

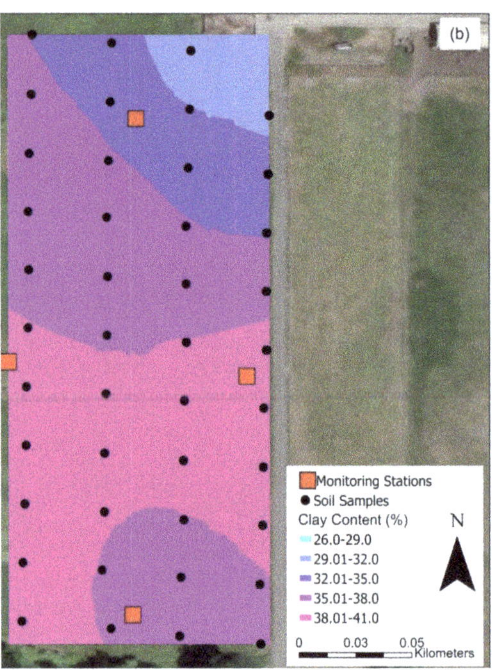

Figure 3. Clay distribution using data from the 43 soil samples and monitoring stations using Ordinary Kriging at the (**a**) 0.2 m depth and (**b**) 0.5 m depth.

3.2. Soil Water Balance

DP was variable among stations and irrigation seasons (Table 1). At the end of the 2020 season, the South station had the largest amount of cumulative *DP* (98 mm), followed by the West (94 mm), East (69 mm), and North (20 mm) stations (Table 1). The amount of *DP* for each station did not appear to be related to the cumulative amount of *TWA* (*IRR* + *P*), as the East station had the largest total *IRR* (280 mm) for 2020 but resulted in the second lowest total *DP* amount (69 mm). In comparison, the West station *IRR* was 214 mm (the smallest of the season) resulting in 94 mm of *DP* (the second largest). For the 2021 irrigation period, the most groundwater recharge through *DP* occurred in the West station (153 mm), followed by the East (101 mm), South (99 mm), and North station (92 mm). The greatest amount of *TWA* during this season corresponded to the South station (391 mm), which had the second lowest amount of *DP* (99 mm). By comparison, the North station *TWA* was 368 mm (smallest of the season), and resulted in 92 mm of *DP*, the lowest of the season. Total *DP* across all stations in 2020 was 281 mm, and *TWA* was 1,006 mm, while in the longer 2021 season, *DP* was 445 mm with a *TWA* of 1,524 mm (Table 1). The statistical analysis showed that the relationships between *DP* and *TWA* alone were weak ($p > 0.05$) for all stations during both irrigation seasons. However, the role of antecedent soil water, as reflected in ΔS, was an important factor in estimating *DP*. The regression analysis conducted for all irrigation applications in each year showed a positive linear relationship ($p \leq 0.05$; R^2 values from 0.72 to 0.99) between *DP* and *TWA* - ΔS for all soil stations in 2020, but not for the East station in 2021. Overall, seasonal *TWA* values were higher than ΔS, with the fraction of ΔS to *TWA* ranging from 0.4% to 0.8%. The exception was the North station in 2020 when ΔS exceeded *TWA*. The monitoring station was toward the bottom of a more pronounced slope in that part of the field. We believe that potential oversaturation of the soil profile at this station may have occurred during the longer 48 h irrigation events used to raise soil water conditions at the beginning of the 2020 season.

Table 1. Water budget component results for each soil monitoring station (North, West, East, and South) during the 2020 and 2021 irrigation seasons. The water budget components are irrigation (*IRR*), precipitation (*P*), total change in soil water (ΔS), evapotranspiration (*AET*), and deep percolation (*DP*). The season-cumulative results for all irrigation applications (n) for each station are shown. All results are reported in mm.

Year	Station	n	IR	P	DS	AET	DP
2020	North	6	250	9	302	42	20
	West	6	214	9	114	45	94
	East	6	280	9	238	52	69
	South	6	253	9	118	45	98
2021	North	13	368	0	202	96	92
	West	13	386	0	150	99	153
	East	13	379	0	233	95	101
	South	13	391	0	285	99	99

3.3. Groundwater Levels and Aquifer Recharge

A gradual rise in groundwater levels, following the onset of the irrigation season, was observed in all four wells during both years. Overall, groundwater levels remained higher (up to 1.2 m) than initial conditions through both irrigation seasons. Groundwater levels in all wells rose from the beginning to the end of the 2020 irrigation season, while in 2021 the North and East wells' groundwater levels rose throughout the summer (Figure 4a,b). A relatively moderate rise and decline in groundwater levels was observed after all six irrigation applications at each well location in 2020. The exception was the North well, following the first irrigation (Figure 4). The more frequent and larger number (n = 13) of irrigation applications in 2021 resulted in relatively sharp groundwater level rises and declines in all wells. The South well appeared to show groundwater level increases even when no direct on-site water applications occurred. This was attributed to the location of the well at the bottom of the pasture, which may have been affected by lateral flow contributions from the irrigation in other parts of the field. The groundwater level rises observed in the West well illustrate the saturated-ponding conditions that occurred during several irrigation events in 2021 when peak water-level-rise plateaued for several hours before receding. During the 2020 irrigation season, groundwater levels in the West well showed a diurnal oscillation that was attributed to water uptake from the riparian vegetation present in a storm discharge channel adjacent to the experimental pasture field. Heavier clay content conditions and a lower terrain were observed at this West well location.

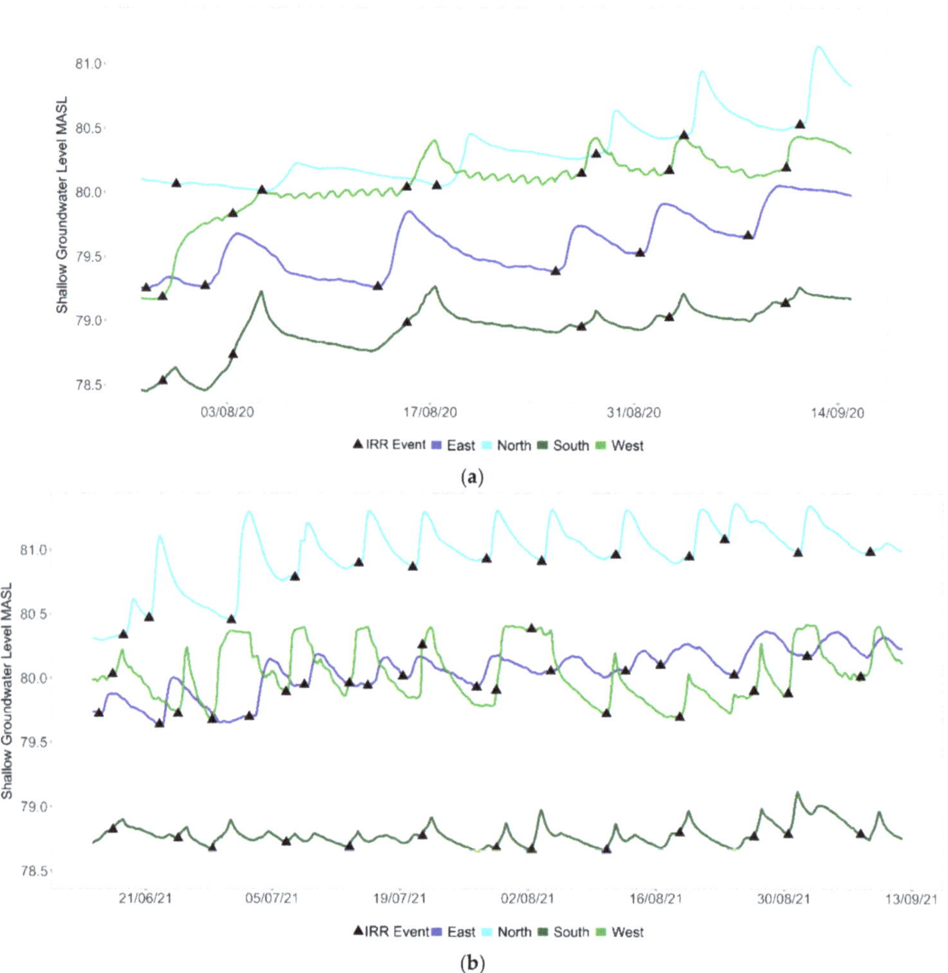

Figure 4. Daily averaged groundwater level in meters-above-sea-level (MASL) at the four monitoring wells on the experimental field during the (**a**) 2020 and (**b**) 2021 irrigation seasons. The irrigation events (*IRR*) for the area near each well location are also illustrated.

The substantially fewer number of *IRR* events at each soil station in 2020 (mean of 449 mm) compared to 2021 (mean of 568 mm) was reflected in the lower *Re* values obtained for each well location. On average, total seasonal aquifer recharge was 132 mm in 2020 and 290 mm in 2021. The *Re* estimates for each monitoring well were less variable in 2020 than in 2021 (Table 2). In 2020, total *Re* ranged from 128 mm in the North well to 137 mm in the East and West wells. In 2021, total *Re* ranged from 190 mm in the West well to 352 mm in the East well. The ANOVA results showed that mean seasonal *Re* was not significantly different ($p > 0.05$) among all four wells in 2020. However, a Kruskal–Wallis One-Way ANOVA on Ranks showed seasonal *Re* was significantly different ($p \leq 0.05$) for West vs. South and North vs. South wells.

Table 2. Irrigation season total groundwater recharge (*Re*) and total irrigation (*IRR*) for each monitoring well location. All measurements are reported in mm.

Station	2020		2021	
	Re	*IRR*	*Re*	*IRR*
North	128	358	340	438
East	137	526	352	460
West	137	385	190	628
South	130	528	278	745

4. Discussion

This investigation explored the connections between surface water and shallow groundwater as they relate to irrigation water transport through the vadose zone into the shallow aquifer. Study results indicated that soil physical attributes observed across the field played an important role in *DP* and aquifer response. A relatively shallow aquifer (<2 m) and preferential flow may have contributed to the rapid transport of irrigation water through the soil profile and into the shallow aquifer. Similar to that reported in other studies [2,24], irrigation frequency and antecedent soil water content were important variables affecting *DP* and shallow aquifer recharge. Differences in the *TWA* and frequency of irrigation were important for the greater *DP* and shallow aquifer recharge estimates observed at the end of the irrigation season in the second year of the experiment. The higher frequency of irrigation events observed during the 2021 irrigation season helped maintain higher soil-moisture levels and, consequently, more *DP* and aquifer recharge than in the previous year. Irrigation frequency and irrigation duration differences between the two seasons likely impacted *DP* distinctly, as research has shown that even an hour's break in irrigation causes the upper part of the soil profile to desaturate and allows for air to enter the soil profile and to fill pores, disrupting *DP* [25]. At the beginning of both seasons, *TWA* was slow to saturate the soil, with most of the water stored in the soil profile, and consequently less *DP* was observed. Antecedent soil water content was important in the *DP* estimates obtained for individual irrigation events. The change in soil water content (ΔS) made up 47% to 85% of total irrigation in 2020, and 39% to 73% in 2021. The effects of *TWA* for individual irrigation events were less apparent, and no strong relationships between irrigation amount and *DP* were obtained. This was partly attributed to the relatively low amounts of water applied (mean = 29 mm) during each irrigation event. At the beginning of both irrigation seasons, groundwater level rises occurred before the soil sensors' profile (upper 0.8 m) reported saturation.

Overall, cumulative *Re* estimates during both irrigation seasons were greater than *DP*. This is different from the results of a previous study [5] conducted in flood-irrigated fields with highly permeable soils in the southwestern USA. The larger *Re* estimates obtained in this study were attributed to irrigation water moving down the soil profile to the shallow aquifer through macropore flow paths, which is a behavior that has been documented in other deep percolation investigations (e.g., [24,26]). We credited the potential presence of macropore flow paths to the soil physical properties of the experimental field. A study conducted by [27] in a field near our study site revealed the presence of macropore flow due to shrinking and swelling of the clayey soil, paths caused by earth worms, and tunnels dug by voles. The transient water-contributions to the shallow aquifer from recently applied irrigations in neighboring sections of the field may have resulted in the larger *Re* values observed when compared to the *DP* values based on soil water estimates. The large differences in *Re* vs. *DP* obtained using different techniques (SWBM vs. WTFM) indicate caution ought to be exercised when assessing irrigation contributions to the shallow groundwater system.

The potential presence of preferential flow paths could also help explain the difference in *DP* and *Re* calculations between the stations during both seasons. For example, both the West (40.6%) and East (43.8%) stations had the highest average clay content but never had the lowest *DP* or *Re* calculations. The higher clay content may have resulted in a higher number of preferential pathways available for water to rapidly reach the shallow aquifer [9] while minimizing the amount of water lost to *ET* and stored in the soil profile. Rapid responses from the water table were observed throughout the season even when the soil reached saturation and cracks were no longer visible in the soil surface, which has been shown to stop preferential flow due to soil swelling [28]. However, previous research [29] has shown that sprinkler irrigation may only partially close soil cracks, maintaining an avenue for rapid water transport. Preferential flow paths may also persist in the deeper levels of a saturated soil subsurface [30] while others may only partially close or remain fully open throughout the soil profile [31]. The presence of preferential flow paths could be determined using lab [32] or field [33] dye experiments. The existence and role of preferential flow paths in soil water and groundwater recharge at this experimental pasture field remain unknown: further investigation may be needed to determine if that is a condition that can help explain the shallow groundwater-level fluctuations and *Re* values obtained in this study.

Study limitations included a later start in the 2020 irrigation season than in 2021. This caused adjustments in the irrigation scheduling during the first year (a smaller number of irrigation events, longer duration) to compensate for the drier soil conditions observed at the beginning of the season. In addition, the type of rotational irrigation system (pod sprinkler) employed did not allow for full irrigation cover across the entire field at the same time. The field was separated in four subsections, and *DP* and *Re* estimated for each individual subsection. Therefore, no direct comparison of specific *DP* events and *Re* was conducted. Instead, we calculated *Re* based on the various distinct water table rises and declines observed during each irrigation season. The soil water content sensors used in this study were not calibrated for site-specific soil characteristics. It is possible that some of the SWBM calculations (i.e., ΔS) may have been affected by potential discrepancies in θ readings.

Beyond improved measurements for the water balance and aquifer recharge methodology employed, results from this investigation contribute towards understanding of the relationships between irrigation, soil water, and shallow aquifer recharge in agricultural fields with fine-textured soils in the Pacific Northwest region of the USA. These findings can be used to aide water managers and farmers in similar agroecological systems, in determining the effect of irrigation on solute transport, groundwater recharge, and possibly streamflow augmentation during critical baseflow periods.

Author Contributions: C.G.O., D.G.G. and D.G. developed the study design and conducted field data collection. All authors contributed to data analyses and the writing of the manuscript. All authors have read and agreed to the published version of the manuscript.

Funding: This research received no external funding.

Institutional Review Board Statement: Not applicable.

Informed Consent Statement: Not applicable.

Acknowledgments: The authors gratefully acknowledge the various graduate and undergraduate students from Oregon State University who helped with field data collection activities related to the results presented here. The study was supported by the Oregon Agricultural Experiment Station.

Conflicts of Interest: The authors declare no conflict of interest.

Appendix A Soil Physical Properties and Textural Classification

Table A1. Soil physical properties by monitoring station: by depth, including mean values and standard error of the three samples collected at each depth to determine dry soil bulk density (ρ_b); soil particle distribution of sand, silt, and clay; and soil texture. Soil particle distribution was calculated using an aggregate of the three samples collected at each depth.

Station	Soil Depth (m)	ρ_b (Mg m^{-3})	Clay (%)	Silt (%)	Sand (%)	Soil Texture
North	0.2	1.5 ± 0.01	30.1	45.9	24.0	clay loam
	0.5	1.5 ± 0.01	30.7	39.9	29.3	clay loam
	0.8	1.4 ± 0.004	36.1	19.3	44.7	clay loam
West	0.2	1.5 ± 0.04	44.9	30.5	24.7	clay
	0.5	1.4 ± 0.02	43.5	43.1	13.3	silty clay
	0.8	1.6 ± 0.01	33.5	57.1	9.3	silty clay loam
East	0.2	1.7 ± 0.06	42.9	35.8	21.3	clay
	0.5	1.7 ± 0.04	44.2	32.5	23.3	clay
	0.8	1.6 ± 0.03	44.2	34.5	21.3	clay
South	0.2	1.0 ± 0.01	21.4	24.6	54.0	sandy clay loam
	0.5	1.1 ± 0.01	34.1	26.6	39.3	clay loam
	0.8	1.2 ± 0.02	39.4	23.3	37.3	clay loam

References

1. Arnold, L.R. *Estimates of Deep-Percolation Return Flow beneath a Flood- and a Sprinkler-Irrigated Site in Weld County, Colorado, 2008–2009*; U.S. Geological Survey Scientific Investigations Report 2011–5001; US Geological Survey: Reston, VA, USA, 2011; 225p.
2. Gutiérrez-Jurado, K.Y.; Fernald, A.G.; Guldan, S.J.; Ochoa, C.G. Surface water and groundwater interactions in traditionally irrigated fields in Northern New Mexico, U.S.A. *Water* **2017**, *9*, 102. [CrossRef]
3. Li, D. Quantifying water use and groundwater recharge under flood irrigation in an arid oasis of northwestern China. *Agric. Water Manag.* **2020**, *240*, 106326. [CrossRef]
4. Fernald, A.G.; Cevik, S.Y.; Ochoa, C.G.; Tidwell, V.C.; King, J.P.; Guldan, S.J. River Hydrograph Retransmission Functions of Irrigated Valley Surface Water–Groundwater Interactions. *J. Irrig. Drain. Eng.* **2010**, *136*, 823–835. [CrossRef]
5. Ochoa, C.G.; Fernald, A.G.; Guldan, S.J.; Tidwell, V.C.; Shukla, M.K. Shallow Aquifer Recharge from Irrigation in a Semiarid Agricultural Valley in New Mexico. *J. Hydrol. Eng.* **2013**, *18*, 1219–1230. [CrossRef]
6. Bethune, M.G.; Selle, B.; Wang, Q.J. Understanding and predicting deep percolation under surface irrigation. *Water Resour. Res.* **2008**, *44*, 1–16. [CrossRef]
7. Bethune, M. Towards effective control of deep drainage under border-check irrigated pasture in the Murray-Darling Basin: A review. *Aust. J. Agric. Res.* **2004**, *55*, 485–494. [CrossRef]
8. Lal, R.; Shukla, M.K. *Principles of Soil Physics*; Marcel Dekker: New York, NY, USA, 2004.
9. Baram, S.; Kurtzman, D.; Dahan, O. Water percolation through a clayey vadose zone. *J. Hydrol.* **2012**, *424–425*, 165–171. [CrossRef]
10. Kurtzman, D.; Scanlon, B.R. Groundwater Recharge through Vertisols: Irrigated Cropland vs. Natural Land, Israel. *Vadose Zone J.* **2011**, *10*, 662–674. [CrossRef]
11. Boyko, K.; Fernald, A.G.; Bawazir, A.S. Improving groundwater recharge estimates in alfalfa fields of New Mexico with actual evapotranspiration measurements. *Agric. Water Manag.* **2021**, *244*, 106532. [CrossRef]
12. Grogan, D.S.; Wisser, D.; Prusevich, A.; Lammers, R.B.; Frolking, S. The use and re-use of unsustainable groundwater for irrigation: A global budget. *Environ. Res. Lett.* **2017**, *12*, 034017. [CrossRef]
13. Scanlon, B.R.; Healy, R.W.; Cook, P.G. Choosing appropriate techniques for quantifying groundwater recharge. *Hydrogeol. J.* **2002**, *10*, 18–39. [CrossRef]
14. Sophocleous, M.A. Combining the soil water balance and water-level fluctuation methods to estimate natural groundwater recharge: Practical aspects. *J. Hydrol.* **1991**, *124*, 229–241. [CrossRef]
15. Healy, R.W.; Cook, P.G. Using groundwater levels to estimate recharge. *Hydrogeol. J.* **2002**, *10*, 91–109. [CrossRef]
16. Nimmer, M.; Thompson, A.; Misra, D. Modeling Water Table Mounding and Contaminant Transport beneath Storm-Water Infiltration Basins. *J. Hydrol. Eng.* **2010**, *15*, 963–973. [CrossRef]
17. Childs, E.C. The nonsteady state of the water table in drained land. *J. Geophys. Res.* **1960**, *65*, 780–782. [CrossRef]
18. Soil Survey Staff, Natural Resources Conservation Service, United States Department of Agriculture. Soil Series Classification Database. Available online: http://websurvey.sc.egov.usda.gov/ (accessed on 6 February 2021).

19. Oregon State University, Corvallis, Oregon, USA—Climate Summary. Available online: https://wrcc.dri.edu/cgibin/cliMAIN. pl?or1862 (accessed on 31 January 2022).
20. Blake, G.R.; Hartge, K.H. Bulk density. In *Methods of Soil Analysis, Part 1—Physical and Mineralogical Methods*; Agronomy Monographs 9; American Society of Agronomy: Madison, WI, USA, 1986; Volume 9.
21. Gee, G.W.; Bauder, J.W. Particle size analysis. In *Method of Soil Analysis, Part 1: Physical and Mineralogical Methods*; Klute, A., Ed.; Soil Science Society of America: Madison, WI, USA, 1986.
22. Bureau of Reclamation. Available online: https://www.usbr.gov/pn/agrimet/cropcurves/PASTcc.html (accessed on 12 January 2022).
23. Dingman, S.L. Ground water in the hydrologic cycle. In *Physical Hydrology*, 2nd ed.; Prentice Hall: Upper Saddle River, NJ, USA, 2002; pp. 325–388.
24. Ochoa, C.G.; Fernald, A.G.; Guldan, S.J.; Shukla, M.K. Deep percolation and its effects on shallow groundwater level rise following flood irrigation. *Trans. ASABE* **2007**, *50*, 73–81. [CrossRef]
25. Wangemann, S.G.; Kohl, R.A.; Molumeli, P.A. Infiltration and percolation influenced by antecedent soil water content and air entrapment. *Trans. ASAE* **2000**, *43*, 1517–1523. [CrossRef]
26. Ochoa, C.G.; Fernald, A.G.; Guldan, S.J.; Shukla, M.K. Water Movement through a Shallow Vadose Zone: A Field Irrigation Experiment. *Vadose Zone J.* **2009**, *8*, 414–425. [CrossRef]
27. Cassidy, J.R. Effect of Burrowing Mammals on the Hydrology of a Drained Riparian Ecosystem. Ph.D. Thesis, Oregon State University, Corvallis, OR, USA, 2002.
28. Römkens, M.J.M.; Prasad, S.N. Rain Infiltration into swelling/shrinking/cracking soils. *Agric. Water Manag.* **2006**, *86*, 196–205. [CrossRef]
29. Qi, W.; Zhang, Z.-y.; Wang, C.; Chen, Y.; Zhang, Z.-m. Crack closure and flow regimes in cracked clay loam subjected to different irrigation methods. *Geoderma* **2020**, *358*, 113978. [CrossRef]
30. Greve, A.K.; Acworth, R.I.; Kelly, B.F.J. Detection of subsurface soil cracks by vertical anisotropy profiles of apparent electrical resistivity. *Geophysics* **2010**, *75*, WA85–WA93. [CrossRef]
31. Greve, A.K.; Andersen, M.S.; Acworth, R.I. Monitoring the transition from preferential to matrix flow in cracking clay soil through changes in electrical anisotropy. *Geoderma* **2012**, *179–180*, 46–52. [CrossRef]
32. Zhang, Y.; Cao, Z.; Hou, F.; Cheng, J. Characterizing preferential flow paths in texturally similar soils under different land uses by combining drainage and dye-staining methods. *Water* **2021**, *13*, 219. [CrossRef]
33. Liu, M.; Guo, L.; Yi, J.; Lin, H.; Lou, S.; Zhang, H.; Li, T. Characterising preferential flow and its interaction with the soil matrix using dye tracing in the Three Gorges Reservoir Area of China. *Soil Res.* **2018**, *56*, 588–600. [CrossRef]

Article

Participatory and Integrated Modelling under Contentious Water Use in Semiarid Basins

Rodrigo Rojas [1,*], **Juan Castilla-Rho** [2,†], **Gabriella Bennison** [3], **Robert Bridgart** [4], **Camilo Prats** [5] and **Edmundo Claro** [3]

[1] CSIRO Land and Water, EcoSciences Precinct, 41 Boggo Road, Dutton Park, Brisbane, QLD 4102, Australia
[2] Faculty of Engineering and Information Technology, School of Information, Systems and Modelling, University of Technology Sydney, Broadway, Sydney, NSW 2007, Australia; juan.castillarho@canberra.edu.au
[3] Fundación CSIRO Chile Research, 4700 Apoquindo Av., Las Condes, Santiago 7500000, Chile; gabriella.bennison@csiro.au (G.B.); edmundo.claro@csiro.au (E.C.)
[4] CSIRO Land and Water, Environmental Informatics Group, Clayton South, VIC 3169, Australia; robert.bridgart@csiro.au
[5] Fundación Atacama Verde, Caldera 1570000, Chile; camiloprats@gmail.com
[*] Correspondence: rodrigo.rojas@csiro.au; Tel.: +61-0738335600
[†] Current address: Institute for Governance and Policy Analysis, Faculty of Business, Governance and Law, University of Canberra, Bruce, Canberra, ACT 2617, Australia.

Abstract: Addressing modern water management challenges requires the integration of physical, environmental and socio-economic aspects, including diverse stakeholders' values, interests and goals. Early stakeholder involvement increases the likelihood of acceptance and legitimacy of potential solutions to these challenges. Participatory modelling allows stakeholders to co-design solutions, thus facilitating knowledge co-construction/social learning. In this work, we combine integrated modelling and participatory modelling to develop and deploy a digital platform supporting decision-making for water management in a semiarid basin under contentious water use. The purpose of this tool is exploring "on-the-fly" alternative water management strategies and potential policy pathways with stakeholders. We first co-designed specific water management strategies/impact indicators and collected local knowledge about farmers' behaviour regarding groundwater regulation. Second, we coupled a node–link water balance model, a groundwater model and an agent-based model in a digital platform (SimCopiapo) for scenario exploration. This was done with constant input from key stakeholders through a participatory process. Our results suggest that reductions of groundwater demand (40%) alone are not sufficient to capture stakeholders' interests and steer the system towards sustainable water use, and thus a portfolio of management strategies including exchanges of water rights, improvements to hydraulic infrastructure and robust enforcement policies is required. The establishment of an efficient enforcement policy to monitor compliance on caps imposed on groundwater use and sanction those breaching this regulation is required to trigger the minimum momentum for policy acceptance. Finally, the participatory modelling process led to the definition of a diverse collection of strategies/impact indicators, which are reflections of the stakeholders' interests. This indicates that not only the final product—i.e., SimCopiapo—is of value but also the process leading to its creation.

Keywords: stakeholder participation; surface water-groundwater interaction; scenario modelling; integrated water management; agent-based modelling; SimCopiapo

Citation: Rojas, R.; Castilla-Rho, J.; Bennison, G.; Bridgart, R.; Prats, C.; Claro, E. Participatory and Integrated Modelling under Contentious Water Use in Semiarid Basins. *Hydrology* **2022**, *9*, 49. https://doi.org/10.3390/hydrology9030049

Academic Editors: Il-Moon Chung, Sun Woo Chang, Yeonsang Hwang and Yeonjoo Kim

Received: 14 February 2022
Accepted: 11 March 2022
Published: 16 March 2022

Publisher's Note: MDPI stays neutral with regard to jurisdictional claims in published maps and institutional affiliations.

1. Introduction

Water resources are fundamental for supporting livelihoods, food production, energy generation and ecosystem services across the globe. Despite their relevance, water systems are under continuous threats, thus undermining water security [1] and promoting water stress [2]. Interdependencies between water, ecological and social systems across multiple scales and dimensions (e.g., water–energy–food–environment nexus [3,4]) continuously challenge the way water resources have been managed [5]. In this regard, Hoff [6] states that " . . . water management and governance have not yet adapted to these cross-scale and cross-sectoral interdependencies and their dynamics and associated uncertainties".

Water management challenges are no longer addressed solely as technical problems but rather have become part of complex policy and decision-making processes, where multiple stakeholders and institutions reflecting an array of diverse values and interests are involved [7–9]. "Integrated" approaches to account for the array of drivers that help to constrain/condition these water management challenges have therefore received a surge of attention in recent years [10–12].

Kelly et al. [13] discuss the term "integration" in the context of integrated assessment and define five levels with multiple loci in the modelling process: (a) integrated treatment of issues, (b) integration with stakeholders, (c) integration of disciplines, (d) integration of processes and models, and (e) integration of scales of consideration. Integration of biophysical and socioeconomic aspects [10,14] and integration across processes/models (e.g., surface water and groundwater interactions [15,16]), as well as integration with stakeholders [17], have all been documented in the water management-related literature. In the context of surface water–groundwater interactions, Barthel and Barnhaz [16] suggest that "integrated modelling" should explore aspects beyond the purely physical coupling process between surface water and groundwater systems and cover multiple scientific domains and disciplines, thus aligning with the level "integration of processes and models" proposed by Kelly et al. [13].

Jakeman et al. [18] suggest that the development and application of integrated modelling stands on several building blocks, with participatory modelling [19,20] and the development of modelling tools and software/hardware technologies considered as key pillars. In the context of policy analysis, more specifically, participatory processes are essential for linking science and policy [21] and to achieve the legitimacy of processes [22], with stakeholder participation and computer-based models regarded as key components of the participatory and collaborative modelling [17]. This society–science–policy interface [23] is usually moulded by different contextual pressures and communication protocols, thus rendering early stakeholder participation critical for successful outcomes in policy making [24].

There is no doubt that stakeholder participation in water resource management has received substantial attention in the last years [25–27]. The popularity of participatory modelling in particular has seen a substantial growth due to its compatibility with environmental paradigms such as Integrated Water Resources Management (IWRM) and Adaptive Management (AM) [28]. An advantage of participatory modelling resides in the potential to integrate meaningful input from decision makers and stakeholders into the modelling process [29]. Based on case studies from Africa, Asia, Europe and Oceania, Penny and Goddard [30] noted however that experimentation and learning beyond the "expert" group (to include non-expert participation) was mostly absent from discussions around model development.

To enable a participatory involvement, Basco-Carrera et al. [17] suggest that developed tools and models in the context of participatory modelling should be built using open source or freeware software where possible to facilitate distribution and use by stakeholders. Similarly, Carmona et al. [31] suggest that decision-making tools for successful stakeholder participation in natural resources management should be transparent, flexible and designed to elicit knowledge from different groups. Transparency and flexibility in the process of model development are also advocated by Bots et al. [21], with the aim of increasing stakeholders' trust by making the usually perceived "black box" model transparent.

Despite the clear need for stakeholder participation in the modelling development process, van Bruggen et al. [32] suggest that limited attention has been given to the model-based exploration and design of policy pathways with stakeholders. They argue that disciplinary fragmentation and the "not-invented-here" academic syndrome ("a negative attitude to knowledge that originates from a source outside the own domain" [33]) are factors hindering the development of modelling with stakeholders.

In arid and semiarid regions, collaborative processes and water governance are usually a major challenge [31,34–36] driven by contentious water use and competing stakeholder interests and values. This situation impacts the successful materialisation of the integration levels described by Kelly et al. [13] and poses challenges to the participatory modelling process as highlighted by Carmona et al. [31]. In particular, in these regions, the interaction between surface water and groundwater plays an important role [16,36,37]. This interaction is often complicated by agricultural and/or mining activities, as they will potentially alter the fragile flow regimes of the coupled water system [37]. As highlighted by Gorelick and Zheng [38], groundwater plays an important role, and its relevance will continue in the coming years, more importantly in arid and semi-arid regions.

This article describes the implementation of a participatory modelling process to develop and deploy an integrated modelling tool and digital platform (SimCopiapo) to support decision making in water management in a semiarid basin under contentious water use. The purpose of this digital platform is to explore alternative water management strategies to support scenario analysis and potential policy pathways with stakeholders, thus contributing to addressing the research need identified by van Bruggen et al. [32].

We build upon the work of Galvez et al. [34] to set up a participatory process in the Copiapó River Basin (CRB), northern Chile. We follow the integrated modelling levels suggested by Kelly et al. [13], and as such we include in the proposed integrated modelling tool surface water–groundwater interactions (*integration of processes and models*); local knowledge and expertise in water operational rules (*integration with stakeholders*); short-, mid- and long-term outputs, as well as sub-daily reservoir operations, daily water balance in irrigation districts and monthly time steps in groundwater assessment and different spatial scales for aquifer sectors and irrigation districts (*integration of scales of consideration*); and an agent-based model (ABM) to account for farmers' compliance against imposed caps on groundwater allocations (*integrated treatment of issues*). As suggested by the literature, we develop the integrated modelling tool and software platform in open source code with constant input from different stakeholder groups (water users, regulators, civil society, academy) [34] for transparency and flexibility [17,31] and to promote the legitimacy of the process [22] and ownership of results. The novelty of this work lies in advancing previous modelling efforts in the CRB [39–41] by improving on the operational rules of critical infrastructure in the CRB and co-designing water management strategies and impact indicators, all of which are designed with continuous input from key stakeholders by employing formal participatory and stakeholder engagement processes. A major feature of the proposed digital platform (SimCopiapo), compared to previous modelling efforts in the CRB, is the ability for users to run "on-the-fly" a loosely coupled [16] node–link water balance model and fully distributed groundwater model during interactive participatory sessions, thus facilitating social learning and knowledge co-creation. This was done in order to address research needs identified in the specialised literature [17,31,42]. Finally, the proposed digital platform (and integrated model) can be seen as a boundary object [43] bridging stakeholders and facilitating mutual understanding and cooperation—a practical exercise that has not been implemented before in the CRB [34].

The remainder of this article is arranged as follows. Section 2 describes the case study and the two-step methodological framework implemented in this work. Results of the integrated modelling process are analysed in Section 3 for a series of water management strategies and a base scenario. Section 4 presents a discussion of these results, and concluding remarks are offered in Section 5.

2. Materials and Methods

2.1. Case Study: Copiapó River Basin

The Copiapó River Basin (CRB) covers an area of 18,700 km^2 and is located in Northern Chile at the southern boundary of the Atacama Desert (Figure 1). The discharge contributions of the main tributaries Pulido, Jorquera and Manflas rivers in the headwater basins are regulated by the Lautaro Reservoir (26 Mm^3). The gauging station "Río Copiapó en La Puerta" shows an average discharge of 2.6 m^3 s^{-1}, whereas average annual precipitation in Copiapó city is 19 mm, reaching up to 500 mm y^{-1} for altitudes over 5000 m above sea level (asl) [44]. The CRB is a clear example of a semiarid basin under sustained water stress originating from both natural and anthropogenic causes, where water management can be regarded as inadequate [40,41,45,46].

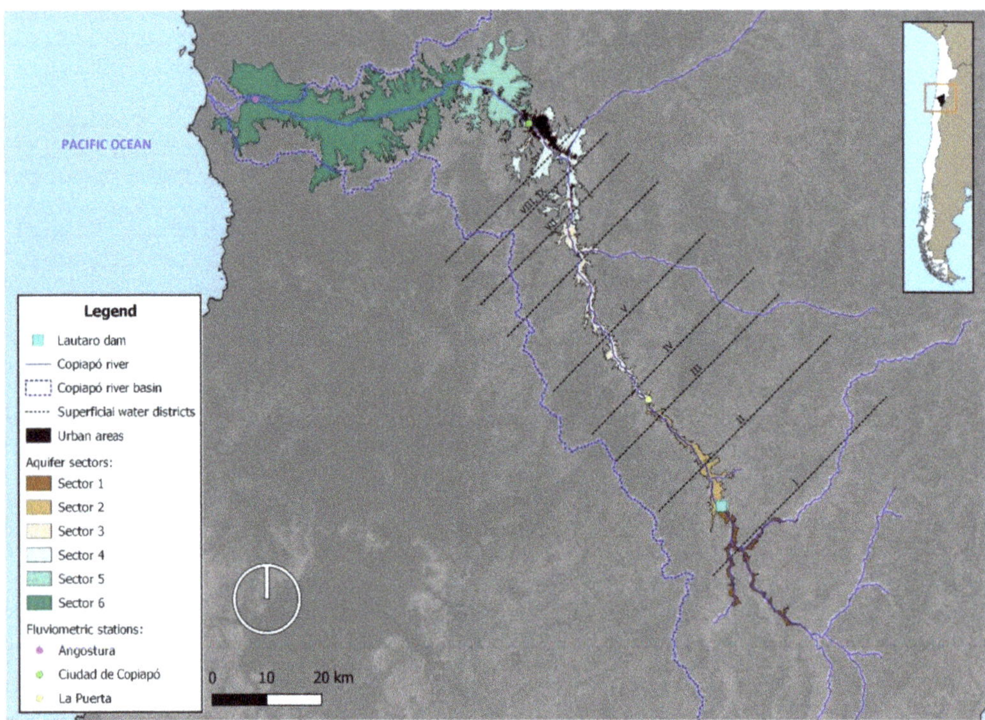

Figure 1. Location of the Copiapó River basin, main groundwater (aquifer) sectors, surface water irrigation districts (I: irrigation district D1, II: irrigation district D2, III: irrigation district D3, IV: irrigation district D4, V: irrigation district D5, VI: irrigation district D6, VII: irrigation district D7, VIII: irrigation district D8, IX: irrigation district D9. Most downstream districts (VIII and IX) are combined into irrigation district D89), and Lautaro Reservoir at the confluence of main tributaries (Jorquera, Manflas and Pulido rivers) to the Copiapó River. Urban areas in black color, coloured circles represent gauging stations (after [34]).

Currently, available surface water in nine irrigation districts (see Figure 1) is fully allocated for consumption, whereas the overexploitation of groundwater has been previously well documented in the literature [34,40,41] and is manifested by deteriorating groundwater quality and persistent deepening of groundwater levels. Administration of groundwater rights/licenses takes place in six groundwater/aquifer sectors (see Figure 1). Around 60% of groundwater demand is used for highly technified irrigation, whereas mining activities account for 30% and drinking water for 10% of the demand [47]. On an average water year, the Copiapó River dries up halfway to the outlet at the Pacific Ocean due to upstream water consumption and zero contributions from lateral intermediate sub-basins [48].

Groundwater rights for consumption total ca. 19 m^3 s^{-1} in the CRB, which contrasts with the estimated average recharge rates of 3 to 4.8 m^3 s^{-1} and effective groundwater demands of 6 to 14 m^3 s^{-1} [47,49]. Therefore, permanent conflicts between water users at different levels (upstream vs downstream users, surface water vs groundwater users) are detrimental factors for effective water resources management in the basin. These conflicts can be typified as Type 2, river basin conflicts, and Type 3, overexploited groundwater systems, by Bauer [50]. Rinaudo and Donoso [45] identified five factors leading to the current over-exploitation of Copiapo's groundwater resources: (i) limited knowledge of groundwater, (ii) legal complexity and political pressure, (iii) poorly-defined water permits, (iv) compliance and enforcement problems and (v) inconsistencies between management of surface water and groundwater.

Despite the peculiarities of the water resource management model in the Copiapó Basin [45,50], this management landscape is likely to reflect similar operational conditions as in other semiarid water-stressed basins around the world, thus providing generality to our findings.

2.2. Methodological Framework

We applied an intertwined two-step methodological framework in this work (Figure 2). First, we implemented a participatory process with existing key stakeholders to define and explore potential water management strategies and impact indicators of interest to stakeholders and to collect data on farmers' behaviours regarding groundwater regulation and operational rules for critical hydraulic infrastructure (c.f. Lautaro Reservoir). This step builds upon the work by Galvez et al. [34], who identified key stakeholders and barriers to collaborative water governance in the CRB and feeds into step 2. The second step consisted of designing and implementing a digital platform (termed SimCopiapo), which hosts a Graphical User Interface (GUI) (see Figure A1 in the Appendix A) with capabilities for stakeholders/modelers to run "on-the-fly" a node–link water balance model, a groundwater model and an agent-based model (ABM) [51,52]. This second step implemented the aspects identified through the participatory process in step 1. During participatory workshops, SimCopiapo was mainly run by stakeholders organised into groups with guidance provided by the research team. The purpose of this digital platform was to collaboratively explore different water management strategies as support for exploratory scenario analysis during participatory decision-making sessions with stakeholders. In the following sections, details for both steps are described.

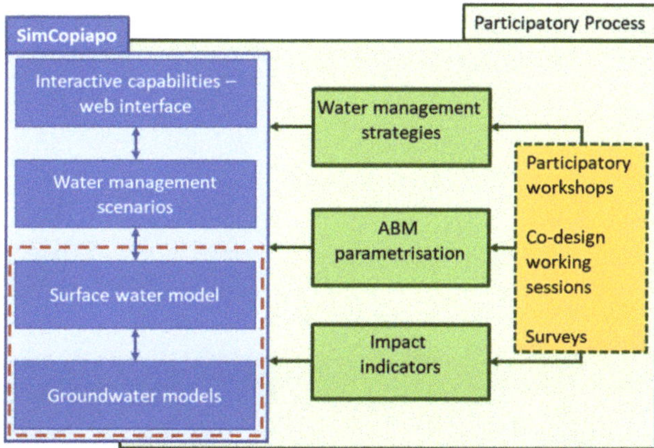

Figure 2. General methodological framework highlighting development of the digital platform SimCopiapo, participatory processes and the integration of the surface water and groundwater models (red dashed line).

2.2.1. Participatory Process

Galvez et al. [34] provide an overview of the stakeholders involved in the participatory process. We implemented 4 plenary workshops with 31 institutions and 5 working sessions with specific groups of stakeholders for data/local knowledge collection. In addition, we implemented two surveys with regional organisations in the study area: Water Resources Regional Advisory Committee (CARRH) (on-line) and Copiapó Exporters and Producers Association (APECO) (on-line). Surveys were used to collect information about farmers' behaviours regarding tolerance towards groundwater regulation (e.g., follow groundwater allocation rules) and the propensity to breach these rules following the social sub-model proposed by Castilla-Rho et al. [51,52]. This information was used to parameterise an agent-based model (ABM) to assess compliance against caps on groundwater use as explained in Section 2.2.3.

2.2.2. Water Management Strategies and Impact Indicators

From the participatory process, we co-designed with stakeholders 13 water management strategies grouped in 4 domains: (a) exchanges of water uses/rights among users, (b) improvement to current hydraulic infrastructure, (c) management of groundwater recharge and (d) management of water demand. These water management strategies are shown and described in detail in Table 1 and were implemented in the SimCopiapo platform.

The participatory process also allowed the definition of a series of key impact indicators of interest to stakeholders. To this end, we followed a similar approach as that proposed by Santos Coelho [53]. The main impact indicators identified were (a) river flows through the Copiapó city (termed as urban flows), (b) environmental flows at the Copiapó basin outlet, (c) storage at Lautaro Reservoir (headwater basin), (d) percentage change in aquifer storage in groundwater sectors 2 to 6 after implementing water management strategies and (e) compliance with the cap on groundwater use. A description of the main impact indicators is presented in Table 2. These and other impact indicators (e.g., water security for individual irrigation districts) are automatically generated and exported by the SimCopiapo platform.

45

Table 1. Water management strategies devised through the participatory process with main stakeholders in the Copiapó River basin.

Water Management Strategy		Description	Anticipated Impacts
Water use/rights exchanges	1.1	**Water use/right exchange between Candelaria—Aguas Chanar** This strategy consists of Aguas Chanar (water utility) decreasing groundwater abstraction in GW sectors 5 and 6 by 175 L/s, which instead is obtained from desalination water available from Candelaria mining company. As compensation, Aguas Chanar grants Candelaria mining company the same volume from the wastewater treatment plant downstream of Copiapó city.	• Groundwater heads recovery in GW Sector 5 • Perceived quality of potable water improves • Water security for urban population increases • Potential for improvements in environmental flows at the basin outlet
	1.2	**Water use/right exchange between Caserones—Ramadilla River** This strategy consists of Caserones mining company stopping groundwater abstraction (200 L/s) in GW sector 2. As compensation, Caserones mining company extracts the same volume as surface water in the headwater basin of Ramadilla River.	• Groundwater heads recovery in GW sector 2 • Decreases in tributary flows to Copiapó River in the headwater basin
	1.3-a	**Water use/right exchange between SW Irrigation districts D8 and D9 (D89)—SW irrigation districts D1-D7** This strategy consists of re-allocating unused surface water rights from irrigation districts D8 and D9 (downstream and combined into a single district, D89) to upstream irrigation districts D1 to D7. (It is worth noting that irrigation districts D8 and D9 are the most limited in areal extent compared to other districts and are located in the outskirts of Copiapó city, thus experimenting a fast land use change from rural to urban areas. See Figure 1 for location of combined district D89).	• Surface water security increases in irrigation districts 1 to 7 • Groundwater recharge increases in GW sectors 2, 3 and 4
	1.3-b	**Water use/right exchange between SW Irrigation districts D8 and D9 (D89)—Aguas Chanar** This strategy consists of re-allocating available surface water rights from irrigation districts D8 and D9 (downstream and combined into a single district, D89) to Aguas Chanar (water utility), which reduces groundwater abstraction in GW sectors 5 and 6 by Aguas Chanar.	• Groundwater heads recovery in GW sectors 5 and 6 • Perceived quality of potable water improves
	1.3-c	**Water use/right exchange between SW Irrigation districts D8 and D9 (D89)—Localised recharge Copiapó River** This strategy consists of using available surface water rights from irrigation districts D8 and D9 to recharge GW sectors 3 and 4 (upstream Copiapó City) through localised recharge in the Copiapó riverbed.	• Groundwater heads recovery in GW sectors 3, 4 and 5 • Improved quality of life in Copiapó city as more frequent surface flows observed through Copiapó city (urban flows)
	1.3-d	**Water use/right exchange between SW Irrigation districts 8 and 9—GW Sector 5/with excess localised recharge Copiapó River** This strategy consists of using available surface water rights from irrigation districts D8 and D9 conveyed by pipe system to farmers of GW sector 5 for irrigation purposes; water surplus as localised recharge through Copiapó riverbed.	• Groundwater heads recovery in GW sector 4 and 5 • Irrigation water security increases for farmers in GW sectors 5 and 6 • Improved quality of life in Copiapó city as more frequent surface flows observed through Copiapó city (urban flows)
	1.3-e	**Water use/right exchange between SW Irrigation districts 8 and 9—GW Sector 5/with excess Managed Aquifer Recharge in GW Sector 5** This strategy consists of using available surface water rights from irrigation districts D8 and D9 conveyed by pipe system to farmers of GW sector 5; water surplus as managed aquifer recharge in GW sector 5.	• Groundwater heads recovery in GW sector 4 and 5 • Irrigation water security increases for farmers in GW sectors 5 and 6

Table 1. *Cont.*

Water Management Strategy		Description	Anticipated Impacts
Hydraulic infrastructure	2.1	**Impermeabilisation Lautaro Reservoir (100% in a 5-year period)** Lautaro Reservoir is the main regulation infrastructure in the basin, and it has a storage capacity of ca. 25 Hm^3. Infiltration losses however are close to 50% of the stored volume. This strategy consists of installing impermeabilisation geotextiles in the Lautaro Reservoir to reduce these infiltration losses.	• Volumes available at Lautaro Reservoir increase • Surface water available for irrigation districts 1 to 9 increases • Water security for farmers in irrigation districts 1 to 9 increases • Recharge rates and groundwater heads in GW sector 2 decrease
	2.2	**Surface water conveyance to irrigation sectors through pipes instead of open channels** Currently there is 42% conveyance losses in the irrigation system in the Copiapó basin. This strategy consists of replacing open-channel systems with pipe systems to reduce conveyance and evaporation losses. This assumes there is no expansion in the irrigated surface.	• Infiltration in irrigation districts 1 to 9 decreases • Surface water available for irrigation districts 1 to 9 increases • Water security for farmers in irrigation districts 1 to 9 increases • Groundwater recharge rates decrease in GW sectors 2, 3 and 4
	2.3	**Operation of desalination plant** This strategy consists of operating the desalination plant designed for the Atacama region. It considers a staged supply plan (90 L/s, 450 L/s, 930 L/s) until providing close to 930 L/s of drinking water after 25 years of operation. Aguas Chanar (water utility) progressively ceases groundwater exploitation in GW sectors 4, 5 and 6.	• Marine ecosystems potentially impacted • Water security for farmers in irrigation districts 1 to 9 increases • Surface water available for irrigation districts 1 to 9 increases • Water security for urban population increases • Perceived quality of potable water • Groundwater heads recovery in GW sectors 4, 5 and 6
Recharge management	3.1	**Managed aquifer recharge along the Copiapó River** This strategy consists of building infiltration ponds along the main river course at specific locations: (a) Nantoco (GW sector 2), (b) upstream Kaukari Park (GW sector 4), (c) downstream Kaukari park (GW sector 5), and (d) Piedra Colgada (GW sector 6). These ponds have been restricted to ca. 4 ha (surface area), 1.5 m depth and assuming a representative infiltration rate of 1 m/d based on [54].	• Groundwater heads recovery in GW sectors 2, 4, 5 • Protection against extreme events (floods) • Water security for urban population increases
Demand management	4.1	**Prorate of groundwater uses in GW sectors 3, 4 and 5** This strategy consists of decreasing groundwater demand in GW sectors 3, 4 and 5 by 40% and aligns with results by DGA-DICTUC [49], Suarez et al. [40] and Hunter et al. [41]. In this strategy we implemented an agent-based model (ABM) to assess the compliance achieved against this demand restriction. Based on cultural parameters surveyed among irrigators in the Copiapó basin, two sub-strategies were analysed: (4.1.a) high level of monitoring and fines by the regulator, and (4.1.b) low level of monitoring and fines by the regulator.	• Groundwater heads recovery across GW sectors 3, 4 and 5 • Compliance rate with demand restriction • Improved capabilities by the water regulator agency to devise monitoring/enforcement strategies based on prorate compliance levels
	5.1	**Greywater reuse/recirculation** This strategy consists of reusing greywater to reduce the demand of drinking water in 20% in urban areas of the Copiapó basin. Water utility decreases groundwater exploitation in GW sectors 5 and 6.	• Groundwater heads recovery across GW sectors 5 and 6

Table 2. Description of impact indicators and expected values co-designed with stakeholders through the participatory process.

Impact Indicator	Description	Expected Value by Stakeholders
Urban flows	Percentage of the simulated period where the simulated discharge at the Copiapó city gauging station is in the range [0; 5000] L/s. Based on the hydraulic design of the river cross section at that station, above 5000 L/s is considered to represent a high risk of flooding.	No. of months 0 < Urban flows < 5000 L/s/Total months simulated
Environmental flows at the outlet of the Copiapó Basin	Percentage of the simulated period where the simulated discharge at the "Copiapó en Desembocadura" gauging station is greater than 50 L/s. This value corresponds to the historical average outflow from the basin.	No. of months environmental flows > 50 L/s/Total months simulated
Storage of Lautaro Reservoir	Percentage of the simulated period where the simulated volume in the Lautaro Reservoir is greater than 50% of its total storage capacity.	No. of months storage Lautaro Reservoir > 50%/Total months simulated
Aquifer Volume GW sectors 2 to 6	Change ratio between final and initial aquifer volume, e.g., >1.0 indicates aquifer volume at end of simulation period is greater than initial volume.	Aquifer storage$_{FINAL}$/Aquifer storage$_{INITIAL}$ > 1.0
Groundwater cap compliance	Compliance level expressed as the percentage of farmers adhering to the groundwater cap imposed by the regulator. This indicator is specific to water management strategy 4.1 only.	Compliance rate > 50%

2.2.3. SimCopiapo Platform

SimCopiapo is a digital platform built mostly in Python, including an API web and front-end HTML/JavaScript (Figure A1 in Appendix A). It includes capabilities to select pre-defined water management strategies devised by stakeholders, run "on-the-fly" a loosely coupled node-link water balance and groundwater models [16,55], display graphs/plots/maps for rapid assessment and export reports summarising the results of management strategies selected by the user. It also contains an ABM associated with water management strategy 4.1 (groundwater demand management) to assess farmers' compliance against caps in groundwater use imposed by the regulator.

Figure 3 shows the interaction between the components of the SimCopiapo platform, including node–link water balance, groundwater and agent-based models. SimCopiapo uses geospatial information (irrigation areas, aquifer sectors, channel network, production wells, etc.), historical hydrological timeseries at the headwater basins for the period 1991–2016 and a series of alternative water management strategies (see Table 1) selected by the user to set up a specific scenario run. SimCopiapo users also have the opportunity to select/input pre-defined alternative hydrological time series driving the simulation (historical 1991–2016, 50% historical, etc.) or to include new time series if required. SimCopiapo is built as an open-source tool to allow continuous, replicable, reproducible and transparent research and improvements by other users [56–58], thus improving on previous efforts developed under proprietary hydrological software [39–41,49].

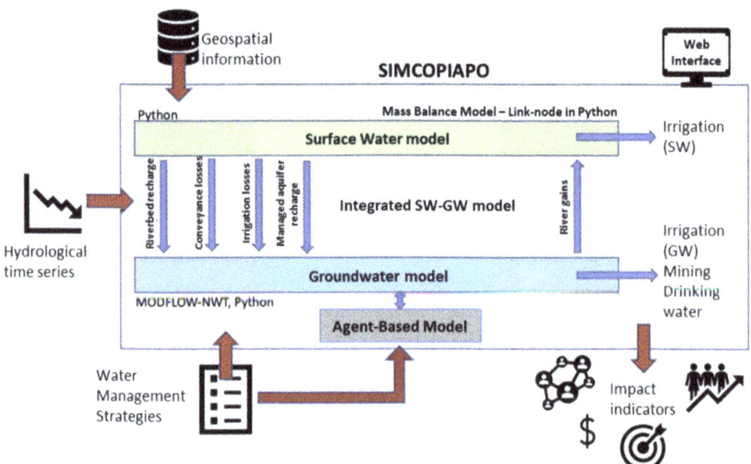

Figure 3. SimCopiapo digital platform and interactions of the integrated surface water and groundwater models and different modules and platform components.

It is worth noting that the objective of the digital platform is to facilitate the interaction among stakeholders in the contentious Copiapó river basin, where competing and conflicting water uses exist and the level of collaboration for water management is limited and generally perceived as inadequate [34]. Therefore, the focus is on providing a research tool able to run basin-scale assessments in order to support rapid appraisal of water management strategies enabling stakeholder discussion, collaboration and decision-making. Under these premises, SimCopiapo is aligned more closely to what Oxley et al. [24] define as a policy-oriented model, where *accurate* process representation is traded for *adequate* process representation and emphasis is focused on addressing practical policy issues. For this purpose, we illustrate how the digital platform can be populated with surface water and groundwater models previously documented and validated by stakeholders (e.g., [39–41,49]).

DGA-HIDROMAS [39] provides the latest surface water model available for the Copiapó basin implemented in the AQUATOOL software—a proprietary generic decision support system for water resources planning [59]. We translated the topology of the AQUATOOL model for the Copiapó basin into a node–link model coded in Python, accounting for the daily mass balance of the surface water system. The conceptual representation of this topology and the operation of the irrigation districts in the Copiapó River Basin is presented in Figure 4. This figure shows the upstream Lautaro reservoir as the main hydraulic infrastructure regulating surface flows to supply irrigation water to nine districts (D1–D9, irrigation districts D8 and D9 are combined into a single district, D89. See Figure 1 for locations of irrigation districts). Available surface water regulated from Lautaro reservoir is equally allocated between districts D1 to D7 (12% each), whereas districts D8 and D9 are allocated 8% each (i.e., 16% combined for D89). Figure 4 also shows the crop sectors (e.g., R2a-XX) belonging to each irrigation district, with some of the irrigation districts including more than one crop sector (e.g., R2a-13 and R2a-14 belong to irrigation district D6). It is worth emphasising that D8 and D9 are the most downstream irrigation districts located at the outskirts of Copiapó city, thus experiencing changes in land use patterns from rural to urban. Upstream of the gauging station "Copiapo @ Mal Paso", most of the available surface water is conveyed through a channel (1000 L/s maximum capacity), leaving just excess water flowing through the natural river course.

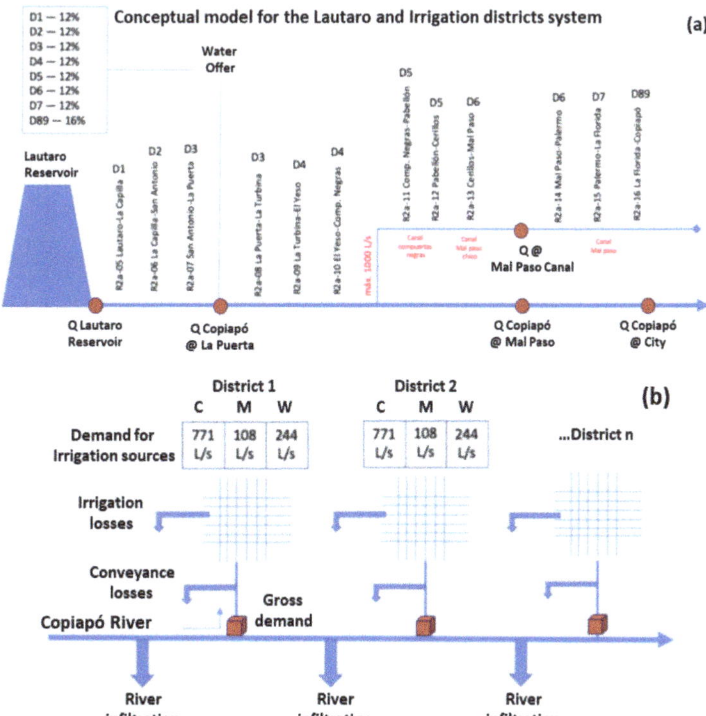

Figure 4. Conceptual model for the (**a**) operation of the Lautaro Reservoir and the (**b**) irrigation districts in the Copiapó River Basin. Values for parameters of the conceptual model are obtained from [39,40] and included in the node–link water balance model. Water sources for each irrigation district are distinguished between C: channel, W: wells and M: mixed source and are based on crop area supplied by that source.

For each irrigation district, information on crop types/surfaces, irrigation and water conveyance efficiencies, water sources, etc. is used to perform an internal supply–demand balance per irrigation area, considering alternative water sources (surface water, groundwater or combination). The water source indicates the water volume supplying crop areas for each irrigation district. The volumes demanded for each irrigation district are obtained from the crop surveys by DGA-DICTUC [49] and the water licenses. In this way, irrigation demands are supplied first by channel source, then mixed source and finally groundwater (wells). Results of the water balance for each irrigation district are spatially coupled to the aquifer sectors implemented in the groundwater model, which together with the infiltration through the riverbed define the main recharge rates for the aquifer sectors (see Figure 1).

This node–link mass balance model was fully coupled with the latest available groundwater model for the Copiapó aquifer developed in MODFLOW-2005 [60] by DGA-HIDROMAS [39]. Using the FloPy Python package [61], we translated this MODFLOW-2005 model for operation in Python and coupled it with the node–link model of the surface water system in the SimCopiapo platform. The daily node–link water balance model was aggregated to a monthly time-step for consistency with the MODFLOW model for the Copiapó aquifer. For full details on the coupling process we refer the reader to [62].

As shown in Figure 3, SimCopiapo also includes an agent-based model (ABM) to assess farmers' compliance against caps imposed on groundwater use in the Copiapó aquifer. This ABM is based on the social sub-model developed by Castilla-Rho et al. [51,52], which represents a social utility function, S, that follows a Cobb–Douglas functional form:

$$S = \text{grid}^m \, (1 - \text{group})^n \tag{1}$$

where m = number of times a farmer reports a neighbour taking groundwater illegally, n = number of times a farmer is seen taken groundwater illegally, and grid-group are categories of the Cultural Theory proposed by Douglas [63]. S (social utility function) represents the loss of social reputation and the social costs to groundwater users when reporting non-compliant behaviour. Using survey data collected from farmers in the Copiapó basin [62] and the four grid-group categories (Egalitarian–Hierarchist–Individualist–Fatalist) proposed by Douglas [63] in Cultural Theory, we were able to parametrise equation 1 and thus farmers' decision-making processes. The user of SimCopiapo can adjust two parameters associated with the ABM model: (a) the percentage of groundwater users monitored by the regulator to check compliance and (b) the severity of the fines (as a percentage of the total farm revenue) if the farmer is caught taking groundwater illegally.

Equation (1) quantifies the loss of social reputation and the social costs to farmers when reporting non-compliant neighbours engaged in illegal extraction of groundwater in the Copiapó basin, and thus impact farmers' future decisions of engaging in non-complaint behaviour (i.e., taking groundwater illegally). Other factors impacting this decision relate to farmers' probability of being monitored by the regulator and the severity of fines if farmers are caught in non-compliant behaviour [48,49]. This social metric is combined with an economic (gross margins from crop enterprise) and institutional (monitoring/monetary fines) score into each farmers' objective function for decision making; i.e., whether to take groundwater illegally or not. For details on this implementation, the reader is referred to [51,52].

2.2.4. Improvements on Previous Integrated Modelling Tools

Suarez et al. [40] presented an integrated model for the CRB using the SIMGEN module of AQUATOOL software [59] (surface water only). More recently, Hunter et al. [41] presented an integrated model for the CRB based on [40] coupling the WEAP (Water Evaluation and Planning system) model [64] and the MODFLOW [60] groundwater model described by DGA-HIDROMAS [39]. Although these works claim the advantages of their corresponding integrated modelling frameworks, both tools rely on proprietary software and are therefore not amenable for rapid modifications by interested stakeholders, have been developed with limited input from key stakeholders in terms of potential water management strategies as well as impact indicators of interest to stakeholders, and concentrate only on demand management strategies. In this work, we improved on several aspects on the integrated modelling framework for the CRB: (1) the groundwater model developed by DGA-HIDROMAS [39] has been checked for spatial and temporal consistency of aquifer contributions to surface water at La Puerta and Angostura gauging stations; (2) evapotranspiration and groundwater demands were revised and activated; for the surface water model, (3) Lautaro Reservoir operational rule for water allocation was completely re-designed and implemented thanks to the advice of the Vigilance Board of the Copiapó River and its Tributaries; and (4) the operation of the irrigation districts was revised and deployed in Python considering (a) controlled discharge from the Lautaro Reservoir and (b) a supply–demand model for the irrigation districts considering allocation volumes, irrigation demands, irrigation losses, conveyance losses and gross water demand. For details on other improvements regarding updates on mining groundwater demands, downstream irrigation districts, drinking water demands and losses in the potable water network, the reader is referred to [62].

3. Results

3.1. Results Participatory Process

A total of 31 organisations representing the civil society (6), regional state agencies (16) and private/productive (9) sectors were engaged in the participatory process. On average and across all participatory workshops and working sessions, stakeholders from the civil society were the least involved (35% of organisations engaged), whereas stakeholders of the private sector were the most engaged, with 52% of the institutions of this sector taking part in the participatory sessions.

Regarding the on-line surveys for parameterising the ABM for groundwater regulation, the CARRH survey showed that civil society stakeholders were the most engaged, with 83% of the institutions of this sector providing responses, whereas only 31% of the institutions of the public sector were engaged in this process. Forty-four percent of private sector stakeholders participated in this survey. The APECO survey targeted 25 farmers of the CRB, with more than 83% concentrated in the upstream aquifer sectors 1 to 3 and the remaining 17% concentrated in downstream aquifer sectors 4 and 5. No responses were obtained from farmers in aquifer sector 6. Although not shown here, results of both surveys indicate a clear trend towards validating the importance of regulating groundwater resources for sustainable use and minimizing impacts to ecosystems and third parties, and enforcing this regulation in practice. Discrepancies among the CARRH stakeholders were observed on justifying the illegal extraction of groundwater on economic (profits) grounds and allocating importance to social costs (loss of reputation) if caught breaching caps on groundwater use imposed by the regulator. This discrepancies can be attributed to the heterogeneity of the stakeholders composing the CARRH [34]. On the contrary, the APECO survey indicated that farmers attributed a higher importance to the loss of social reputation (individual) if caught breaching the imposed cap on groundwater use and allocated a higher importance to collective enforcement policy such as effective monitoring of groundwater use. For details about the outcomes of the surveys and the implementation in the ABM model, the reader is referred to [50,51,60].

3.2. Validation of Node–Link Water Balance Approach to Surface Water Modelling in SimCopiapo

As the main driver controlling surface water flows and the recharge to aquifers is the operation of the Lautaro Reservoir, we focused on reproducing the observed discharges measured at the gauging station immediately downstream of the Lautaro Reservoir (Figure 5). For the simulated period implemented in SimCopiapo (1991–2016), we observe a much better correspondence between observed and simulated discharges compared to the original surface water models implemented by [40,41]. In general, peak releases are properly simulated with the exceptions of hydrological years 1997/1998 and 2001/2002, where both the original surface water model and the node–link model implemented in SimCopiapo show difficulties in capturing the peak behaviour.

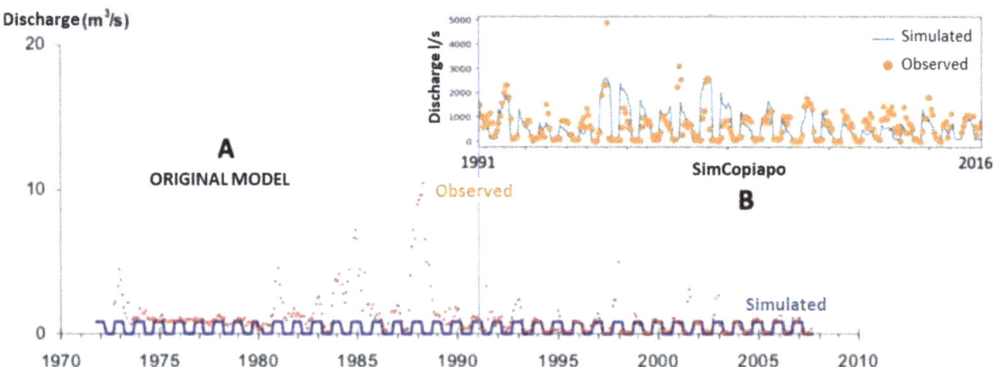

Figure 5. Release discharge from Lautaro Reservoir for the (**A**) original model developed by Hunter et al. [41] and Suarez et al. [40] and (**B**) adapted operational rule implemented in SimCopiapo.

Similarly, Figure 6 shows the simulated and observed discharges for the period 1991–2016 in La Puerta and Copiapó City gauging stations (see Figure 1 for locations). This figure shows that the operation of the Lautaro Reservoir and the improved supply–demand model for the irrigation districts implemented in the node–link water balance model in the SimCopiapo platform can reproduce the observed discharges in a reasonable manner, preserving the long-term trend and capturing relevant peaks in 1998/1999 and 2003/2004. It is worth noting that "Rio Copiapó en Ciudad" (Figure 6b) is a gauging station located downstream of all irrigation districts, and as such reflects excess volumes after irrigation use located upstream of Copiapó city.

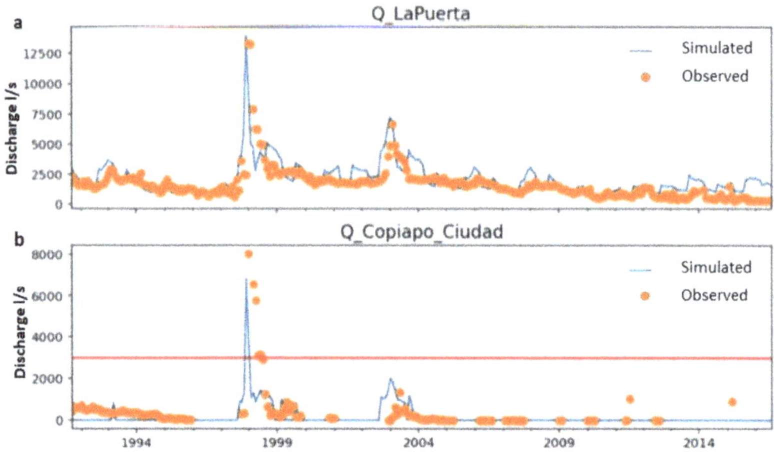

Figure 6. Simulated versus observed discharges in (**a**) "Rio Copiapó en La Puerta" and (**b**) "Rio Copiapó en Ciudad" gauging stations. Horizontal red line represents the average controlled discharge from Lautaro Reservoir (3020 L/s).

Closing the water balance obtained from loosely-coupled surface water and groundwater models has been identified as a drawback of the integration process [16,55]. La Puerta gauging station in the CRB is the main control point to verify that the coupling process of both the daily node–link mass balance model (implemented in Python) and the MODFLOW model (implemented in FloPy) is correct. In this sector, the Copiapó valley shows an important reduction in its cross section, and basement rocks are uplifted, thus substantially constraining the aquifer cross section, resulting in substantial contributions from the aquifer to the Copiapó River [39,49]. In general, discharges at La Puerta gauging station are influenced by the groundwater throughflows from groundwater sector 2 and the infiltration rates from the Lautaro Reservoir (see Figure 1). The infiltration losses in the Lautaro Reservoir were therefore adjusted between 500 L/s and 3500 L/s (as a function of the stored volume) until a reasonable match between the daily node–link mass balance model and groundwater model outputs at La Puerta was obtained. Figure 7 shows the match between both the node–link water balance and MODFLOW models at La Puerta gauging station. In general, we observe a good match between river gains from groundwater simulated through the drain package of MODFLOW and the node balance at La Puerta gauging station. We observe a good fit when simulating the temporality and magnitude of the time series, with a range between 1000 and 2500 L/s at "Rio Copiapó en La Puerta". Few discrepancies are observed at the start of the simulation period, most likely attributed to the stabilisation of parameters in the node–link model (warm-up period) and due to the aggregation of the monthly time steps into 6 month stress periods in MODFLOW.

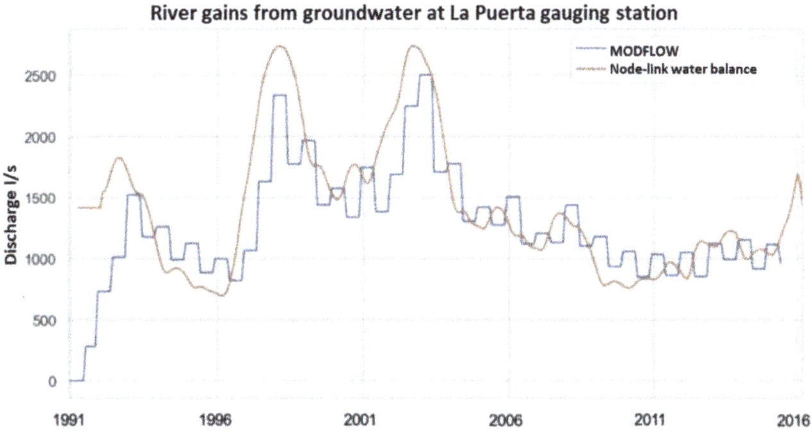

Figure 7. Validation of coupling process for daily surface water mass balance models and groundwater model at "Rio Copiapó en La Puerta" gauging station.

3.3. Results for Individual Water Management Strategies

To assess the results of individual (and combined) water management strategies we defined a base scenario reflecting hydraulic infrastructure, water demands, crop types and land uses corresponding to year 2018. This base scenario reflects a business-as-usual (BAU) approach. Both the base scenario and water management strategies are assessed for a 25-year period (2018–2042) in order to isolate the marginal impacts of implementing such strategies, using the observed hydrology for the period 1991–2016 as forcing data.

Table A1 in the Appendix A shows the individual results for each water management strategy described in Section 2.2.2. Individual water management strategies show spatially bounded impacts and marginal cumulative impacts at the end of the simulation period. Strategy 3.1, for example, shows an increase in infiltration flows in the Copiapó river of less than 3% the potential recharge volume due to constraints in the size of the recharge ponds and the available surface flows for infiltration. In the next sections, we analyse a selected group of results for individual strategies.

3.3.1. Water Uses/Rights Exchanges

For strategies promoting water uses/rights exchanges among users, the most attractive corresponds to strategy 1.3-c in Table 1, which promotes urban flows increases up to 263 L/s on average for the 25-year simulated period. Figure 8 shows the monthly frequency of the occurrence of urban flows for the base scenario and strategy 1.3-c. Stakeholders have defined the occurrence of urban flows through Copiapó city as an important indicator of quality of life, and results show a substantial increase in the occurrence of urban flows from 19% to 96% by implementing strategy 1.3-c.

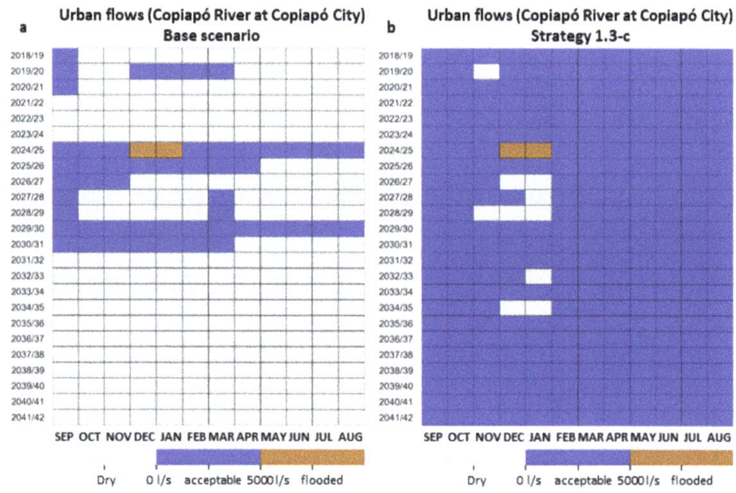

Figure 8. Monthly frequency of occurrence of urban flows at the Copiapó City gauging station for (**a**) base scenario and (**b**) strategy 1.3-c, for dry, acceptable, and flooded thresholds. Blue cells record the occurrence of average monthly discharges greater than 0 L/s and less than 5000 L/s at Copiapó city.

3.3.2. Improvement to Current Hydraulic Infrastructure

In terms of water management strategies promoting improvements to current hydraulic infrastructure, strategy 2.1 (impermeabilisation of Lautaro Reservoir) shows substantial impacts (positive and negative). Figure 9 shows the water balance for the Lautaro Reservoir for both the base scenario and strategy 2.1. For this figure, we observe that releases from the Lautaro Reservoir become regular and over 2000 L/s, with 18 out of 20 water years using the spillway to regulate the reservoir's capacity. After fully implementing the impermeabilisation of the inundated surface by year 5, infiltration losses become 0, and the reservoir volume is above 50% its capacity for 14 out of the 20 remaining years. Although not shown here, this results in increases in water security for irrigation districts no. 6, 7, 8 and 9, which are most closely located downstream of the reservoir. In addition, by implementing strategy 2.1, an increase in the frequency of occurrence of urban flows through Copiapó city from 18% to 30% of the months in the simulation period 2018–2042 is observed.

Despite the positive impacts from the surface water perspective, a negative impact is observed for the groundwater sector 2, located immediately downstream of the Lautaro Reservoir. Groundwater sector 2 is a narrow tube-like aquifer recharged mainly through upstream groundwater throughflows originating from upstream aquifers and, most importantly, infiltration losses from the Lautaro Reservoir. Figure 10 shows the groundwater levels of representative observation wells located in groundwater sector 2. After implementing the impermeabilisation of the Lautaro Reservoir, a sustained decreasing trend is observed in the mid- (MT) and long-term (LT), reaching average values of −0.8 m/y. It is worth noting that these decreasing trends concentrate in the upstream half of groundwater sector 2, whereas groundwater levels around La Puerta remain stable or increase given the constriction of the aquifer section explained in sections above.

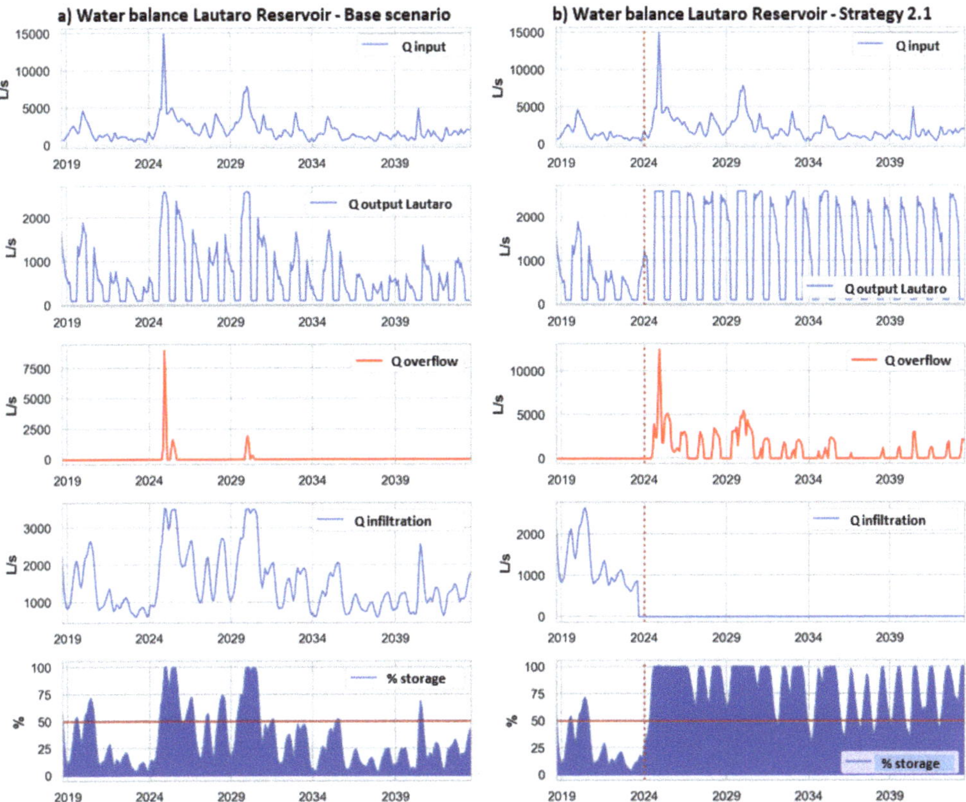

Figure 9. Water balance for the Lautaro Reservoir for (**a**) base scenario and (**b**) strategy 2.1, reflecting the impermeabilisation of 100% inundated surface after a 5 year period (vertical red dotted line).

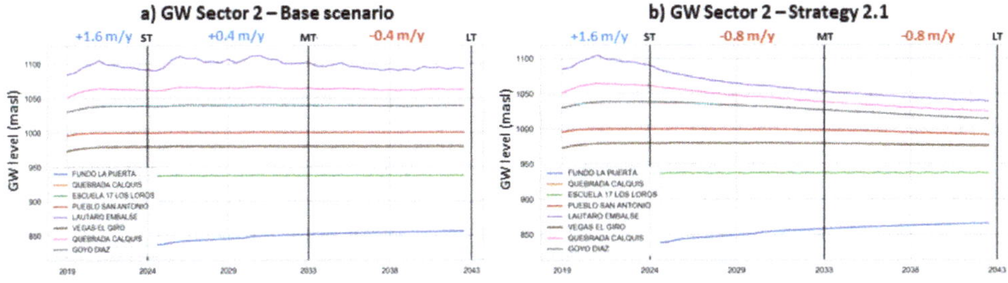

Figure 10. Time series of groundwater levels in representative observation wells located in the groundwater sector 2 for (**a**) base scenario, and (**b**) Strategy 2.1. ST: short-term, MT: mid-term, LT: long-term. Increasing average trend for each period for all observation wells in blue. Decreasing average trend for each period for all observation wells in red.

3.3.3. Demand Management

For the strategies promoting management of the groundwater demand, strategy 4.1 (a proportional reduction of 40% of groundwater use across groundwater sectors 3, 4 and 5) shows substantial impacts. This strategy was implemented through the ABM and assessed the level of compliance of the imposed cap on groundwater use and the impact on groundwater level/balance. Based on our previous experience, both monitoring of groundwater users and fine levels are strong deterrents when dealing with non-compliant behaviour in groundwater management [48,49]. Figure 11 shows two levels of enforcement tested in this strategy: (a) monitoring of 90% of users in groundwater sectors 3, 4 and 5 and substantial fine levels (90% gross profit from farm enterprise) if farmers are caught breaching the cap, thus defining a strong enforcement policy; and (b) lax monitoring (20% of users in groundwater sectors 3, 4 and 5) and fine levels (20% gross profit from farm enterprise), thus defining a weak enforcement policy. Figure 11a shows that for groundwater sectors 3, 4, 5 and 6, implementing the cap on groundwater use together with a strong enforcement policy brings storage volumes in these aquifer sectors back to values that are better than the initial state of the base scenario. On the contrary, when the enforcement policy is weak (Figure 11b), the impacts are limited to aquifer sector 5 and to a lesser extent in aquifer sector 6 given the proportional volume of groundwater for irrigation in these sectors. This indicates that at least 20% of the groundwater users of aquifers sectors 3, 4 and 5 need to be monitored if a cap on groundwater extraction is imposed by the regulator. For other impact indicators, implementing strategy 4.1 has a limited impact.

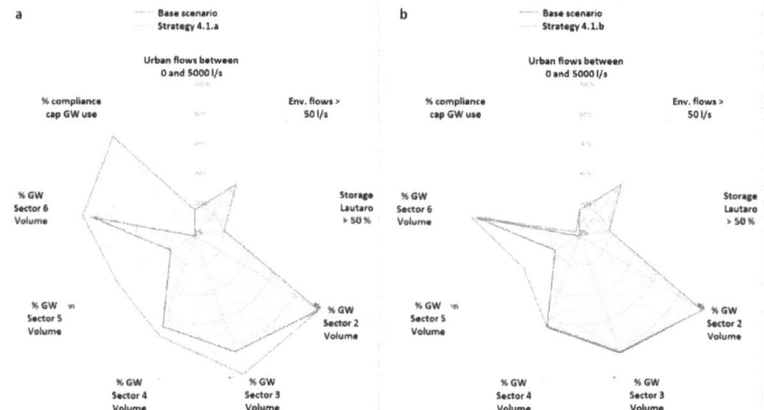

Figure 11. Impact indicators for strategy 4.1 (cap on groundwater use implemented through ABM) for two enforcement strategies: (**a**) high level of monitoring (90% of groundwater users) and fines (90% of revenue if farmer caught breaching the cap) and (**b**) low level of monitoring (20% of groundwater users) and fines (20% revenue if farmer caught breaching the cap).

3.4. Results for Combined Water Management Strategies

Any single water management strategy cannot address the basin-scale water management challenges identified in the CRB by different authors (see e.g., [34,40,41,45]). Impacts of individual strategies are sectoral and, in some cases, spatially and temporally constrained. It then seems appropriate to combine alternative water management strategies to assess multiple stakeholders' interests/perspectives. Based on the participatory process, a prioritised combination of water management strategies attractive to stakeholders of the CRB was defined by simultaneously implementing strategies 1.3-c, 2.1, 2.3, and 4.1a (see Table 1). Strategy 2.3, which corresponds to operating a desalination plant to supply 90 L/s (first

5 years), 450 L/s (next 5 years) and 930 L/s (last 15 years), has been included as this strategy is already in early operation in the CRB.

Figure 12 shows the results for impact indicators when combining these strategies. We found a substantial increase in the occurrence of urban flows from 20% to ca. 100% of the simulated period (25 years), thus enhancing the quality of life perceived by stakeholders of the CRB; marginal increases in the environmental flows at the basin outlet, thus promoting a healthy habitat for the wetland at the Copiapó River mouth; substantial increases in the storage volume of the Lautaro Reservoir from 20% to 72% of the simulated time with volumes greater than 50% its maximum storage capacity, thus impacting water security for irrigation districts; recoveries in heads and stored volumes in groundwater sectors 3, 4, 5 and 6, thus decreasing pumping costs to users and contributing to groundwater sustainability in the long-term; high levels of compliance (>80% groundwater users) to caps on groundwater use supported by a robust enforcement policy formulated around high monitoring rates and substantial fines.

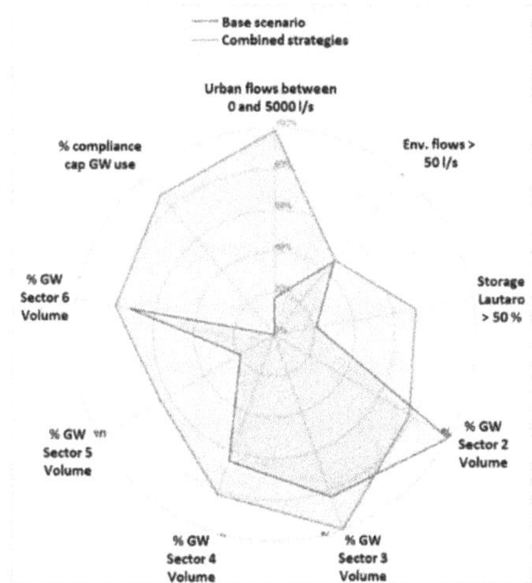

Figure 12. Impact indicators for the combinatory of water management strategies prioritised through the participatory process.

All these positive impacts also carry a negative impact, which is the detrimental impact on groundwater heads and stored volumes in the upper section of the aquifer sector 2, immediately downstream of the Lautaro Reservoir. Figure 13 shows that after implementing the combination of water management strategies, decreases in groundwater heads in the upper section of aquifer sector 2 can reach up to 40 m compared to the base scenario. Long-term increases in groundwater heads in sectors 3 and 4 can reach between 30 m and 40 m in aquifers immediately downstream of La Puerta gauging station and around 10 m in aquifer sector 5 downstream of Copiapó City gauging station.

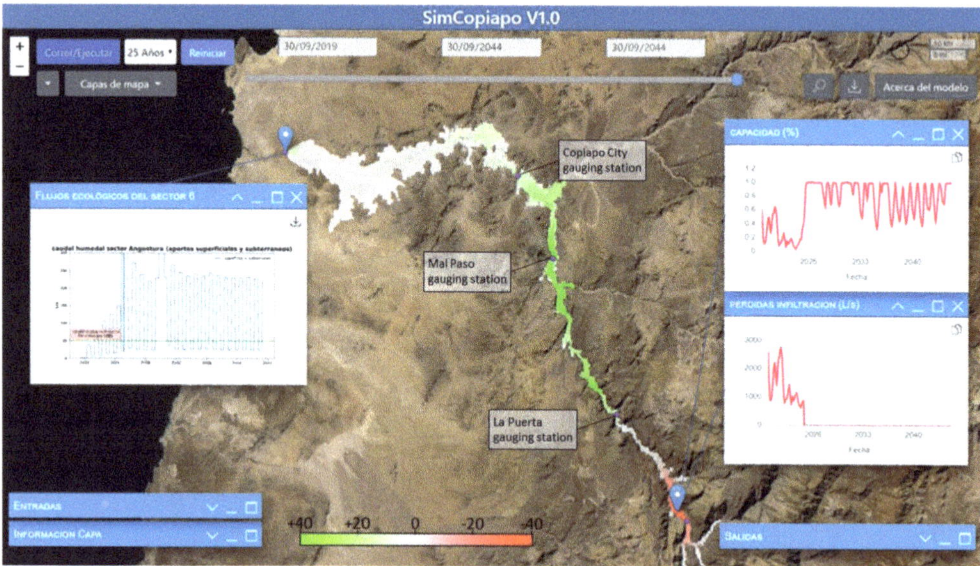

Figure 13. Groundwater head difference (m) between base scenario and combined water strategies. Inside panels show results of Lautaro Reservoir water balance (storage (% maximum volume) and infiltration losses (L/s)) and river flows (L/s) at the outlet of the CRB.

4. Discussion

The situation in the CRB indicates human and environmental vulnerabilities [38] stemming from the sustained exploitation of groundwater resources. Several authors have tried to explain the factors driving the water crisis in the CRB from economic, regulatory and management perspectives [34,45,46,48], whereas others have suggested technical solutions such as basin-scale or sectorial groundwater use restrictions [40,41,49]. Evidence by Wurl et al. [65] in a similar context (arid overexploited aquifer in Mexico) suggests that water management problems can no longer be addressed purely as technical problems and should consider a wide range of stakeholders' perspectives to strengthen the resilience of water resources. In this work, we have contributed towards this by devising potential water management strategies (i.e., technical solutions) and relevant impact indicators through a bottom-up participatory process driven by key stakeholders in the CRB.

Our approach directly addresses one of the tasks Rinaudo and Donoso [45] suggest the regulator should implement as part of a groundwater management model; i.e., implement an efficient enforcement strategy, e.g., by proposing minimum monitoring coverage and fine levels to achieve compliance on cap reductions based on ABM results. Although not explicitly addressed by our integrated modelling approach, the remaining tasks defined by Donoso and Rinuado (i.e., calculate sustainable groundwater abstraction limits, defining sharing rules, reallocation of water use rights and rules to adjust volume of water use rights) can be assessed implementing minor modifications to the SimCopiapo tool (e.g., test rules to adjust volumes of groundwater rights in different aquifer sectors).

SimCopiapo can be classified as a policy-oriented model [24] where adequate process representation, addressing practical policy issues and supporting decision-making with stakeholder participation are regarded as key features [42]. One of the purposes of the participatory modelling was social learning and acceptance of model improvements through direct participation in designing the conceptual model, water management strategies and impact indicators for discussion. Following the classification of Hare [42] the participatory modelling exercise implemented in SimCopiapo aligns with a Front and Back-End (FABE) category, where stakeholder involvement concentrates on early (conceptual model design, definition of operational rules, impact indicators and water management strategies) and later stages (assessment of water management strategies, discussion of potential policy pathways) of the modelling process. The effectiveness of the methods used in this participatory process is yet to be asserted through a follow-up process with decision and policy-makers of the CRB. A promising research direction is a post-hoc assessment using boundary objects attributes (credibility, salience, legitimacy) as suggested by Falconi and Palmer [43] to assess the success of SimCopiapo as a participatory model.

Our integrated approach and proposed digital platform (SimCopiapo) contribute to addressing the challenges identified in the use of models to operationalise IWRM by Badham et al. [66]. First, the bottom-up participatory process contributed to addressing a difficult problem in water policy by streamlining multiple pressures, conflicting values, competing goals and limited resources in a transparent way; and second, it helped in handling the human element in IWRM by reconciling potential conflictive agendas by stakeholders. The latter has been recognised as an important research avenue in water management [46,67,68].

Results show that no single strategy is able to provide definite long-term solutions to the water management challenges observed in the CRB nor to capture the multiplicity of stakeholders' interests expressed through the impact indicators identified. DGA-DICTUC [49], Suarez et al. [40] and Hunter et al. [41] proposed basin-wide or sectoral reductions in groundwater use (demand management) by values between 20% and 50% on the basis of cost analysis or a multi-dimensional measure of sustainability. While useful, assuming monetary motivations are central to water management and a key driver of behavioural change in groundwater users ignores the role that social, ecological and cultural values might have in this regard, thus constraining the assessment of water management strategies [69,70]. SimCopiapo contributes to equilibrating the assessment by transparently assessing physical, ecological and social aspects of the water management strategies devised, thus counter-balancing the bias towards exclusively cost-based assessments observed in the literature (e.g., [71–73]).

Our results indicate that management of groundwater use is one of the most critical water strategies to recover the aquifer sectors in the CRB. However, there needs to be a clear enforcement policy to trigger the minimum momentum required to achieve social acceptance of this policy. This is fully aligned to one of the drivers suggested by Rinaudo and Donoso [45] triggering the water crisis in the CRB. Results indicate that there seems to be a middle point between lax and strong enforcement policies to achieve this reduction in demand in a sustainable way. The definition of where this middle point lies is beyond the scope of this article, but our results bring a first approximation to this; i.e., between 20% and 90% monitoring coverage and between lax (fines accounting for 20% revenue) and strong (fines accounting for up to 90% revenue) fine levels for breaching the imposed cap on groundwater use. These results are fully aligned with findings by Castilla-Rho et al. [51,52] for other aquifers around the globe.

Future research avenues might consider including crop choices in the supply–demand and ABM models, implementation of other water management strategies and impact indicators, optimising the level of monitoring and fines to achieve a target compliance in different aquifer sectors or groundwater user groups (e.g., mining, industry) and testing multi-level ABM parametrisations for time-varying water management policies as in Du et al. [74].

5. Conclusions

In this work, we demonstrated the value of combining participatory modelling and integrated modelling to develop a digital platform tool (SimCopiapo) supporting social learning and knowledge co-construction for water management in a semiarid basin under contentious water use. We have contributed to transparently positioning a policy-based model as a boundary object to bridge a diverse group of stakeholders with individual and competing interests and perspectives.

Current water management in the Copiapó River Basin (CRB) is not providing the required solutions for resource sustainability, with previous research to date proposing (optimised) cost-based technical solutions focusing solely on groundwater demand reductions. This narrow perspective however does not seem to hold for complex water management problems with no single/simple solutions that act and depend on values and priorities by multiple stakeholders. Early stakeholder engagement and participation for social learning and knowledge co-construction are therefore essential steps in this process.

Our results suggest that management of groundwater demand in the CRB together with a portfolio of strategies including water rights exchanges, improvements to hydraulic infrastructure and robust enforcement policies are best suited to capture the diversity of stakeholders' interests and perspectives when addressing the water management challenges observed in the CRB. This diversity is expressed through a series of impact indicators, which, directly or indirectly, are a reflection of not only available groundwater resources for use but also ecological (e.g., basin outlets) and social (e.g., urban flows) aspects of relevance to stakeholders. An important aspect to manage groundwater demand is the establishment of an efficient enforcement policy to monitor caps imposed on groundwater use. In the absence of a clear policy and an institutional/legal framework to achieve this, water users' behaviours will continue to be non-cooperative, therefore leading to unsustainable groundwater use in the long-term.

Finally, we can conclude that including stakeholders in the participatory modelling process has led to the definition of a rich and diverse collection of water management strategies and ways to assess these strategies, which are a good reflection of stakeholders' interests and visions. This indicates that not only the final product—i.e., the SimCopiapo digital platform—is of value but also the process leading to its creation.

Author Contributions: Conceptualisation, R.R., J.C.-R. and G.B.; methodology, R.R. and J.C.-R.; software, R.B. and J.C.-R.; validation, R.R., J.C.-R., G.B. and C.P.; formal analysis, R.R. and J.C.-R.; investigation, R.R., J.C.-R., G.B. and C.P.; resources, E.C.; data curation, R.R., J.C.-R., G.B., R.B. and C.P.; writing—original draft preparation, R.R.; writing—review and editing, R.R., J.C.-R., G.B., R.B., E.C. and C.P.; visualisation, R.R. and J.C.-R.; supervision, R.R. and E.C.; project administration, E.C. and G.B.; funding acquisition, R.R., G.B., J.C.-R. and E.C. All authors have read and agreed to the published version of the manuscript.

Funding: This research was funded by the Atacama Regional Government (Chile) through the "Fondo de Innovación para la Competitividad", FIC-R 2016, código BIP N° 30486475.

Institutional Review Board Statement: Not applicable.

Informed Consent Statement: Informed consent was obtained from all subjects involved in the study, and clearance from the CSIRO Ethics Committee was obtained for this research (097/18).

Data Availability Statement: All data collected and necessary to reproduce this research are available from CSIRO Research Data Planner (rdp.csiro.au) and the corresponding project website https://research. csiro.au/gestion-copiapo/en/projects-simcopiapo-participative-modelling-for-water-management/, accessed on 14 March 2022.

Acknowledgments: We are grateful to the many regional/local stakeholders who took part in the participatory sessions throughout this research. Authors also thank Alexis Medina and Victor Galvez for comments made to improve earlier versions of this manuscript.

Conflicts of Interest: The authors declare no conflict of interest.

Appendix A

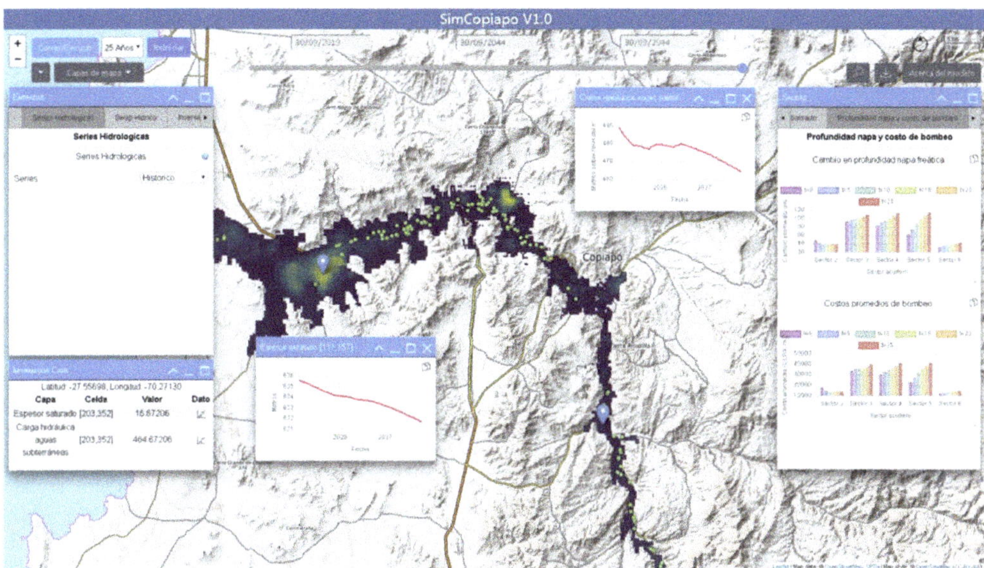

Figure A1. Graphical user interface for SimCopiapo platform v1.0 (in Spanish).

Table A1. Summary of impacts for water management strategies implemented in SimCopiapo.

Water Management Strategy		Description	Simulated Impact
Water use/rights exchanges	1.1	Water use/right exchange between Candelaria–Aguas Chanar	(o+) Groundwater heads/volumes increase in GW sectors 5 and 6
	1.2	Water use/right exchange between Caserones–Ramadilla River	(o−) Stored volumes in Lautaro Reservoir decrease (o−) Groundwater heads/volumes decrease in GW sectors 3 and 4 (o−) Urban flows through Copiapó city decrease (o−) Irrigation security decreases in districts 1, 7, 8 and 9
	1.3-a	Water use/right exchange between SW Irrigation districts 8 and 9–SW irrigation districts 1–7	(++) Urban flows through Copiapó city increase (o+) Groundwater heads/volumes increase in GW sectors 3, 4 and 5 (o+) Irrigation security increases in district 6 (o−) Irrigation security decreases in district 7
	1.3-b	Water use/right exchange between SW Irrigation districts 8 and 9–Aguas Chanar	(o+) Groundwater heads/volumes increase in GW sectors 4 (o+) Groundwater heads/volumes increase in GW sectors 3 and 5 (oo) No substantial impact detected in irrigation districts
	1.3-c	Water use/right exchange between SW Irrigation districts 8 and 9–localised recharge Copiapó River	(++) Urban flows through Copiapó city increase (o+) Groundwater heads/volumes increase in GW sectors 3, 4, 5 and 6 (oo) No substantial impact detected in irrigation districts
	1.3-d	Water use/right exchange between SW Irrigation districts 8 and 9–GW Sector 5/with excess localised recharge Copiapó River	(o+) Groundwater heads/volumes increase in GW sectors 3, 4, 5 and 6 (o+) Environmental flows at the outlet of the basin increase (oo) No substantial impact detected in irrigation districts
	1.3-e	Water use/right exchange between SW Irrigation districts 8 and 9–GW Sector 5/with excess Managed Aquifer Recharge in GW Sector 5	(o+) Groundwater heads/volumes increase in GW sectors 3, 4 and 5 (oo) No substantial impact detected in irrigation districts

Table A1. *Cont.*

Water Management Strategy		Description	Simulated Impact
Hydraulic infrastructure	2.1	Impermeabilisation Lautaro Reservoir (100% in a 5 year period)	(++) Stored volumes in Lautaro Reservoir increase (o+) Urban flows through Copiapó city increase (o+) Groundwater heads/volumes increase in GW sectors 3, 4, 5 and 6 (o+) Irrigation security increases in districts 6, 7, 8 and 9 (− −) Groundwater heads/volumes decrease in GW sector 2
	2.2	Surface water conveyance to irrigation sectors through pipes instead of open channels	(++) Irrigation security increases in all irrigation districts (o+) Groundwater heads/volumes increase in GW sectors 3 and 4
	2.3	Operation of desalination plant	(o+) Groundwater heads/volumes increase in GW sectors 5 and 6 (oo) No substantial impact detected in irrigation districts
Management of recharge	3.1	Managed aquifer recharge along the Copiapó River	(o+) Groundwater heads/volumes increase in GW sector 4 (oo) No substantial impact detected in irrigation districts
Demand management	4.1-a	Prorate of groundwater uses in GW sectors 3, 4 and 5 (high enforcement level: monitoring and fines at 90%)	(++) Groundwater heads/volumes increase in GW sectors 3 and 5 (++) Compliance with caps in groundwater use (o+) Groundwater heads/volumes increase in GW sectors 4 and 6 (oo) No substantial impact detected in irrigation districts
	4.1-b	Prorate of groundwater uses in GW sectors 3, 4 and 5 (low enforcement level: monitoring and fines at 20%)	(++) Groundwater heads/volumes increase in GW sector 5 (o+) Compliance with caps in groundwater use (o+) Groundwater heads/volumes increase in GW sector 6 (oo) No substantial impact detected in irrigation districts
	5.1	Greywater reuse/recirculation	(oo) No substantial impact detected

(oo): no decreases/increases over the simulation period with respect to base scenario; (− −): decreases over the simulation period more than 20% with respect to base scenario; (o−): decreases over the simulation period less than 20% with respect to base scenario; (o+): increases over the simulation period less than 20% with respect to base scenario; (++): increases over the simulation period more than 20% with respect to base scenario.

References

1. Vörösmarty, C.J.; McIntyre, P.B.; Gessner, M.O.; Dudgeon, D.; Prusevich, A.; Green, P.; Glidden, S.; Bunn, S.E.; Sullivan, C.A.; Liermann, C.R.; et al. Global threats to human water security and river biodiversity. *Nature* **2010**, *467*, 555–561. [CrossRef] [PubMed]
2. Vörösmarty, C.J.; Green, P.; Salisbury, J.; Lammers, R.B. Global water resources: Vulnerability from climate change and population growth. *Science* **2000**, *289*, 284–288. [CrossRef] [PubMed]
3. Biggs, E.M.; Bruce, E.; Boruff, B.; Duncan, J.M.A.; Horsley, J.; Pauli, N.; McNeill, K.; Neef, A.; Van Ogtrop, F.; Curnow, J.; et al. Sustainable development and the water-energy-food nexus: A perspective on livelihoods. *Environ. Sci. Policy* **2015**, *54*, 389–397. [CrossRef]
4. Ringler, C.; Bhaduri, A.; Lawford, R. The nexus across water, energy, land and food (WELF): Potential for improved resource use efficiency? *Curr. Opin. Environ. Sustain.* **2013**, *5*, 617–624. [CrossRef]
5. United Nations. *The United Nations World Water Development Report 2018—Nature-Based Solutions for Water*; UNESCO: Paris, France, 2018.
6. Hoff, H. Global water resources and their management. *Curr. Opin. Environ. Sustain.* **2009**, *1*, 141–147. [CrossRef]
7. Buytaert, W.; Zulkafli, Z.; Grainger, S.; Acosta, L.; Alemie, T.C.; Bastiaensen, J.; De Bièvre, B.; Bhusal, J.; Clark, J.; Dewulf, A.; et al. Citizen science in hydrology and water resources: Opportunities for knowledge generation, ecosystem service management, and sustainable development. *Front. Earth Sci.* **2014**, *2*, 1–21. [CrossRef]
8. Paul, J.D.; Buytaert, W.; Allen, S.; Ballesteros-Cánovas, J.A.; Bhusal, J.; Cieslik, K.; Clark, J.; Dugar, S.; Hannah, D.M.; Stoffel, M.; et al. Citizen science for hydrological risk reduction and resilience building. *WIREs Water* **2018**, *5*, 1262. [CrossRef]

9.   Warren, A. Collaborative Modelling in Water Resources Management. Master's Thesis, Delft University of Technology, TUDelft, Delft, The Netherlands, 2016.
10. Zare, F.; Elsawah, S.; Iwanaga, T.; Jakeman, A.J.; Pierce, S.A. Integrated water assessment and modelling: A bibliometric analysis of trends in the water resource sector. *J. Hydrol.* **2017**, *552*, 765–778. [CrossRef]
11. GWP. *Integrated Water Resources Management*; Global Water Partnership: Stockholm, Sweden, 2012; ISBN 9163092298.
12. Biswas, A.K. Integrated water resources management: A reassessment: A water forum contribution. *Water Int.* **2004**, *29*, 248–256. [CrossRef]
13. Kelly, R.A.; Jakeman, A.J.; Barreteau, O.; Borsuk, M.E.; ElSawah, S.; Hamilton, S.H.; Henriksen, H.J.; Kuikka, S.; Maier, H.R.; Rizzoli, A.E.; et al. Selecting among five common modelling approaches for integrated environmental assessment and management. *Environ. Model. Softw.* **2013**, *47*, 159–181. [CrossRef]
14. Rojas, R.; Feyen, L.; Watkiss, P. Climate change and river floods in the European Union: Socio-economic consequences and the costs and benefits of adaptation. *Glob. Environ. Chang.* **2013**, *23*, 1737–1751. [CrossRef]
15. Barthel, R. HESS Opinions "Integration of groundwater and surface water research: An interdisciplinary problem?". *Hydrol. Earth Syst. Sci.* **2014**, *18*, 2615–2628. [CrossRef]
16. Barthel, R.; Banzhaf, S. Groundwater and Surface Water Interaction at the Regional-scale—A Review with Focus on Regional Integrated Models. *Water Resour. Manag.* **2016**, *30*, 1–32. [CrossRef]
17. Basco-Carrera, L.; Warren, A.; van Beek, E.; Jonoski, A.; Giardino, A. Collaborative modelling or participatory modelling? A framework for water resources management. *Environ. Model. Softw.* **2017**, *91*, 95–110. [CrossRef]
18. Jakeman, A.J.; El Sawah, S.; Guillaume, J.H.A.; Pierce, S.A. Making progress in integrated modelling and environmental decision support. *IFIP Adv. Inf. Commun. Technol.* **2011**, *359*, 15–25. [CrossRef]
19. Voinov, A.; Bousquet, F. Modelling with stakeholders. *Environ. Model. Softw.* **2010**, *25*, 1268–1281. [CrossRef]
20. Voinov, A.; Kolagani, N.; McCall, M.K.; Glynn, P.D.; Kragt, M.E.; Ostermann, F.O.; Pierce, S.A.; Ramu, P. Modelling with stakeholders—Next generation. *Environ. Model. Softw.* **2016**, *77*, 196–220. [CrossRef]
21. Bots, P.W.G.; Bijlsma, R.; von Korff, Y.; van der Fluit, N.; Wolters, H. Supporting the constructive use of existing hydrological models in participatory settings: A set of "rules of the game". *Ecol. Soc.* **2011**, *16*, 1–16. [CrossRef]
22. Seidl, R. A functional-dynamic reflection on participatory processes in modeling projects. *Ambio* **2015**, *44*, 750–765. [CrossRef]
23. Voinov, A.; Gaddis, E.J.B. Lessons for successful participatory watershed modeling: A perspective from modeling practitioners. *Ecol. Modell.* **2008**, *216*, 197–207. [CrossRef]
24. Oxley, T.; McIntosh, B.S.; Winder, N.; Mulligan, M.; Engelen, G. Integrated modelling and decision-support tools: A Mediterranean example. *Environ. Model. Softw.* **2004**, *19*, 999–1010. [CrossRef]
25. Pahl-Wostl, C.; Craps, M.; Dewulf, A.; Mostert, E.; Tabara, D.; Taillieu, T. Social learning and water resources management. *Ecol. Soc.* **2007**, *12*, 1–19. [CrossRef]
26. Mostert, E.; Craps, M.; Pahl-Wostl, C. Social learning: The key to integrated water resources management? *Water Int.* **2008**, *33*, 293–304. [CrossRef]
27. Carr, G.; Blöschl, G.; Loucks, D.P. Evaluating participation in water resource management: A review. *Water Resour. Res.* **2012**, *48*, 11401. [CrossRef]
28. Voinov, A.; Gaddis, E.B. Values in Participatory Modeling: Theory and Practice. In *Environmental Modeling with Stakeholders: Theory, Methods, and Applications*; Gray, S., Paolisso, M., Jordan, R., Gray, S., Eds.; Springer: Cham, Switzerland, 2017; pp. 47–63. ISBN 9783319250533.
29. Robinson, K.F.; Fuller, A.K. Participatory Modelling and Structured Decision Making. In *Environmental Modeling with Stakeholders: Theory, Methods, and Applications*; Gray, S., Paolisso, M., Jordan, R., Gray, S., Eds.; Springer: Cham, Switzerland, 2017; pp. 83–101. ISBN 9783319250533.
30. Penny, G.; Goddard, J.J. Resilience principles in socio-hydrology: A case-study review. *Water Secur.* **2018**, *4–5*, 37–43. [CrossRef]
31. Carmona, G.; Varela-Ortega, C.; Bromley, J. Participatory modelling to support decision making in water management under uncertainty: Two comparative case studies in the Guadiana river basin, Spain. *J. Environ. Manag.* **2013**, *128*, 400–412. [CrossRef] [PubMed]
32. van Bruggen, A.; Nikolic, I.; Kwakkel, J. Modeling with stakeholders for transformative change. *Sustainability* **2019**, *11*, 825. [CrossRef]
33. Grosse Kathoefer, D.; Leker, J. Knowledge transfer in academia: An exploratory study on the Not-Invented-Here Syndrome. *J. Technol. Transf.* **2012**, *37*, 658–675. [CrossRef]
34. Galvez, V.; Rojas, R.; Bennison, G.; Prats, C.; Claro, E. Collaborate or perish: Water resources management under contentious water use in a semiarid basin. *Int. J. River Basin Manag.* **2020**, *18*, 421–437. [CrossRef]
35. Xu, S.; Ou, J. Good Water Governance for the Sustainable Development of the Arid and Semi-arid Areas of Northwest China. *IOP Conf. Ser. Earth Environ. Sci.* **2018**, *199*, 1–8. [CrossRef]
36. Tian, Y.; Xiong, J.; He, X.; Pi, X.; Jiang, S.; Han, F.; Zheng, Y. Joint Operation of Surface Water and Groundwater Reservoirs to Address Water Conflicts in Arid Regions: An Integrated Modeling Study. *Water* **2018**, *10*, 1105. [CrossRef]
37. Tian, Y.; Zheng, Y.; Wu, B.; Wu, X.; Liu, J.; Zheng, C. Modeling surface water-groundwater interaction in arid and semi-arid regions with intensive agriculture. *Environ. Model. Softw.* **2015**, *63*, 170–184. [CrossRef]

38. Gorelick, S.M.; Zheng, C. Global change and the groundwater management challenge. *Water Resour. Res.* **2015**, *51*, 3031–3051. [CrossRef]
39. DGA-HIDROMAS. *Actualización de la Modelación Integrada y Subterránea del Acuífero de la Cuenca del río Copiapó*; Direccion General de Aguas: Santiago, Chile, 2013.
40. Suárez, F.; Muñoz, J.F.; Fernández, B.; Dorsaz, J.M.; Hunter, C.K.; Karavitis, C.A.; Gironás, J. Integrated water resource management and energy requirements for water supply in the Copiapó River basin, Chile. *Water* **2014**, *6*, 2590–2613. [CrossRef]
41. Hunter, C.; Gironás, J.; Bolster, D.; Karavitis, C.A. A dynamic, multivariate sustainability measure for robust analysis of water management under climate and demand uncertainty in an arid environment. *Water* **2015**, *7*, 5928–5958. [CrossRef]
42. Hare, M. Forms of participatory modelling and its potential for widespread adoption in the water sector. *Environ. Policy Gov.* **2011**, *21*, 386–402. [CrossRef]
43. Falconi, S.M.; Palmer, R.N. An interdisciplinary framework for participatory modeling design and evaluation—What makes models effective participatory decision tools? *Water Resour. Res.* **2017**, *53*, 1625–1645. [CrossRef]
44. DGA. *Atlas del Agua*; Direccion General de Aguas: Santiago, Chile, 2016; Volume 1.
45. Rinaudo, J.D.; Donoso, G. State, market or community failure? Untangling the determinants of groundwater depletion in Copiapó (Chile). *Int. J. Water Resour. Dev.* **2019**, *35*, 283–304. [CrossRef]
46. Blanco, E.; Donoso, G. Drivers for collective groundwater management: The case of Copiapo, Chile. In *Global Water Security Issues (GWSI) 2020 Theme: The Role of Sound Groundwater Resources Management and Governance to Achieve Water Security*; UNESCO: Paris, France, 2020; pp. 1–18.
47. DGA-HIDRICA-ERIDANUS. *Plan Estratégico de Gestión Hídrica en la Cuenca de Copiapó*; Direccion General de Aguas: Santiago, Chile, 2020.
48. Bitran, E.; Rivera, P.; Villena, M.J. Water management problems in the Copiapó Basin, Chile: Markets, severe scarcity and the regulator. *Water Policy* **2014**, *16*, 844–863. [CrossRef]
49. DGA-DICTUC. *Cuenca Del Río Copiapó Informe Final—Tomo I*; Direccion General de Aguas: Santiago, Chile, 2010; Volume 1.
50. Bauer, C.J. Water conflicts and entrenched governance problems in chile's market model. *Water Altern.* **2015**, *8*, 147–172.
51. Castilla-Rho, J.C.; Rojas, R.; Andersen, M.S.; Holley, C.; Mariethoz, G. Social tipping points in global groundwater management. *Nat. Hum. Behav.* **2017**, *1*, 640–649. [CrossRef] [PubMed]
52. Castilla-Rho, J.C.; Rojas, R.; Andersen, M.S.; Holley, C.; Mariethoz, G. Sustainable groundwater management: How long and what will it take? *Glob. Environ. Chang.* **2019**, *58*, 101972. [CrossRef]
53. Santos Coelho, R.; Lopes, R.; Coelho, P.S.; Ramos, T.B.; Antunes, P. Participatory selection of indicators for water resources planning and strategic environmental assessment in Portugal. *Environ. Impact Assess. Rev.* **2022**, *92*, 106701. [CrossRef]
54. Bekele, E.B.; Donn, M.; Barry, K.; Vanderzalm, J.; Kaksonen, A.; Puzon, G.; Wylie, J.; Miotlinkski, K.; Cahill, K.; Walsh, T.; et al. *Managed Aquifer Recharge and Recycling Options: Understanding Clogging Processes and Water Quality Impacts*; Australian Water Recycling Centre of Excellence: Brisbane, Australia, 2015.
55. Haque, A.; Salama, A.; Lo, K.; Wu, P. Surface and groundwater interactions: A review of coupling strategies in detailed domain models. *Hydrology* **2021**, *8*, 35. [CrossRef]
56. Ince, D.C.; Hatton, L.; Graham-Cumming, J. The case for open computer programs. *Nature* **2012**, *482*, 485–488. [CrossRef]
57. Nature Editorial. Code share. *Nature* **2014**, *514*, 536. [CrossRef] [PubMed]
58. Nature Geoscience Editorial. Towards Transparency. *Nat. Geosci.* **2014**, *7*, 777. [CrossRef]
59. Andreu, J.; Capilla, J.; Sanchis, E. AQUATOOL, a generalized decision-support system for water-resources planning and operational management. *J. Hydrol.* **1996**, *177*, 269–291. [CrossRef]
60. Harbaugh, A.W. *MODFLOW-2005, the U.S. Geological Survey Modular Ground-Water Model—The Ground-Water Flow Process*; US Department of the Interior, US Geological Survey: Reston, VA, USA, 2005.
61. Bakker, M.; Post, V.; Langevin, C.D.; Hughes, J.D.; White, J.T.; Starn, J.J.; Fienen, M.N. Scripting MODFLOW Model Development Using Python and FloPy. *Groundwater* **2016**, *54*, 733–739. [CrossRef]
62. Bennison, G.; Rojas, R.; Castilla-Rho, J.C.; Prats, C.; Bridgart, R. *SimCopiapó: Modelación Participativa Para la Gestión del Agua*; Fundacion CSIRO Chile Research: Santiago, Chile, 2019.
63. Douglas, M. *A History of Grid and Group Cultural Theory*; University of Toronto: Toronto, ON, Canada, 2007.
64. Yates, D.; Sieber, J.; Purkey, D.; Huber-Lee, A. WEAP21—A demand-, priority-, and preference-driven water planning model. Part 1: Model characteristics. *Water Int.* **2005**, *30*, 487–500. [CrossRef]
65. Wurl, J.; Gámez, A.E.; Ivanova, A.; Imaz Lamadrid, M.A.; Hernández-Morales, P. Socio-hydrological resilience of an arid aquifer system, subject to changing climate and inadequate agricultural management: A case study from the Valley of Santo Domingo, Mexico. *J. Hydrol.* **2018**, *559*, 486–498. [CrossRef]
66. Badham, J.; Elsawah, S.; Guillaume, J.H.A.; Hamilton, S.H.; Hunt, R.J.; Jakeman, A.J.; Pierce, S.A.; Snow, V.O.; Babbar-Sebens, M.; Fu, B.; et al. Effective modeling for Integrated Water Resource Management: A guide to contextual practices by phases and steps and future opportunities. *Environ. Model. Softw.* **2019**, *116*, 40–56. [CrossRef]
67. Suárez, F.; Leray, S.; Sanzana, P. Groundwater resources. In *Water Resources in Chile*; Fernandez, B., Gironas, J., Eds.; Springer: Cham, Switzerland, 2021; Volume 8, pp. 93–127. ISBN 9783030569013.

68. Li, X.; Zhang, L.; Zheng, Y.; Yang, D.; Wu, F.; Tian, Y.; Han, F.; Gao, B.; Li, H.; Zhang, Y.; et al. Novel hybrid coupling of ecohydrology and socioeconomy at river basin scale: A watershed system model for the Heihe River basin. *Environ. Model. Softw.* **2021**, *141*, 105058. [CrossRef]
69. Douglas, M.M.; Jackson, S.; Canham, C.A.; Laborde, S.; Beesley, L.; Kennard, M.J.; Pusey, B.J.; Loomes, R.; Setterfield, S.A. Conceptualizing Hydro-socio-ecological Relationships to Enable More Integrated and Inclusive Water Allocation Planning. *One Earth* **2019**, *1*, 361–373. [CrossRef]
70. Heinrichs, D.H.; Rojas, R. Cultural values in water management and governance: Where do we stand? *Water* **2022**, *14*, 803. [CrossRef]
71. Medellín-Azuara, J.; Howitt, R.E.; Harou, J.J. Predicting farmer responses to water pricing, rationing and subsidies assuming profit maximizing investment in irrigation technology. *Agric. Water Manag.* **2012**, *108*, 73–82. [CrossRef]
72. Smidt, S.J.; Haacker, E.M.K.; Kendall, A.D.; Deines, J.M.; Pei, L.; Cotterman, K.A.; Li, H.; Liu, X.; Basso, B.; Hyndman, D.W. Complex water management in modern agriculture: Trends in the water-energy-food nexus over the High Plains Aquifer. *Sci. Total Environ.* **2016**, *566–567*, 988–1001. [CrossRef]
73. Delgado-Galván, X.; Izquierdo, J.; Benítez, J.; Pérez-García, R. Joint stakeholder decision-making on the management of the Silao-Romita aquifer using AHP. *Environ. Model. Softw.* **2014**, *51*, 310–322. [CrossRef]
74. Du, E.; Tian, Y.; Cai, X.; Zheng, Y.; Han, F.; Li, X.; Zhao, M.; Yang, Y.; Zheng, C. Evaluating Distributed Policies for Conjunctive Surface Water-Groundwater Management in Large River Basins: Water Uses Versus Hydrological Impacts. *Water Resour. Res.* **2022**, *58*, e2021WR031352. [CrossRef]

 *hydrology*

Article

Inferring Hydrological Information at the Regional Scale by Means of $\delta^{18}O$–δ^2H Relationships: Insights from the Northern Italian Apennines

Federico Cervi [1,*] and Alberto Tazioli [2]

[1] Independent Researcher, 42122 Reggio Emilia, Italy
[2] Dipartimento di Scienze e Ingegneria della Materia, dell'Ambiente ed Urbanistica, Università Politecnica delle Marche, 60131 Ancona, Italy; a.tazioli@staff.univpm.it
* Correspondence: fd.cervi@gmail.com

Abstract: We compared five regression approaches, namely, ordinary least squares, major axis, reduced major axis, robust, and Prais–Winsten to estimate $\delta^{18}O$–δ^2H relationships in four water types (precipitation, surface water, groundwater collected in wells from lowlands, and groundwater from low-yield springs) from the northern Italian Apennines. Differences in terms of slopes and intercepts of the different regressions were quantified and investigated by means of univariate, bivariate, and multivariate statistical analyses. We found that magnitudes of such differences were significant for water types surface water and groundwater (both in the case of wells and springs), and were related to robustness of regressions (i.e., standard deviations of the estimates and sensitiveness to outliers). With reference to surface water, we found the young water fraction was significant in inducing changes of slopes and intercepts, leading us to suppose a certain role of kinetic fractionation processes as well (i.e., modification of former water isotopes from both snow cover in the upper part of the catchments and precipitation linked to pre-infiltrative evaporation and evapotranspiration processes). As final remarks, due to the usefulness of $\delta^{18}O$–δ^2H relationships in hydrological and hydrogeological studies, we provide some recommendations that should be followed when assessing the abovementioned water types from the northern Italian Apennines.

Keywords: stable water isotopes; young water fraction; global meteoric water line; northern Italian Apennines

Citation: Cervi, F.; Tazioli, A. Inferring Hydrological Information at the Regional Scale by Means of $\delta^{18}O$–δ^2H Relationships: Insights from the Northern Italian Apennines. *Hydrology* **2022**, *9*, 41. https://doi.org/10.3390/hydrology9020041

Academic Editors: Il-Moon Chung, Sun Woo Chang, Yeonsang Hwang and Yeonjoo Kim

Received: 6 January 2022
Accepted: 16 February 2022
Published: 21 February 2022

Publisher's Note: MDPI stays neutral with regard to jurisdictional claims in published maps and institutional affiliations.

1. Introduction

Oxygen (^{18}O and ^{16}O) and hydrogen isotopes (^2H and ^1H) of water are commonly used in surface and subsurface hydrology [1–3]. They are considered environmental tracers in the form of $\delta^{18}O$ and δ^2H, where $\delta(‰) = \left[\left(\frac{R_{sample}}{R_{standard}} - 1 \right) \times 1000 \right]$ and R is the corresponding isotopic ratio ($^{18}O/^{16}O$ or $^2H/^1H$) in the water sample or in the standard (usually V-SMOW, i.e., the Vienna Standard Mean Oceanic Water). If a multitude of water samples is collected from the same source (a rain gauge, a river, a spring, or a well), the corresponding $\delta^{18}O$–δ^2H pairs in a Cartesian graph will be aligned along a regression line in the form of y = mx + q, where y is δ^2H, x is $\delta^{18}O$, m is the slope, and q the intercept. This fact was first noted by [4] when reporting several $\delta^{18}O$ and δ^2H values from precipitation waters worldwide, which allowed definition of the so-called "global meteorological water line" (GMWL) equal to $\delta^2H(‰) = 8.0 \cdot \delta^{18}O + 10.0$. The authors of [5] have substantially confirmed this slope and intercept by processing more recent data. Although this relationship is valid everywhere, more accurate regression lines (with different slope and intercept) can be obtained by selecting $\delta^{18}O$ and δ^2H values from restricted areas. This is due to the specific isotopic fractionation processes (i.e., vapour pressure and temperature conditions) controlling precipitation over each area. Moreover, since further post-condensation and temperature-driven processes such as evaporation and evapotranspiration could act prior to infiltration and/or during runoff, $\delta^{18}O$–δ^2H regressions from rivers (river water lines, RWLs) and groundwater (groundwater lines, GWLs) may also differ from that of the precipitation occurring in their recharge areas (meteoric water lines, MWLs). In fact, evaporation and evapotranspiration lead to a fractionation between the different isotopologues of water, with lighter water molecules ($^1H_2{}^{16}O$) vaporising faster than heavier ones ($^2H_2{}^{18}O$) and inducing an enrichment of the latter into the liquid residual. In this case, as a water parcel evaporates, its isotopic composition usually evolves with a $\delta^{18}O$–δ^2H regression line whose slope is lower than those of MWLs.

It is evident that such changes in slopes from RWLs and GWLs concerning those of MWLs can be used to infer information on the hydrological processes occurring at the slope and the catchment scales. As an example, and without claiming to be exhaustive, the following studies highlighted that a change in slope from RWLs and GWLs concerning those of MWL can be used to:

- Display the role of the riparian zone in feeding base flow in low-relief and forested catchments [6];
- Highlight pre-infiltrative evaporation/evapotranspiration that acted by modifying the waters before their infiltration towards the aquifer and, subsequently, to the base flow of rivers [7];
- Demonstrate that groundwater may be fossil (related to other climate recharge condition) or actually recharged by losses from the streambed or even a mixing among these two components [8];
- Reveal instream or lake evaporation in a nonarid environment [9,10];
- Elucidate which component (precipitation or losses from the streambeds) is exclusive or prevalent in recharging alluvial aquifers [11].

Starting with the pioneering works by [4,12] regarding meteoric water and up to this day, $\delta^{18}O$–δ^2H regression lines are usually carried out by means of the ordinary least squares (OLS) method, i.e., an approach that minimises the sum of the squared vertical distances between the y data values and the corresponding y values on the fitted line (the predictions). Thus, the OLS design assumes that there is no variation in the independent variable (x) and is considered as the simplest method among the several available linear regression models. By focusing on the aforementioned isotopes of water, we should also take care of the variations associated with variable (x) as the same or different isotopic fractionation processes, which may have developed even at different rates and may have affected both δ-values.

For this reason, [13] proposed a more complex linear regression approach for obtaining MWLs lines that consider associated errors with both dependent (y) and independent variables (x), such as the reduced major axis (RMA) and the major axis (MA). In the end, it is found that MA approaches usually led to the smallest discrepancies between the estimated and predicted values (a measure of goodness of fit usually described with the well-known coefficient of determination R^2, i.e., smallest discrepancies are identified by higher values of R^2) and larger slopes in MWLs calculation than those obtained with RMA and OLS [14]. The recent attempt made by [15] on several water types (river water, groundwater, soil water) confirmed that RWLs and GWLs were also characterised by larger slopes in the case of MA regression while the lowest values were noted when using OLS. The same authors found that in all cases (MWLs, RWLs, GWLs), the higher the R^2 between $\delta^{18}O$ and δ^2H values, the smaller were the differences in the slopes obtained by MA, RMA, and OLS. This was due to the different sensitiveness of the linear regression approaches to outliers and large measurement errors rather than temperature-driven post-condensation processes.

This study aims to verify whether such discrepancies in slopes and intercepts from different regression methods are present (thus significant) or not in four water types (precipitation, surface water, groundwater collected in wells from the lowlands, groundwater from low-yield springs) from the northern Italian Apennines. For this reason, we exploited datasets already published in the literature, e.g., [7,11,16–18], and we carried out visual inspection (heat maps) and statistical comparison of results from the three aforementioned approaches (OLS, RMA, MA) already tested in [15]. With reference to OLS approach, we further verified whether preliminary weighting of the isotopic data to the corresponding values of discharge or precipitation may have induced changes to our results or not. In addition, we tested two methods, namely, Prais–Winsten (PW) and robust (R), to investigate possible influence on the final $\delta^{18}O$–δ^2H alignments of nonstationary processes (here, we recall that the OLS, RMA, and MA approaches are based on the assumption that data are not serially correlated, thus a $\delta^{18}O$–δ^2H pair from a determined time period is not correlated with the earlier one, while properties as mean, variance, and autocorrelation are constant over time, i.e., stationary, while recent papers in the literature highlighted the possible nonstationary behaviour of such series of isotopic data [19]) or even outliers (i.e., anomalous isotopic values) within the series of isotopes.

Furthermore, we provided a possible explanation for the geographic and climatic factors (i.e., catchment descriptors) influencing the several regressions and finally we made some considerations concerning their applicability in the context of mountainous areas such as the northern Italian Apennines.

2. Overview of the Climatic, Geomorphological, and Hydrogeological Features of the Study Area

The study area is located in the northern Italian Apennines and belongs to the Emilia–Romagna Region (Figure 1). It includes nine catchments between the Trebbia River and the Savio River, with river gauges (in which the samples were collected) located close to the foothills of the mountain chain. The area has an overall extension of 6261 km^2. Maximum altitudes are in the southern sectors, where the main watershed divide lies (with main peaks showing elevations higher than 2000 m a.s.l., such as Mt. Cimone with its 2165 m a.s.l.) is the southern border of the Emilia–Romagna Region. Elevation decreases towards the NE direction to approximately 40 m a.s.l. at the Savio River gauge.

All the nine rivers originate from the main watershed divide and flow towards the NE. Six rivers (namely, Trebbia, Nure, Taro, Enza, Secchia, and Panaro) are tributaries of the Po River while the other three (Reno, Lamone, and Savio) enter the Adriatic Sea. Catchment areas are between 193 km^2 (Lamone) and 1300 km^2 (Secchia), while flow lengths range from 28 km (Enza) to 85.2 km (Secchia). Mean annual discharges during the period 2006–2016 are included between 8.4 m^3 s^{-1} (Savio) and 30.4 m^3 s^{-1} (Secchia).

From a hydrogeological point of view, a report [20] grouped bedrock outcrops in the aforementioned catchments into six main classes (or hydrogeological complexes, namely: clay, marl, flysch, foreland flysch, ophiolite, and limestone). Those composed of poorly permeable or impermeable materials (clay, marl, and flysch hydrogeological complexes; see Figure 1) are the most represented in terms of areal coverage, leading to a runoff response of rivers that closely follows the rainfall distribution during the year (pluvial discharge regime). Rivers originating from the most elevated parts of the main watershed divide (Secchia, Panaro) are characterised by a nival–pluvial discharge regime as they are influenced by the melting of snow cover accumulated during the winter months in the upper parts of their catchments [21].

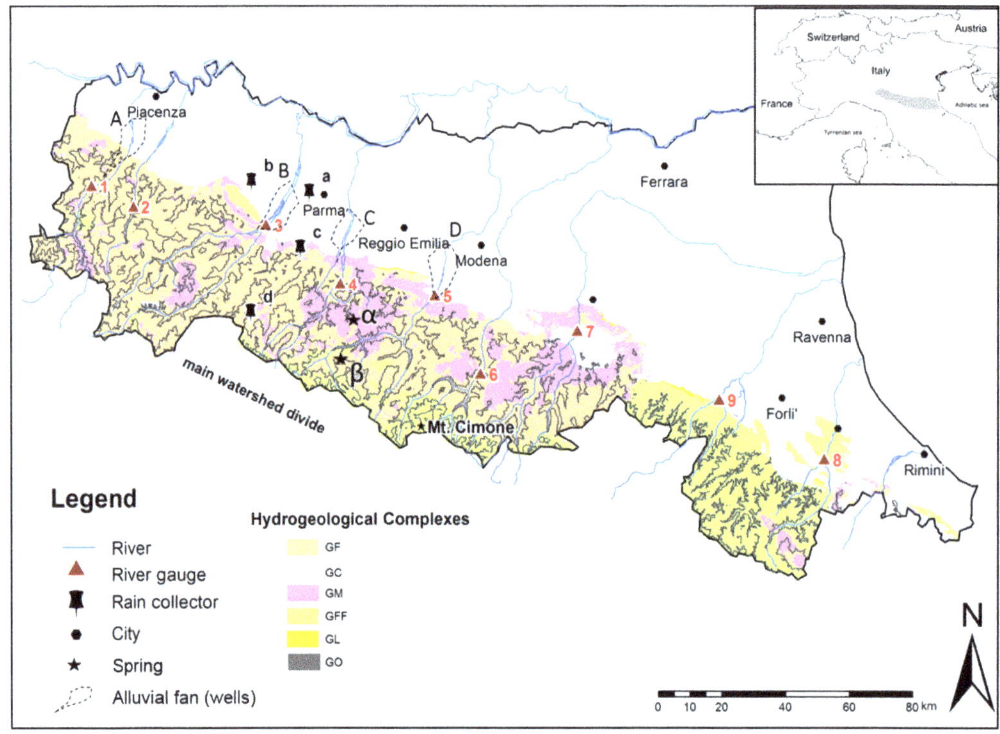

Figure 1. Sketch map of the area (modified after [7]) with locations where sampling has been carried out by previous studies for precipitation (rain gauges with letters a to d), surficial water (rivers numbered 1 to 9), groundwater from springs (Greek letters α and β), and groundwater from wells (located in the alluvial fans with capital letters A to D). Hydrogeological complexes are reported following [20]; G_C: clay; G_M: marl; G_F: flysch; G_{FF}: foreland flysch; G_L: limestone; G_O: ophiolite. For further details, see Table 1.

Table 1. Main features of the sampling points from which isotopic data were derived. For the corresponding map locations, readers are referred to Figure 1.

Location	Type	Code	Number of Samples	Timing of Sampling	Length of Time	References
Parma	Precipitation	a	41	Monthly	3 January–6 December	[7,16]
Lodesana	Precipitation	b	18	Monthly	3 December–5 May	[7,16]
Langhirano	Precipitation	c	18	Monthly	3 December–5 May	[7,16]
Berceto	Precipitation	d	14	Monthly	4 September–5 October	[7,16]
Trebbia	Surface water	1	36	Monthly	5 January–7 December	[16]
Nure	Surface water	2	24	Monthly	6 January–7 December	[16]
Taro	Surface water	3	36	Monthly	5 January–7 December	[16]
Enza	Surface water	4	24	Monthly	6 January–7 December	[16]
Secchia	Surface water	5	24	Monthly	6 January–7 December	[16]
Panaro	Surface water	6	36	Monthly	5 January–7 December	[16]

Table 1. *Cont.*

Location	Type	Code	Number of Samples	Timing of Sampling	Length of Time	References
Reno	Surface water	7	24	Monthly	6 January–7 December	[16]
Lamone	Surface water	8	24	Monthly	6 January–7 December	[16]
Savio	Surface water	9	36	Monthly	5 January–7 December	[16]
Trebbia	Groundwater from wells	A	66	Four-Monthly	4 January–7 December	[16]
Taro	Groundwater from wells	B	23	Four-Monthly	4 January–7 December	[16]
Enza	Groundwater from wells	C	23	Four-Monthly	4 January–7 December	[16]
Secchia	Groundwater from wells	D	33	Four-Monthly	4 January–7 December	[16]
Pietra di Bismantova	Groundwater from springs	α	32	Monthly Two-Monthly Three-Monthly	14 January–15 December	[17]
Montecagno	Groundwater from springs	β	21	Monthly Two-Monthly Three-Monthly	14 March–15 December	[18]

In the vicinity of the foothills (therefore close to the corresponding river gauges), several wells drilled in the alluvial fans of the Trebbia, Taro, Enza, and Secchia rivers continuously pump groundwater for both agricultural and drinking purposes. As previously highlighted by [11], by exploiting water stable isotopes, wells pumping water from confined aquifers in Trebbia and Taro alluvial fans are also likely to be recharged by zenithal precipitation infiltrating through gravels and sands that outcrop at the foothills of the northern Apennines (i.e., apical part of the alluvial fans). On the contrary, an important quota of recharge also occurs from streambed dispersion (focused on the apical part of their alluvial fans, see [22]) seems to affect wells located in the alluvial fans of the Enza and Secchia rivers. Two groups of low-yield springs from the Secchia River catchment (namely, Pietra di Bismantova and Montecagno) were also considered. These springs should be considered as representative of the common ones in the northern Italian Apennines, whose discharges are closely related to the rainfall pattern while outflows are strongly reduced (often in the order of 1 L·s^{-1} or less) at the end of the summer periods (shallow groundwater flow paths; for more details see [20,23,24]).

From the climatic point of view, and as already reported in [25], the mean annual rainfall distribution during the period 1990–2015 exceeds 2200 mm/y near the main watershed divide and progressively decreases to about 900 mm/y in the foothills. The rainfall distribution during the year is characterised by a marked minimum in the summer season and two maxima during autumn (the main one) and spring. Close to the main watershed divide, the cumulative annual snow cover can reach 2–3 m at the end of the winter season [21]. Potential evapotranspiration ranges from about 500 up to 650 mm/y in the lowlands and is mainly active during the summer months [25].

3. Methodology

The methodology used in this study consists of five steps involving data inspection and statistical comparison on datasets, including rainfall (4 rain gauges, namely: Parma, Lodesana, Langhirano, Berceto), surface water (9 rivers, namely: Trebbia, Nure, Taro, Enza, Secchia, Panaro, Reno, Lamone, Savio) groundwater from wells (aggregated isotopic values from wells belonging to 4 alluvial fans, namely: Trebbia—4 wells, Taro—5 wells, Enza—3 wells, and Secchia—5 wells), and groundwater from springs (aggregated isotopic values from springs belonging to 2 areas from the Secchia catchment, namely: Pietra di Bismantova and Montecagno). Firstly, a check on the assumption of stationary behaviour of each series of stable isotopes was carried out by means of conventional statistical tests coupled with inspection of standardised residuals.

Secondly, slopes and intercepts from δ^2H–$\delta^{18}O$ alignments were obtained by using 5 different regression approaches. Moreover, we further considered 2 different regressions applied to δ-values from rain gauges and rivers that had been preliminary weighted to the corresponding quota of precipitation and discharge, respectively. Thirdly, slopes and intercepts were visually inspected by means of heat maps to identify discrepancies among the several regression methods. Fourthly, slopes and intercepts were compared through bivariate (correlation matrices) and multivariate analyses (hierarchic clustering, i.e., dendrograms) to identify linear correlations and similarities. Fifthly, and with reference to the only surface water, we made a comparison between differences in slopes and intercepts with some selected catchments to verify linear or nonlinear correlations among these variables.

For convenience (see Figure 1 for location of the sampling points and Table 1 for further details), we report below rain gauges signified by letters (from a to d: "a"—Parma, "b"—Lodesana, "c"—Langhirano, "d"—Berceto); surface water locations as numbers (from 1 to 9: "1"—Trebbia, "2"—Nure, "3"—Taro, "4"—Enza, "5"—Secchia, "6"—Panaro, "7"—Reno, "8"—Lamone, "9"—Savio); groundwater from wells as capital letters ("A"—Trebbia, "B"—Taro, "C"—Enza, "D"—Secchia); and groundwater from springs as Greek letters ("α"—Pietra di Bismantova, "β"—Montecagno).

3.1. Isotopic Datasets

The dataset from 4 rain gauges and 9 rivers consists of monthly isotopic data while 17 water wells were characterised by grabbed four-monthly samples. All the data are derived from [7,16]. With reference to rain gauges, isotopic datasets lasted over the period from January 2003 to December 2006. Isotopic data from surface water and groundwater covered the period from January 2004 to December 2007. Further monthly, two-monthly, and three-monthly isotopic data from 2 groups of nearby low-yield springs were considered. These data were published in [17,18] and included the period January 2014 to December 2016.

The final dataset consists of 553 isotopic values, of which 91 are from precipitation, 264 from surface water, 145 from groundwater collected in wells, and 53 from groundwater collected in springs. Precipitation and river discharge that were used for further weighting procedures (see Section 3.3, "Linear Regression Types") come from [26].

As reported in the previous works of [7,11,16], the isotopic analyses were carried out by using isotope ratio mass spectrometry (IRMS) while instrument precision (1σ) was on the order of ± 0.05‰ for $\delta^{18}O$ and ± 0.7‰ for δ^2H. With reference to groundwater from springs [17,18], the corresponding isotopic analyses were carried out by mixed technique involving IRMS for $\delta^{18}O$ and cavity ring-down spectroscopy (CRDS) for δ^2H. Instrument precision (1σ) was assessed as ± 0.1‰ for $\delta^{18}O$ and ± 1.0‰ for δ^2H.

3.2. Verifying the Stationary Behaviour of Isotopic Data Series

As anticipated in the introduction, three of the five linear regression approaches are considered in this study, namely, ordinary least squares (OLS), reduced major axis (RMA) and major axis (MA), which are based on the assumption that a series of stable isotopes are stationary [27,28]. This means that statistical properties as mean and variance remain constant along each $\delta^{18}O$–δ^2H alignment, i.e., modelling errors are normally distributed and homoscedastic. Moreover, the closest pairs of $\delta^{18}O$–δ^2H must not be affected by autocorrelation phenomena as the modelling errors (i.e., residuals) from the regressions must be independent [27].

The presence of outliers or heteroscedasticity (i.e., modelling errors have not the same variance over the alignment) or autocorrelation lead the assumption of stationarity to be violated, thus slopes and intercepts from the abovementioned regression may not be meaningful. In this work, we exploited conventional statistical tests for verifying multivariate normality (i.e., presence of outliers inducing non-normality; Doornik–Hansen test [28,29]) heteroscedasticity (Breusch–Pagan test [30]), and autocorrelation (Durbin–Watson [31]).

All of the abovementioned tests are based on a comparison of the corresponding statistics' *p*-value results with a threshold value (level of significance α set at 0.01) to decide whether the null hypotheses have to be rejected ($p < 0.01$) or not ($p > 0.01$). The following null hypotheses were selected: multivariate normality (Doornik–Hansen), homoscedasticity (Breusch–Pagan), no autocorrelation (at a lag of 1) in the residuals (Durbin–Watson). With reference to the Durbin–Watson test, it must be specified that values are included between 0 and 4. Values close to 0 indicate almost total positive autocorrelation while results in the proximity of 4 indicate a total negative autocorrelation. Values between 1.5 and 2.5 show there is no autocorrelation in the data.

As suggested by [27], in the case of rejection of the null hypothesis of multivariate normality or homoscedasticity, we further verified the standardised residuals for identifying outliers (i.e., those values for which standardised residuals fall outside the interval from -4 to 4). Moreover and always following [27], if residuals were found to be serially correlated, we made autocorrelation functions of the standardised residuals to further confirm the lag length of autocorrelation.

3.3. Linear Regression Types

In this work, 5 types of linear regressions were tested, namely, ordinary least squares (OLS), reduced major axis (RMA), major axis (MA), robust (R), and Prais–Winsten (PW). To avoid excessive mathematical details, we provide a cursory examination of the methods and the reader is referred to more specific literature [14,15,27,32]. For convenience, we report all the corresponding equations by replacing δ^{18}O with x and δ^{2}H with y, while n is the number of samples and r is the Pearson correlation coefficient.

The OLS regression assumes that the x values are fixed (i.e., it is commonly used when *x* values have very few associated errors) and finds the line that minimises the squared errors in the y values. The slope of the linear regression (slope$_{OLS}$) is calculated as follows:

$$slope_{OLS} = r \times \frac{sd_y}{sd_x} \tag{1}$$

where *sd* represents standard deviations calculated for *x* variables (sd_x) and *y* variables (sd_y). Unlike OLS, RMA and MA try to minimise both the *x* and the *y* errors [33]. In the case of RMA, the corresponding slope$_{RMA}$ can be obtained with:

$$slope_{RMA} = sign[r] \times \frac{sd_y}{sd_x} \tag{2}$$

where *sign[r]* is the algebraic sign of the Pearson coefficient. The slope$_{MA}$ is calculated by:

$$slope_{MA} = -A + \frac{\sqrt{r^2 + A^2}}{r} \tag{3}$$

where *A* can be obtained as:

$$A = 0.5 \times \left(\frac{sd_x}{sd_y} - \frac{sd_y}{sd_x} \right) \tag{4}$$

As anticipated in introduction, the PW regression [34] has never been used for developing δ^{18}O–δ^{2}H alignments, as series of isotopic data have always been considered to the present as time-invariant (i.e., stationary). Recently, [19,35] highlighted that multiannual series of such isotopic data may be affected by nonstationary processes (such as, for example, trends in the means or presence of far-off values). In this case (nonstationary multiannual series of δ^{18}O–δ^{2}H pairs), the use of common regression methods such as OLS, RMA, and MA could induce residuals to be larger and characterised by stronger serial correlations.

PW is commonly used for data with serially correlated residuals of the estimates [36]. As a matter of fact, this approach takes into account AR1 (i.e., autoregression of the first order) serial correlation of the errors in a linear regression model. The procedure recursively estimates the coefficients and the error autocorrelation of the model until sufficient convergence is reached. All the estimates are obtained by the abovementioned OLS.

As in the case of PW approach, the R method has also never been tested for $\delta^{18}O$–δ^2H regressions. This method is an advanced Model I (in which x is always the independent variable) regression which is less sensitive to outliers than OLS estimates. Having less restrictive assumptions, R is recognised to provide much better regression coefficient estimates than OLS when outliers are present in the data. In particular, this approach has proven to be successful in the case of "almost" normally distributed errors but with some far-off values. This happens as outliers usually violate the assumption of normally distributed residual in OLS method. The algorithm is "least trimmed squares" reported by [37], in which the method consists of finding that subset of x–y pairs whose deletion from the entire dataset would lead to the regression having the smallest residual sum of squares. As in the case of PW approach, estimates of each subset are calculated owing to the OLS method. It must be added that, depending on the size and number of outliers, R regression conducts its own residual analysis and downweight or even these x–y pairs; this fact deserves an accurate inspection of the outliers made by the operator prior to any removal in order to decide whether these x–y pairs have to be considered or not.

For all 5 different regression approaches, the corresponding intercept is obtained with the following:

$$intercept = \frac{\sum_{i=1}^{n} y_i}{n} - slope\frac{\sum_{i=1}^{n} x_i}{N} \tag{5}$$

In all the abovementioned linear regressions, each observation has an equal influence of the orientation of the fitted line. As a matter of fact, it is well recognised that some isotopic data may be more important than others as related to a higher amount of water (for example, a flood in the case of a river or a high discharge event of a spring or high rainfall amount during a storm event). In this case, greater influence in the regression should be given to these isotopic data. In order to also take this effect into account, OLS were applied to isotopic datasets that had been previously weighted on the corresponding monthly amount of precipitation (rainwater) and discharge (freshwater and river water) by means of two different methods, i.e., the classical one that simply involves multiplying each y_i by the water amount (see [28] for further details; hereafter called W) and as reported in [38] (see [7] for the formulation; hereafter called B).

3.4. Comparison among Regressions

Initially, slopes and intercepts from all the regressions were compared by means of heat maps. The heat maps are matrices of fixed cell size showing the magnitude of difference among values with a selected binary colour ramp (in our case from red to green, respectively, indicating the lowest value and highest value within the isotopic dataset), in which the colour intensities provide visual cues to the reader about discrepancies between the data. The goal was to verify any presence of clusters to be further investigated by means of bivariate and hierarchical cluster analyses.

As a bivariate analysis, we carried out scatterplot matrices to determine if linear correlation between multiple variables (slopes and intercepts obtained by the different regression approaches) were present or not. Tests were carried out highlighting the level of significance, which was set as $p < 0.01$.

Furthermore, a hierarchical cluster analysis was performed to identify similarities among the series of slopes and intercepts from the whole datasets. Clustering was done according to the unweighted pair–group average (or centroid) method, in which each group consisted of slopes (or intercepts) from a determined regression approach. The method was based on a step-by-step procedure in which series of slopes (or intercepts) were grouped into branched clusters (dendrogram) based on their similarities to one another. As a result, the two most similar series of slopes (or intercepts) were selected and linked based on the smallest average distance among the values of all slopes (or intercepts). Progressively more dissimilar series were linked at greater distances; in the end, they all were joined to one single cluster. The cophenetic coefficient was used as a measure of similarity between each pair of clusters; more than 2 time series being analysed, the dendrogram was supported by a cophenetic distance matrix. Further details on this method can be found in [28].

3.5. A Focus on the Differences among Regression Approach from River Water: Comparison with Catchment Characteristics

As suggested by [15], we investigated whether some selected catchment characteristics (also called descriptors) could have affected differences among values of slopes and intercepts as obtained by the different approaches reported in Section 3.3. In order to make all slopes and intercepts comparable, we followed the approach proposed by [15] that consisted of prior computed differences in the slopes (as $slope_{OLS}$–$slope_{RMA/MA/R/PW/W/B}$) and intercepts (as $intercept_{OLS}$–$intercept_{RMA/MA/R/PW/W/B}$). Then, and following again the procedure reported in [15], we applied the Spearman ranking correlation matrix in which the abovementioned differences in slopes and intercepts were compared with 9 catchment characteristics. It should be added that this approach is a nonparametric measure of rank correlation that provides a statistical dependence between the rankings of two variables at a time. Unlike the Pearson coefficient, the Spearman ranking assesses how well the relationship between two variables can be described using a monotonic function, even if their relationship is not linear [27]. In particular, several authors have highlighted that many hydrological processing occurring at both slope and catchment scales are nonlinear (see for instance [38,39]) and such behaviour was in turn seen in some descriptors calculated by means of time series of stable isotopes of water [7,40–42].

In order to take into account the linearity among the variables, we also considered the linear correlation by providing the Pearson correlation coefficients. Here, we recall that Pearson and Spearman matrices reflect the magnitude of similarity among the parameters by means of r (the Pearson correlation coefficient described in Section 3.3) and r_s coefficients, respectively. Both correlation coefficients (r and r_s) describe the strength and direction between the two variables and return a closer value to 1 (or -1) when the two different datasets have a strong positive (or negative) relationship. The significance probability (*p*-value) for both the Spearman and Pearson correlation coefficients calculated in this study was set at 0.01, meaning that *p*-values lower than 0.01 represented statistically significant relationships. Readers are referred to [28] for further details on statistical formulations.

The 9 catchment characteristics (or descriptors, see Table 2 and Supplementary Materials Table S1 for further details) were those already considered by [7], namely: catchment area (A); elevation (H); precipitation (P); flow length (F); specific mean annual runoff (q); specific river runoff exceeded for 95% of the observation period (q95; this is a well-known low flow index that is used worldwide for the regionalisation procedure and can be estimated even from a relatively short time series of daily runoffs [43]); and the young water fraction (F_{yw} proposed by [42]; this is considered the percentage proportion of catchment outflow younger than approximately 2–3 months and was estimated from the amplitudes of seasonal cycles of stable water isotopes in precipitation and stream flow that had been already calculated in [7]).

It should be noted that 4 descriptors (P, q, q95, F_{yw}) were obtained by processing daily precipitation (42 rain gauges homogeneously distributed over the study area for P), discharges (for both q and q95), and water isotopes time series (for F_{yw}) lasting over the same time period. Moreover, flow length (F) was derived from a 5×5 m gridded digital terrain model created by the digitalisation and linear interpolation of contour lines represented in the regional topography map at a scale of 1:5000.

Table 2. The 9 catchment characteristics included in the analysis. For further details on the catchment characteristics from a single catchment, see Table S1 in Supplementary Materials.

Acronym	Variable	Units	Minimum	Mean	Maximum
A	Catchment area	km^2	193	696	1303
H_{min}	Altitude of stream gauge	m	43	171	421
H_{max}	Maximum altitude	m	1158	1784	2165
H_{mean}	Mean altitude	m	526	754	944
P	Precipitation	mm	924	1090	1304
F	Flow length	km	20.9	55.5	85.2
q	Specific mean annual runoff	$L\,s^{-1}\,km^{-2}$	2.2	15.0	36.3
q95	Specific runoff exceeded for 95% of the time	$L\,s^{-1}\,km^{-2}$	0.0	1.0	1.7
F_{yw}	Young water fraction	%	9.3	13.7	22.9

4. Results

4.1. Stationary Behaviour of Isotopic Data Series

Table 3 summarises the results from the three statistical tests (Doornik–Hansen for multivariate normality, Breusch–Pagan for homoscedasticity, and Durbin–Watson for autocorrelation) used for assessing the compliance with the stationary assumption. Isotopic series from rivers were those mainly affected by problems of non-normal behaviour (rivers "5,6,9") and autocorrelation at a lag of 1 (rivers "1,4,6,7"). The latter were positive (we recall here that values closer to 0 identify positive autocorrelation phenomena) and more intense in the case of rivers "4,6". Moreover, the Breusch–Pagan test suggested residual homoscedasticity for river "8". By considering the plots of standardised residuals (see Figure S1 in Supplementary Materials), the presence of outliers was further confirmed for rivers "5,6,9" (river "5", 3 outliers; river "6", 4 outliers; river "9", 1 outlier) as well as the increase of variance of standardised residuals along estimates for river "1" (i.e., heteroscedasticity). Autocorrelation functions carried on standardised residuals (see Figure S2 in Supplementary Materials) allowed for demonstrating the presence of serial correlations, although with different lag lengths (river "1", 2 lags; river "4", 2 lags; river "6", 3 lags; river "7", 2 lags).

4.2. Slopes and Intercepts

The $\delta^{18}O$–δ^2H relationships are summarised in in Supplementary Materials containing slopes (a; see Table S2), intercepts (b; see Table S3), standard deviation of the estimates (c; see Table S4), standard deviations of the estimates and coefficient of determinations (d: see Table S5) coefficient of determinations R^2. By viewing all the results reported in the form of heat maps (see Figure S3 in Supplementary Materials), the substantial invariance of slopes (from 6.9 to 13.1) and intercepts (from 7.2 to 8.4) from rain gauges located at lower altitudes (a, b, c), with high performance of the regression (R^2 always close to 0.99) was noticed. These results are in agreement with the GMWL (we recall that this line is characterised by slope and intercept equal to 8.0 and 10.0, respectively; see Figure S4 in Supplementary Materials) with no evidence of outliers. When the two weighting approaches were taken into account, no changes among the unweighted values of slope and intercept were found with the exception of intercepts in the rain gauge "a" (intercepts remarkably lower in the case of weighting procedures in the order of +4.0).

Table 3. Results from the three statistical tests aimed at verifying the compliance with the stationary assumption, namely: Doornik–Hansen (multivariate normality), Breusch–Pagan (homoscedasticity), and Durbin–Watson (autocorrelation). * Null hypotheses rejected as $p < 0.01$.

Location	Type	Code	Doornik–Hansen	Breusch–Pagan	Durbin–Watson
Parma	Precipitation	a	1.81	0.17	1.57
Lodesana	Precipitation	b	1.54	2.40	3.65 *
Langhirano	Precipitation	c	1.45	0.76	1.43
Berceto	Precipitation	d	3.83	0.46	0.84
Trebbia	Surface water	1	5.44	7.84 *	1.07 *
Nure	Surface water	2	1.50	0.36	1.41
Taro	Surface water	3	3.23	0.09	1.31
Enza	Surface water	4	1.94	0.90	0.90 *
Secchia	Surface water	5	10.30 *	0.19	1.21
Panaro	Surface water	6	8.98 *	0.86	0.69 *
Reno	Surface water	7	5.89	0.15	1.02 *
Lamone	Surface water	8	7.58	6.95 *	1.44
Savio	Surface water	9	41.42 *	0.00	2.30
Trebbia	Groundwater from wells	A	1.30	1.14	2.05
Taro	Groundwater from wells	B	3.67	0.18	2.28
Enza	Groundwater from wells	C	8.21	3.74	2.42
Secchia	Groundwater from wells	D	6.15	0.41	1.41
Pietra di Bismantova	Groundwater from springs	α	7.98	2.61	2.17
Montecagno	Groundwater from springs	β	7.76	0.37	1.21

With reference to the rain gauge "d", i.e., that located near the main watershed divide, the R approach provided remarkably higher values of both slope (8.1) and intercept (11.5) than those obtained with the other regression approach (we recall that all values of intercepts from "d" were negatives). It must be highlighted that the standard deviations of the estimates are slightly higher than those obtained with the other regressions (see Table S2 in Supplementary Materials).

By considering the surface water, the RMA and MA approaches almost provided slightly higher values of slopes and intercepts (up to 13.0 and 55.2, respectively, in the case of river "5"). In the case of rivers "1,2,3,4,7,8", values of slopes are in the range of those obtained by weighting procedures. On the contrary, intercepts showed a larger variability among the regression methods investigated. It should be highlighted that in the case of rivers "5,6,9" the values of slopes remarkably varied as well, in particular if MA and R were used. As in the case of the abovementioned rain gauge "d", the $\delta^{18}O$–δ^2H alignments from "5,6,9" were characterised by the lowest values of R^2 and the larger values of standard deviations of the estimates (see Table S4 in Supplementary Materials).

It must be noted that the discrepancies reported for these points (i.e., "5,6,9") affected water with the presence of several outliers and/or serial correlations of residuals (see Figures S2 and S3 in Supplementary Materials and Section 4.1), which violated the stationary assumption.

Akin to the cases of rain gauges and rivers, RMA and MA approaches carried out on groundwater from wells and springs were characterised by larger values of both slopes and intercepts than in the case of OLS. With the exception of groundwater from "C" and "D", the R and PW approaches induced larger variations in both slopes and intercepts, which were particularly marked in the case of groundwater from β (i.e., water whose $\delta^{18}O$–δ^2H alignment was characterised by low values of R^2 and large standard deviations of the estimates, see Table S2 in Supplementary Materials).

In Table 4, the matrix reporting correlation coefficients between pairs of slopes indicated that the largest degree of association was found ($p < 0.01$) between OLS–PW and W–B. High values of correlation (with slightly lower degree of association but still $p < 0.01$) also characterised the following relations: OLS–W, OLS–RMA, OLS–B, RMA–PW, RMA–MA, and PW–W. A significant degree of association ($p < 0.01$) was also found for PW–B. It should be highlighted that in several cases regarding R and MA, the degree of associations was very low and, in some cases, even negative (i.e., an increase in the value of slope obtained with a regression corresponds to a decrease in the series obtained with RMA).

Table 4. Correlation matrix reporting associations among the slopes from different regression approaches considered in this study (namely: OLS, RMA, MA, R, W, B). Progressively darker green colour is associated with a higher correlation coefficient. * Significant as $p < 0.01$.

	OLS	RMA	MA	R	PW	W
RMA	0.86 *					
MA	0.35	0.77 *				
R	0.50	0.29	−0.06			
PW	0.99 *	0.87 *	0.38	0.53		
W	0.87 *	0.29	−0.54	0.39	0.79 *	
B	0.84 *	0.31	−0.47	0.27	0.74 *	0.97 *

By considering the intercepts (Table 5), the degree of associations already highlighted for slopes was further confirmed with the exception of OLS–W and OLS–B (here not significant as $p > 0.01$). It should be noted that almost all correlations were slightly lower than the corresponding ones from the slopes.

Table 5. Correlation matrix reporting associations among the intercepts from different regression approaches considered in this study (namely: OLS, RMA, MA, R, W, B). Progressively darker green colour is associated with a higher correlation coefficient. * Significant as $p < 0.01$.

	OLS	RMA	MA	R	PW	W
RMA	0.84 *					
MA	0.27	0.74				
R	0.44	0.23	−0.13			
PW	0.94 *	0.83 *	0.31	0.46		
W	0.59	0.60	−0.02	0.62	0.71 *	
B	0.61	0.63	−0.01	0.63 *	0.71 *	0.98 *

The hierarchic cluster analysis (see Figure S5 in Supplementary Materials) among the slope series from the several regression approaches (reported as different branches composing the dendrogram) demonstrated that MA was associated with none of the other regression methods, while two main group of pairs were clearly separated: the first is represented by OLS–PW while the second by W–B. The aforementioned first and second group were associated with each other while longer branches further linked them to R and RMA.

With reference to intercepts, the dendrogram confirmed the nonassociation of MA with the other regression approaches. Moreover, the two closest series were still those of OLS and PW, which were in turn associated to W. Contrary to the case of slopes, B series was strictly associated to RMA.

4.3. Comparison between the Differences in the Slopes and Intercepts with Catchment Characteristics and Statistics

By taking into account only $\delta^{18}O$–δ^2H regressions from surface water, the Pearson rank correlation matrix (we recall that Pearson assesses linear relationships) comparing differences in slopes with catchment characteristics (see Table 6) did not provide significant ($p < 0.01$) correlations. If the Spearman rank correlation matrix (i.e., assessment of nonlinear relationships) was considered, we found positive and significant ($p < 0.01$) correlations with F_{yw} ($Slope_{OLS-RMA}$ and $Slope_{OLS-MA}$) while negative ones with H_{min} ($Slope_{OLS-W}$, $Slope_{OLS-B}$).

Table 6. Matrix of the Pearson (in grey) and Spearman (in green) rank correlation (values as r and r_s, respectively; r value evidenced in grey while r_s in green) between the differences in the slopes and the selected catchment characteristics considered for the 9 rivers. Relationship between differences in the slopes and coefficient of determination R^2 from $\delta^{18}O$–δ^2H linear regressions are also reported. R^2 values from regressions were calculated starting from signed values of differences in slopes. * Significant as $p < 0.01$.

Descriptor	OLS–RMA		OLS–MA		OLS–R		OLS–PW		OLS–W		OLS–B	
H_{min} (m asl)	0.03	−0.03	0.05	−0.03	0.37	0.57	0.00	−0.01	0.41	−0.61 *	−0.40	−0.65 *
A (kmq)	−0.19	−0.12	−0.22	−0.12	−0.02	−0.01	−0.04	−0.12	0.35	0.15	0.56	0.32
H_{max} (m asl)	−0.41	−0.42	−0.35	−0.42	0.04	−0.01	−0.02	−0.12	0.08	0.12	0.313	0.05
H_{mean} (m asl)	−0.05	−0.28	−0.02	−0.28	0.38	0.28	−0.09	−0.18	−0.14	−0.19	−0.10	−0.19
q (Ls^{-1}/km^2)	−0.57	−0.30	−0.57	−0.31	0.05	0.11	0.04	0.05	0.18	−0.05	0.12	−0.03
q95 (Ls^{-1}/km^2)	−0.14	−0.40	−0.10	−0.40	0.24	0.05	−0.21	−0.29	−0.04	0.10	0.00	−0.04
P (mm)	0.09	0.01	0.08	0.01	0.00	−0.01	0.00	0.01	−0.15	0.11	−0.14	0.02
F (km)	−0.11	0.05	−0.13	−0.05	−0.23	0.25	−0.01	0.02	0.26	0.34	0.42	0.56
F_{yw} (%)	0.41	0.70 *	0.32	0.70 *	−0.09	−0.10	0.00	0.03	0.00	0.00	0.00	0.01
R^2	0.98 *	0.98 *	0.96 *	0.96 *	0.01	0.02	0.10	0.18	−0.46	0.03	−0.32	−0.06
$\delta^{18}O$ range	0.16	0.45	0.13	0.46	−0.38	0.27	0.04	0.00	0.05	0.29	0.15	0.19
δ^2H range	0.00	0.03	−0.26	0.03	0.25	−0.28	−0.34	−0.34	0.25	0.46	0.36	0.45
n° of samples	0.04	0.03	0.04	0.03	0.17	0.12	−0.35	−0.27	−0.03	0.00	−0.03	−0.01

In the case of the Pearson rank correlation matrix applied to differences in intercepts, we did not find significant correlations (see Table 7). On the contrary, and as in the case of slopes, significant ($p < 0.01$) nonlinear relationships were found for F_{yw} ($Slope_{OLS-RMA}$ and $Slope_{OLS-MA}$) and H_{min} (negative correlation for $Slope_{OLS-B}$).

Table 7. Matrix of the Pearson (in grey) and Spearman (in green) rank correlation (values as r and r_s, respectively; r value evidenced in grey while r_s in green) between the differences in the intercepts and the selected catchment characteristics considered for the 9 rivers. Relationship between differences in the slopes and coefficient of determination R^2 from $\delta^{18}O$–δ^2H linear regressions are also reported. R^2 values from regressions were calculated starting from signed values of differences in intercepts. * Significant as $p < 0.01$.

Descriptor	OLS–RMA		OLS–MA		OLS–R		OLS–PW		OLS–W		OLS–B	
H_{min} (m asl)	0.04	−0.03	0.05	−0.03	0.36	0.56	0.00	−0.01	−0.40	−0.53	−0.48	−0.62 *
A (kmq)	−0.19	−0.12	−0.22	−0.12	−0.02	−0.01	−0.04	−0.07	0.35	0.14	0.42	0.28
H_{max} (m asl)	−0.38	−0.42	−0.35	−0.42	0.04	−0.01	−0.02	−0.11	0.10	0.16	0.04	0.06
H_{mean} (m asl)	−0.04	−0.28	−0.02	−0.28	0.36	0.28	−0.08	−0.20	−0.12	−0.14	−0.18	−0.18
q (Ls^{-1}/km^2)	−0.53	−0.30	−0.56	−0.30	0.05	0.11	0.04	0.06	0.20	−0.05	0.10	−0.03
q95 (Ls^{-1}/km^2)	−0.12	−0.40	−0.08	−0.40	0.24	0.05	−0.20	−0.28	−0.02	−0.06	−0.05	0.03
P (mm)	0.10	0.01	0.09	0.01	0.00	−0.01	0.00	0.01	−0.13	−0.12	−0.08	−0.02
F (km)	−0.32	−0.05	0.12	−0.05	−0.22	0.25	−0.01	0.00	0.27	0.30	0.46	0.53
F_{yw} (%)	0.36	0.70 *	0.30	0.70 *	−0.09	−0.10	0.00	0.01	0.01	0.00	0.00	0.00
R^2	0.98 *	0.98 *	0.96 *	0.96 *	0.01	0.00	0.02	0.15	−0.57	−0.12	−0.67 *	0.21
$\delta^{18}O$ range	0.13	0.45	0.12	0.45	−0.41	0.27	−0.04	0.01	0.03	0.27	0.14	0.18
δ^2H range	0.00	0.03	0.00	0.03	−0.30	−0.28	−0.34	0.23	0.22	0.45	0.34	0.46
n° of samples	0.04	0.03	0.04	0.03	0.15	0.12	0.31	0.27	0.03	0.00	0.03	0.01

In all cases, the largest statistical performance (again significant as $p < 0.01$) was found for coefficient of determination R^2 from $\delta^{18}O-\delta^2H$ linear regressions ($Slope_{OLS-RMA}$ and $Slope_{OLS-MA}$; $Intercept_{OLS-RMA}$ and $Intercept_{OLS-MA}$).

5. Discussion

We did not find significant discrepancies in the slopes and intercepts computed by the different regression methods in the case of precipitation data. On the contrary, marked variations were detected in the case of river water and groundwater (both from springs and wells in lowland aquifers) obtained using specific methods. Among others, such discrepancies were somehow reduced in the case of OLS, RMA, and PW. Because of different values of river and spring discharges and the corresponding changes in isotopic content of water during the year, weighting procedures (W, B) were characterised by diverse values of slopes and intercepts rather than the aforementioned OLS, RMA, and PW. Moreover, and as highlighted by both heat maps (Figure S3 in Supplementary Materials) and correlation matrices (Tables 4 and 5) and dendrograms (Figure S5 in Supplementary Materials), slopes and intercepts from the MA and R approaches were not comparable to others (nonsignificant statistical associations). Regardless of water type, the aforementioned discrepancies were promoted when $\delta^{18}O-\delta^2H$ regressions were characterised by weak statistical performances (low values of R^2 and larger values of standard deviations of the estimates). With reference to rivers, the weak statistical performances were linked to the presence of outliers and/or serial correlation of the residuals violating the stationary assumption of OLS, MA, and RMA approaches.

The investigation carried out on the data solely from rivers highlighted that the magnitude of the differences in the slopes and intercepts was related in all cases (with the exception of R and PW) to the coefficient of determination R^2 characterising $\delta^{18}O-\delta^2H$ linear regressions. The largest values of Pearson coefficients (see Tables 6 and 7) led us to consider R^2 as the main causal factor for such differences in slopes and intercepts.

In particular, the larger the correlations between $\delta^{18}O$ and δ^2H, the smaller the differences among slopes and intercepts detected by RMA, MA, W, and B within the specific sampling point (river, well, or spring). This is in agreement with the results reported by [15] and corroborated the hypothesis that statistical performance of the regression was the main driver of these slope and intercept variations. In any case, despite finding highly statistical significance with R^2 in our investigated dataset, no relations between differences of slopes (and intercepts) and the ranges in $\delta^{18}O$ (and δ^2H) along with the number of samples composing the dataset were noticed, thus indicating that extreme values of $\delta^{18}O$ (and δ^2H) were not significant causal factors.

With reference to RMA and MA, the Spearman rank correlation matrices involving differences in slopes and intercepts and catchment descriptors allowed us to find a significant nonlinear association with F_{yw} (we recall here that F_{yw} is the percentage of water younger than 2–3 months). In both cases (RMA and MA) the association (reported also as plots in Figure 2) indicated that the magnitude of the differences in the slopes and in the intercepts decreased along with the quota of young water.

This means that rivers showing low values of F_{yw} are likely to be more affected by differences in slopes and intercepts computed by different regression approaches. By examining the plots reported in Figure 2, it can be evidenced that nonlinearity is driven by two catchments (namely, the Secchia River "5" and Panaro River "6"). As already anticipated in Section 2, these two rivers ("5,6") were the only ones characterised by nival–pluvial discharges due to the melting of the snow cover in the upper part of the catchments during the spring months. Moreover, [7] stated that there were evidences of sublimation in several water samples collected in rivers "5,6" from January 2017 to April 2017. This was further confirmed by the remarkable number of snowfall events that occurred between December 2013 and April 2014 over the highest part of the catchments "5,6". Such events have allowed the consequent snowpack development alternating with partial snowmelt for a snow water equivalent higher than 600 mm [21].

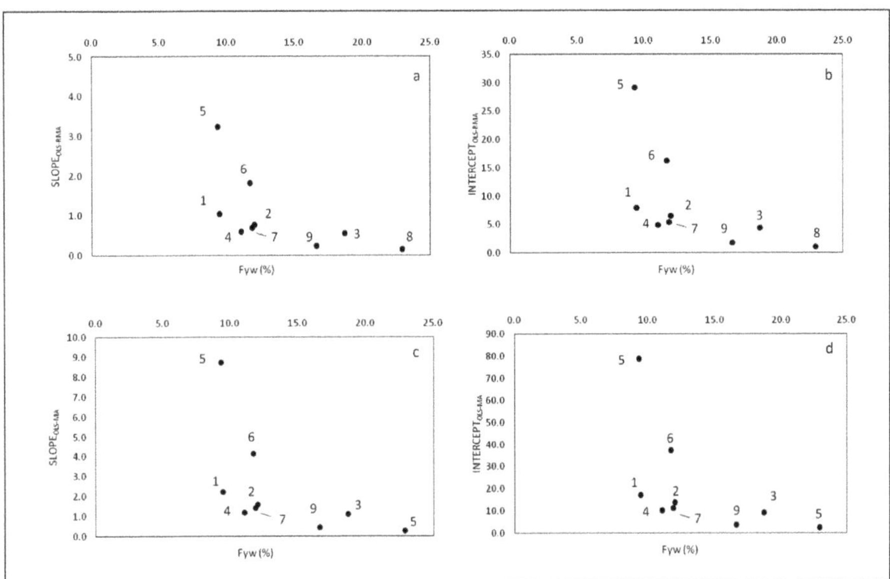

Figure 2. Differences in slopes (a: OLS–RMA: c: OLS–MA) and intercepts (b: OLS–RMA; d: OLS–MA) for OLS–RMA and OLS–MA. Values of differences in slopes and intercepts (y-axes) are reported in modulus form. Codes for rivers (from 1 to 9) are also reported (for further details on river codes, see Table 1).

There, we recall that sublimation occurring during sunny days can modify the former isotopic composition of the superficial snow layers, allowing the release of a vapour phase from the solid skeleton to the atmosphere. In this case, the final snow cover does not preserve the isotopic composition of the original snowfall from which it was derived [44,45], a fact that also led differences in slopes and intercepts from $\delta^{18}O$–δ^2H regressions to be enhanced. In detail, sublimation acting on a snow cover can lead to an enrichment of heaviest isotopes (such as ^{18}O and ^2H) within the solid skeleton and can induce a similar $\delta^{18}O$–δ^2H pattern of that charactering the residual liquid subjected to evaporation (slope decrease of the $\delta^{18}O$–δ^2H alignments for snowpack samples if compared to the water meteoric line (see for field and experimental studies: [46,47])). In particular, the slope decrease can be much more intense if the only late-season snowpack samples are considered (a value of 3.7 was found by [48]).

In case the two rivers "5,6" are removed from the analysis, it is still possible to confirm such alignments, although linear, between y F_{yw} and differences in slope and intercept pairs. In this sense, such relations, still identifying an inverse association between differences in the slopes (and in the intercepts) and quotas of young water, may also be related to other hydrological processes taking place at the catchment scale. As already pointed out by [7], by checking both slopes (river water showed slightly lower values than those characterising rainwater; see Figure S6 in Supplementary Materials) and intercepts (that were negative compared to those from rainwater), all the river water considered underwent evaporation/evapotranspiration processes prior to their infiltration towards the aquifer.

Albeit to a lesser extent, these variations also affected groundwater from wells (which were also fed by streambed dispersion and therefore by water isotopes already modified in the river water; see [11] and low-yield springs (these potentially characterised by pre-infiltrative modification as slopes from $\delta^{18}O$–δ^2H alignments were slightly lower than those obtained from precipitation water; see Figure S7 in Supplementary Materials)). On the contrary, the nonvariability of slopes and intercepts observed in the different $\delta^{18}O$–δ^2H

alignments from precipitation was somehow expected, as these waters were unlikely to be affected by evaporative/sublimation processes once they had entered the rain gauges.

With reference to the effective role of young water fraction F_{yw} in influencing the differences in intercepts and slopes, we believe that further efforts have to be made for such catchments from the hilly part of the northern Italian Apennines and dominated by higher quotas of young water (i.e., $F_{yw} > 25\%$; not analysed in this study for lack of isotopic data). The latter are characterised by intermittent discharge and wide outcrops of low permeable soils and bedrocks (prevailing clayey and marls materials; G_C and G_M in Figure 1). Further investigations should be isotopic-based in order to verify also the role of pre-infiltrative evaporation in isotopic deviation and, above all, in the change of slopes and intercepts.

Following [7], these processes were likely to be promoted in the clay-rich bedrock, where water molecules composing the soil moisture were slowed in percolation and thus kinetic fractionation processes were enhanced.

However, we can provide some preliminary recommendations for use of the different regression approaches for the four water types (precipitation, surface water, groundwater from wells, and low-yield springs) from the northern Italian Apennines:

(i) In the case of $\delta^{18}O$–δ^2H alignments from precipitation, and as no remarkable discrepancies were detected among the several investigated methods, the OLS approach should be preferred.

(ii) For precipitation and surface water, slopes and intercepts from the two weighting procedures W and B were similar. Moreover, there was no evidence of remarkable changes among results obtained from such weighting procedures with those from unweighted OLS. The latter confirms the convenience of using the OLS approach even if, during the year, rainfall or discharge amounts (and isotopic content too) are different between the seasons.

(iii) For surface water and groundwater, the MA and R approaches should not be used in any case as they seem to provide unrealistic values for both slopes and intercepts. The reason has to be searched in the fact that these two approaches are more sensitive to the statistical performance of the regressions (i.e., standard deviations), especially if outliers are present. MA demonstrated to be more sensitive to the statistical performance of the regressions (i.e., standard deviations), especially if outliers are present. Although the R approach was selected to verify its behaviour in the case of outliers, it did not induce improvements in standardised residuals. Moreover, it was also demonstrated that kinetic fractionation processes acting on these water types lead to increase the differences in slopes and intercepts (see, for instance, relationships between differences in intercepts and slopes with young water fraction Fyw). Slopes and intercepts from OLS and PW were the closest, with lower standard deviations sometimes associated to PW regressions. In addition, and with reference to the surface water, PW results were not affected by the kinetic fractionation processes (see Tables 6 and 7).

(iv) Surface water may be affected by nonstationary processes induced by both nonmultivariate normality and serial correlations of the residuals. Thus, prior to carrying out OLS regression on $\delta^{18}O$–δ^2H data from surface water (and groundwater), the presences of outliers, heteroscedasticity, and autocorrelation must be carefully detected by means of both conventional statistical tests and inspections of standardised residuals. In the case of outliers, their importance on the whole data series composing the regression should be evaluated (as an instance, in the case of $\delta^{18}O$–δ^2H pairs from surface water collected during the late summer through the beginning of the autumn period, the strong reduction in discharges may induce their removal from the dataset prior to carrying out the regression). In the event of dealing with time series of stable isotopes affected by autocorrelation, we believe it is convenient to use the PW approach, which, in our case, has proven to solve the serial correlations of residuals.

6. Conclusions

We presented the comparison of five different regression approaches applied to $\delta^{18}O$–δ^2H data from four different water types collected in the northern Italian Apennines. We found that all the tested approaches converged towards similar values of slopes and intercepts for only stable water isotopes from precipitation. Conversely, differences in slopes and intercepts from surface water and groundwater (collected from wells and low-yield springs) were often significant and related to the robustness of the regressions (i.e., standard deviations of the estimates) and their sensitiveness to outliers and autocorrelation. Moreover, and with reference to surface water, we found evidence of a relationship between young water fraction and the magnitudes in differences of slopes and intercepts, suggesting the control of kinetic fractionation processes (mainly related to sublimation acting on snow cover and, secondary, to active pre-infiltrative evaporation and evapotranspiration processes) on such discrepancies. These results allowed us to provide some recommendations for hydrological and hydrogeological studies involving $\delta^{18}O$–δ^2H from the abovementioned water types collected in the northern Italian Apennines. Firstly, as no discrepancies were noticed between slopes and intercepts from all the methods applied to precipitation, the OLS approach is preferred. Secondly, and with reference to the other water types (surface water and groundwater from wells and springs), we warmly suggest carrying out conventional statistical tests coupled with inspection of standardised residuals for a preliminary check on the presence of outliers and autocorrelation phenomena. In the case of managing outliers, the MA and R approaches should be avoided as they are more sensitive to the statistical performance of the regressions and often provide unrealistic values of both slopes and intercepts. Thirdly, for surface water and groundwater, the OLS and PW approaches still showed the highest degree of robustness and produced the closest values of slopes and intercepts, thus resulting as the methods preferable for $\delta^{18}O$–δ^2H regressions. PW would be more reliable in the presence of serial correlations of the residuals (which, in our case, often affected surface water). In the case of managing outliers, the possibility of removing them will have to be considered (as an example in the case of $\delta^{18}O$–δ^2H pairs from marked low-flow periods).

Lastly, despite the presence of marked differences in the amounts of rainfall and their isotopic contents during the year, the convenience of using weighing approaches before applying OLS was not found.

Supplementary Materials: The following are available online at https://www.mdpi.com/article/10.3390/hydrology9020041/s1, Table S1: Catchment characteristics from the 9 rivers considered in this study, Table S2: Slopes from $\delta^{18}O$–δ^2H regressions, Table S3: Intercepts from $\delta^{18}O$–δ^2H regressions, Table S4: Standard deviations of the estimates from $\delta^{18}O$–δ^2H regressions, Table S5: Coefficient of determinations from $\delta^{18}O$–δ^2H regressions, Figure S1: Plots of standardised residuals for surface water affected by nonstationary processes, Figure S2: Autocorrelation functions of standardised residuals for surface water affected by serial correlations, Figure S3: Heat maps reporting slopes and intercepts values, Figure S4: $\delta^{18}O$–δ^2H pairs from rain gauges, Figure S5: Dendrogram for slopes and intercepts series, Figure S6: $\delta^{18}O$–δ^2H pairs from surface water, Figure S7: $\delta^{18}O$–δ^2H pairs from groundwater.

Author Contributions: Conceptualization, F.C.; methodology, F.C. and A.T.; software, F.C.; validation, F.C.; formal analysis, F.C.; investigation, F.C.; data curation, F.C.; writing—original draft preparation, F.C.; writing—review and editing, F.C. and A.T. All authors have read and agreed to the published version of the manuscript.

Funding: This research received no external funding.

Data Availability Statement: The data presented in this study are freely available thorough Supplementary Materials.

Acknowledgments: Authors express their gratitude to two anonymous reviewers as their thoughtful and detailed comments allowed the early version of the manuscript to be greatly improved.

Conflicts of Interest: The authors declare no conflict of interest.

References

1. Clark, I.; Fritz, P. *Environmental Isotopes in Hydrogeology*; CRC Press Lewis Publishers: Boca Raton, FL, USA, 1997.
2. Kendall, C.; McDonnell, J.J. *Isotope Tracers in Catchment Hydrology*; Elsevier: Amsterdam, The Netherlands, 1998; p. 839.
3. Leibundgut, C.; Maloszewski, P.; Kulls, C. *Tracers in Hydrology*; John Wiley & Sons: New York, NY, USA, 2011.
4. Craig, H. Isotopic variations in meteoric waters. *Science* **1961**, *133*, 1702–1703. [CrossRef] [PubMed]
5. Rozanski, K.; Araguás-Araguás, L.; Gonfiantini, R. Isotopic patterns in modern global precipitation. *GMS* **1993**, *78*, 1–36.
6. Klaus, J.; McDonnell, J.J.; Jackson, C.R.; Du, E.; Griffiths, N.A. Where does streamwater come from in low-relief forested watersheds? A dual-isotope approach. *Hydrol. Earth Syst. Sci.* **2015**, *19*, 125–135. [CrossRef]
7. Cervi, F.; Dadomo, A.; Martinelli, G. The analysis of short-term dataset of water stable isotopes provides information on hydrological processes occurring in large catchments from the northern Italian Apennines. *Water* **2015**, *11*, 1360. [CrossRef]
8. Fontes, J.C.; Andrews, J.N.; Edmunds, W.M.; Guerre, A.; Travi, Y. Paleorecharge by the Niger river (Mali) deduced from groundwater geochemistry. *Water Resour. Res.* **1991**, *27*, 199–214. [CrossRef]
9. Reckerth, A.; Stichler, W.; Schmidt, A.; Stumpp, C. Long-term data set analysis of stable isotopic composition in German rivers. *J. Hydrol.* **2017**, *552*, 718–731. [CrossRef]
10. Marchina, C.; Natali, C.; Bianchini, G. The Po River water isotopes during the drought condition of the year 2017. *Water* **2019**, *11*, 150. [CrossRef]
11. Martinelli, G.; Dadomo, A.; Cervi, F. An Attempt to Characterize the Recharge of Alluvial Fans Facing the Northern Italian Apennines: Indications from Water Stable Isotopes. *Water* **2020**, *12*, 1561. [CrossRef]
12. Dansgaard, W. Stable isotopes in precipitation. *Tellus* **1964**, *16*, 436–468. [CrossRef]
13. International Atomic Energy Agency. *Statistical Treatment of Data on Environmental Isotopes in Precipitation*; Technical Reports Series No. 331; IAEA: Vienna, Austria, 1992.
14. Crawford, J.; Hughes, C.E.; Lykoudis, S. Alternative least squares methods for determining the meteoric water line, demonstrated using GNIP data. *J. Hydrol.* **2014**, *519*, 2331–2340. [CrossRef]
15. Marchina, C.; Zuecco, G.; Chiogna, G.; Bianchini, G.; Carturan, L.; Comiti, F.; Engel, M.; Natali, C.; Borga, M.; Penna, D. Alternative methods to determine the $\delta2H$-$\delta18O$ relationship: An application to different water types. *J. Hydrol.* **2020**, *587*, 124951. [CrossRef]
16. Martinelli, G.; Chahoud, A.; Dadomo, A.; Fava, A. Isotopic features of Emilia-Romagna region (North Italy) groundwaters: Environmental and climatological implications. *J. Hydrol.* **2014**, *519*, 1928–1938. [CrossRef]
17. Deiana, M.; Mussi, M.; Ronchetti, F. Discharge and environmental isotope behaviours of adjacent fractured and porous aquifers. *Environ. Earth Sci.* **2017**, *76*, 595. [CrossRef]
18. Deiana, M.; Mussi, M.; Pennisi, M.; Boccolari, M.; Corsini, A.; Ronchetti, F. Contribution of water geochemistry and isotopes ($\delta 18$ O, $\delta 2 H$, 3 H, 87 Sr/86 Sr and $\delta 11 B$) to the study of groundwater flow properties and underlying bedrock structures of a deep landslide. *Environ. Earth Sci.* **2020**, *79*, 30. [CrossRef]
19. Kirchner, J.W. Aggregation in environmental systems—Part 2: Catchment mean transit times and young water fractions under hydrologic nonstationarity. *Hydrol. Earth Syst. Sci.* **2016**, *20*, 299–328. [CrossRef]
20. Cervi, F.; Blöschl, G.; Corsini, A.; Borgatti, L.; Montanari, A. Perennial springs provide information to predict low flows in mountain basins. *Hydrolog. Sci. J.* **2017**, *62*, 2469–2481. [CrossRef]
21. ARPAE-EMR. Regional Agency for Environmental Protection in Emilia-Romagna Region: Bollettini Neve. 2020. Available online: https://www.arpae.it/it/temi-ambientali/meteo/report-meteo/bollettini-innevamento/ (accessed on 31 January 2022).
22. Cervi, F.; Tazioli, A. Quantifying Streambed Dispersion in an Alluvial Fan Facing the Northern Italian Apennines: Implications for Groundwater Management of Vulnerable Aquifers. *Hydrology* **2021**, *8*, 118. [CrossRef]
23. Corsini, A.; Cervi, F.; Ronchetti, F. Weight of evidence and artificial neural networks for potential groundwater spring mapping: An application to the Mt. Modino area (Northern Apennines, Italy). *Geomorphology* **2009**, *111*, 79–87. [CrossRef]
24. Tazioli, A.; Cervi, F.; Doveri, M.; Mussi, M.; Deiana, M.; Ronchetti, F. Estimating the isotopic altitude gradient for hydrogeological studies in mountainous areas: Are the low-yield springs suitable? Insights from the northern Apennines of Italy. *Water* **2019**, *11*, 1764. [CrossRef]
25. Cervi, F.; Nistor, M.M. High resolution of water availability for Emilia-Romagna region over 1961–2015. *Adv. Meteorol.* **2018**, 2489758. [CrossRef]
26. ARPAE-EMR. Regional Agency for Environmental Protection in Emilia-Romagna Region: Annali Idrologici. 2019. Available online: https://www.arpae.it/sim/ (accessed on 31 October 2020).
27. Sheather, S. *A Modern Approach to Regression with R*; Springer Science & Business Media: New York, NY, USA, 2009.
28. Davis, J.C. *Statistics and Data Analysis in Geology*; John Wiley & Sons: New York, NY, USA, 2001.
29. Doornik, J.A.; Hansen, H. An omnibus test for univariate and multivariate normality. *Oxf. B. Econ. Stat.* **2008**, *70*, 927–939. [CrossRef]
30. Breusch, T.S.; Pagan, A.R. A Simple Test for Heteroscedasticity and Random Coefficient Variation. *Econometrica* **1979**, *47*, 1287. [CrossRef]
31. Durbin, J.; Watson, G.S. Testing For Serial Correlation in Least Squares Regression: I. *Biometrika* **1950**, *37*, 409–428. [PubMed]
32. Helsel, D.R.; Hirsch, R.M. *Statistical Methods in Water Resources*; US Geological Survey: Reston, VA, USA, 2002.

33. Warton, D.I.; Wright, I.J.; Falster, D.S.; Westoby, M. Bivariate line-fitting methods for allometry. *Biol. Rev.* **2006**, *81*, 259–291. [CrossRef] [PubMed]
34. Prais, S.J.; Winsten, C.B. *Trend Estimators and Serial Correlation*; Cowles Commission Discussion Paper; Cowles Commission: Chicago, IL, USA, 1954.
35. Kirchner, J.W.; Knapp, J.L. Calculation scripts for ensemble hydrograph separation. *Hydrol. Earth Syst. Sci.* **2020**, *24*, 5539–5558. [CrossRef]
36. Wooldridge, J.M. Introductory Econometrics—A Modern Approach, 5th ed.South-Western Cengage Learning: Boston, MA, USA, 2012.
37. Rousseeuw, P.J.; van Driessen, K. *1999. Computing LTS Regression for Large Data Sets*; Institute of Mathematical Statistics Bulletin: Hayward, CA, USA, 1999.
38. Bergraann, H.; Sackl, B.; Maloszewski, P.; Stichler, W. Hydrological investigations in a small catchment area using isotope data series. In Proceedings of the 5th International Symposium on Underground Water Tracing, Athens, Greece, 22–27 September 1986; Institute of Geology and Mineral Exploration: Athens, Greece, 1986; pp. 255–272.
39. McDonnell, J.J. Where does water go when it rains? Moving beyond the variable source area concept of rainfall runoff response. *Hydrol. Processes* **2003**, *17*, 1869–1875. [CrossRef]
40. Tromp-van Meerveld, H.J.; McDonnell, J.J. Threshold relations in subsurface stormflow: 2. The fill and spill hypothesis. *Water Resour. Res.* **2006**, *42*. [CrossRef]
41. Gallart, F.; Freyberg, J.V.; Valiente, M.; Kirchner, J.W.; Llorens, P.; Latron, J. An improved discharge sensitivity metric for young water fractions. *Hydrol. Earth Syst. Sci.* **2020**, *24*, 1101–1107. [CrossRef]
42. Kirchner, J.W. Aggregation in environmental systems—Part 1: Seasonal tracer cycles quantify young water fractions, but not mean transit times, in spatially heterogeneous catchments. *Hydrol. Earth Syst. Sci.* **2016**, *20*, 279–297. [CrossRef]
43. Laaha, G.; Blöschl, G. Low flow estimates from short stream flow records—A comparison of the methods. *J. Hydrol.* **2005**, *306*, 264–286. [CrossRef]
44. Stichler, W.; Schotterer, U.; Froehlich, K.; Ginot, P.; Kull, C.; Gaeggeler, H.; Pouyaud, B. Influence of sublimation on stable isotope records recovered from high-altitude glaciers in the tropical Andes. *J. Geophys. Res.* **2001**, *106*, 22613–22620. [CrossRef]
45. Sokratov, S.A.; Golubev, V.N. Snow isotopic content change by sublimation. *J. Glaciol.* **2009**, *55*, 823–828. [CrossRef]
46. Earman, S.; Campbell, A.R.; Phillips, F.M.; Newman, B.D. Isotopic exchange between snow and atmospheric water vapor: Estimation of the snowmelt component of groundwater recharge in the southwestern United States. *J. Geophys. Res.* **2006**, *111*, D09302. [CrossRef]
47. Lee, J.H.; Feng, X.H.; Posmentier, E.S.; Faiia, A.M.; Taylor, S. Stable isotopic exchange rate constant between snow and liquid water. *Chem. Geol.* **2009**, *260*, 57–62. [CrossRef]
48. Lechler, A.R.; Niemi, N.A. The influence of snow sublimation on the isotopic composition of spring and surface waters in the southwestern United States: Implications for stable isotope–based paleoaltimetry and hydrologic studies. *Geol. Soc. Am. Bull.* **2012**, *124*, 318–334. [CrossRef]

 *hydrology*

Article

Using Heat as a Tracer to Detect the Development of the Recharge Bulb in Managed Aquifer Recharge Schemes

Esteban Caligaris , Margherita Agostini and Rudy Rossetto *

Institute of Life Sciences, Scuola Superiore Sant'Anna, 56122 Pisa, Italy;
estebanrafael.caligaris@santannapisa.it (E.C.); margherita.agostini@santannapisa.it (M.A.)
* Correspondence: rudy.rossetto@santannapisa.it

Abstract: Managed Aquifer Recharge (MAR), the intentional recharge of aquifers, has surged worldwide in the last 60 years as one of the options to preserve and increase water resources availability. However, estimating the extent of the area impacted by the recharge operations is not an obvious task. In this descriptive study, we monitored the spatiotemporal variation of the groundwater temperature in a phreatic aquifer before and during MAR operations, for 15 days, at the LIFE REWAT pilot infiltration basin using surface water as recharge source. The study was carried out in the winter season, taking advantage of the existing marked difference in temperature between the surface water (cold, between 8 and 13 °C, and in quasi-equilibrium with the air temperature) and the groundwater temperature, ranging between 10 and 18 °C. This difference in heat carried by groundwater was then used as a tracer. Results show that in the experiment the cold infiltrated surface water moved through the aquifer, allowing us to identify the development and extension in two dimensions of the recharge plume resulting from the MAR infiltration basin operations. Forced convection is the dominant heat transport mechanism. Further data, to be gathered at high frequency, and modeling analyses using the heat distribution at different depths are needed to identify the evolution of the recharge bulb in the three-dimensional space.

Keywords: Managed Aquifer Recharge; groundwater tracer; heat transport; surface–ground-water interactions; infiltration basin; groundwater hydrology

Citation: Caligaris, E.; Agostini, M.; Rossetto, R. Using Heat as a Tracer to Detect the Development of the Recharge Bulb in Managed Aquifer Recharge Schemes. *Hydrology* **2022**, *9*, 14. https://doi.org/10.3390/hydrology9010014

Academic Editors: Il-Moon Chung, Sun Woo Chang, Yeonsang Hwang and Yeonjoo Kim

Received: 12 December 2021
Accepted: 11 January 2022
Published: 12 January 2022

Publisher's Note: MDPI stays neutral with regard to jurisdictional claims in published maps and institutional affiliations.

1. Introduction

Freshwater resources are suffering from increasing pressure worldwide. Their contamination and overexploitation are compromising access to safe water [1–3]. This situation pushes towards the search for innovative ways to preserve and increase freshwater resources availability, focusing on sustainable water management techniques. Managed Aquifer Recharge (MAR), the intentional recharge of aquifers potentially using water from various sources, has surged worldwide in the last 60 years as one of these options [4–7].

Measurements of infiltration rates and groundwater levels variations, together with the estimation of the groundwater flows generated during recharge in MAR schemes, are used to evaluate the performance in terms of recharge volumes and the extension of the recharge plume [8,9]. Different groundwater monitoring techniques are usually implemented for this purpose, where the use of sensors to measure groundwater pressure head, electrical conductivity, temperature, and soil moisture is normally accompanied to groundwater sampling for chemical analyses and numerical modeling [10–12].

Ganot et al. [11] assessed the relation between the infiltration and the development of the groundwater mound in MAR using desalinated seawater in an infiltration pond. In their study, the saturated zone of the aquifer was monitored through two groundwater observation wells instrumented with pressure head and electrical conductivity loggers. These measurements were later used in a lumped model where the infiltration dynamics was analyzed to assess the temporal and spatial variation of the recharge. Likewise, the changes in recharge from a river into an aquifer as a result of the implementation of a Riverbank Filtration MAR scheme were evaluated by Rossetto et al. [12] by means of a multidisciplinary approach using hydrodynamics, hydrochemical, and modeling methods, following intensive sensors application [13].

New innovative methodologies to estimate the extension and development of the plume of recharged water in the aquifer are also being proposed. These methodologies apply geophysical methods and can range from the use of electrical resistivity [14–17] up to time-lapse gravity measurements [18,19].

The use of vertical electrical conductivity profiles for the estimation of the saturated hydraulic conductivity and the van Genuchten parameters under an infiltration pond was studied by Mawer et al. [14]. Similarly, Nenna et al. [15] used electrical resistivity probes with the objective of mapping and monitoring the recharge plume from an infiltration pond. By monitoring the temporal variation of the vertical electrical resistivity of different points located under and around the infiltration pond, the temporal variation of the water table could be estimated together with the hydraulic gradients. These data can be used later to estimate the fate of the recharged water. Haaken et al. [16] assessed the use of Electrical Resistivity Tomography (ERT) measurements for characterizing groundwater dynamics under a Soil Aquifer Treatment scheme. Zones with different hydraulic properties were identified by analyzing the temporal variations of these measurements. Likewise, García-Menéndez et al. [17] used ERT to evaluate the effectiveness of MAR in a coastal aquifer. With this technology, the extension and shape of the recharge plume could be identified. This was completed after the joint interpretation of the ERT images with Electrical Conductivity logs from boreholes, and with geological and hydrogeological information of the site. The use of time-lapse gravity surveys was assessed by Davis et al. [18] for the monitoring of an Aquifer Storage and Recovery (ASR) scheme. The use of this geophysical technology was applied successfully during the injection of water into the aquifer for the detection of the general distribution and movement of the injected water. With a similar approach, Chapman et al. [19] used high-precision gravity measurements for the monitoring of another ASR pilot system. In their study, the high-precision gravity surveys were carried before, during, and after two infiltration cycles. The detection of the formation of a mound of recharged groundwater during the recharge cycles was possible with the analysis of the collected data.

The fundamentals of the use of heat as a tracer in groundwater have been previously studied [20]. Groundwater temperature may be measured by lowering a thermometer down a borehole, and the wide availability of waterproof temperature loggers makes this parameter easily accessible [20,21]. Various experimental applications using heat carried by groundwater as a tracer to monitor different aspects of MAR operations have been investigated by diverse authors. For instance, a Fiber Optic Distributed Temperature Sensing technique was used to estimate infiltration rates from recharge basins [22,23]. Similarly, heat was also used as a tracer for the estimation of recharge rates at infiltration ponds [24], and for the estimation of travel time in bank filtration systems [25]. Likewise, the vertical fluxes in heterogeneous aquifers can be estimated using heat [26].

In this study, we monitored the spatio-temporal variation of the groundwater temperature in a phreatic aquifer before and during MAR operations, for 15 days, at a pilot infiltration basin. This change in groundwater temperature is being used to identify the development and extension of the resulting recharge plume following recharge operations at the LIFE REWAT MAR infiltration basin [27].

2. Materials and Methods

In this section, the study site and the MAR scheme are presented alongside the methodology used for monitoring the groundwater temperature changes. The operations at the LIFE REWAT MAR infiltration basin with its different components are also briefly described.

2.1. Study Site

The study site is located in the municipality of Suvereto (Tuscany, Italy) in the alluvial plain of the Cornia River (Figure 1). The Cornia plain hosts a Holocene coastal aquifer constituted by alluvial and swamp-lagoonal deposits. The deposits, largely influenced by the Cornia River dynamics, include gravel, sand, silt, and clay in different proportions and distributions. The stratigraphy of the aquifer under investigation is well presented in Barazzuoli et al. [28]. New drillings allow us to obtain new information confirming the previous hypotheses and work. A large proportion of the aquifer is composed of a gravel lithology in a silty–sandy matrix, possessing a prevalent permeability by interstitial porosity. This layer outcrops the surface or is covered by a layer of silt as a result of fluvial overflows. The aquifer is unconfined in the area of the infiltration basin. Large surface water–groundwater exchanges occur between the River Cornia and the aquifer.

Figure 1. Study area location and measured points.

Figure 2 presents the stratigraphies at points REW_10 (in the center of the infiltration basin), REW_12 and REW_6 (north of the infiltration basin). A relatively thin layer of agricultural soil covers an alternate layer of gravels with different size distribution in silty matrix in the vicinity of the infiltration basin up to about 15 m from soil surface. Some thin lenses of gravels in a clayey matrix can also be found at different depths. As such, the experimental area shows up to a depth of about 15 m from the soil surface, the presence of a gravel-dominated environment, in a matrix variable from silt to sand.

The River Cornia is the main hydrologic feature in the area. The high hydraulic conductivity of the riverbed provides high hydraulic connectivity between the surface water and the aquifer. This enhances surface and groundwater exchanges in the areas near to the river. Hence, the groundwater heads are controlled by the water level of the river, and, locally, by the presence of pumping wells. Because of this, values of electrical conductivity in the aquifer slightly differ from those of surface water. As such, the parameter electrical conductivity cannot be easily used to trace the recharged water. The main groundwater natural flow is directed towards the West, resulting from river recharge and inflows from adjoining hilly areas, with an average hydraulic gradient of 0.2% (Figure 3). From the regional hydrology point of view the area is a recharge area.

Additionally, the study area is characterized by the presence of an important hydrothermal system, which contributes to the recharge of the superficial aquifer by means of upward groundwater flow, causing some thermal and geochemical anomalies [28,29].

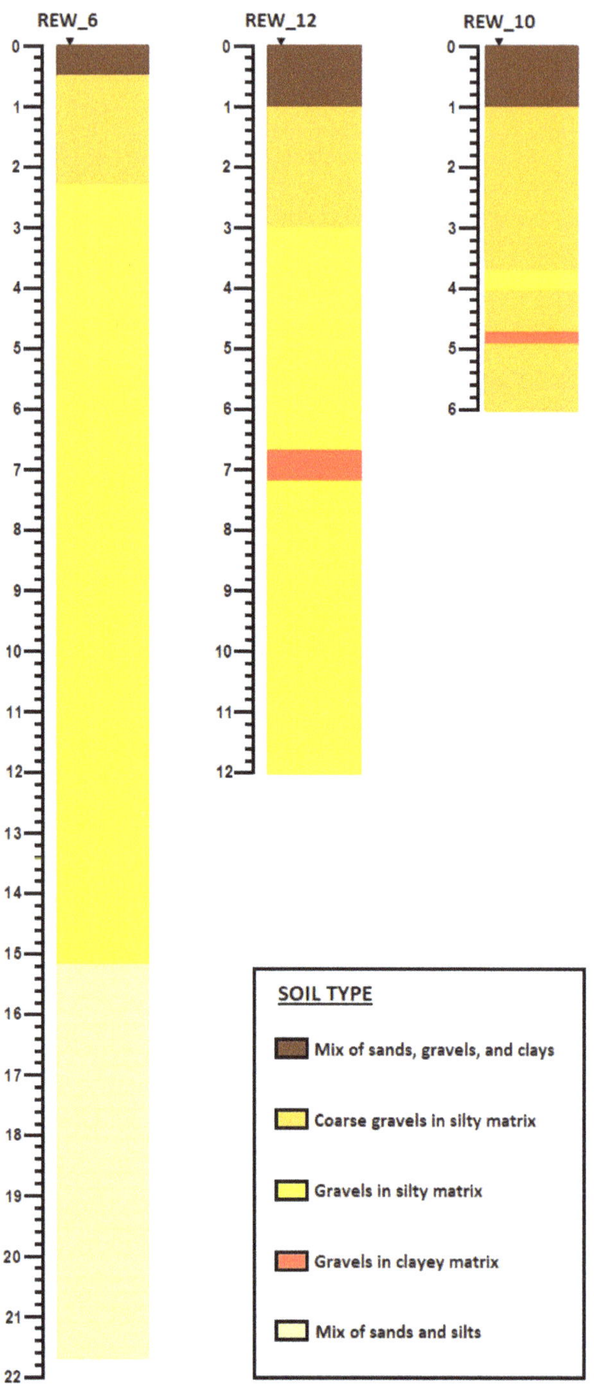

Figure 2. Stratigraphy of three piezometers near the infiltration basin. Information obtained from the analysis of the soil cores during the construction of these piezometers.

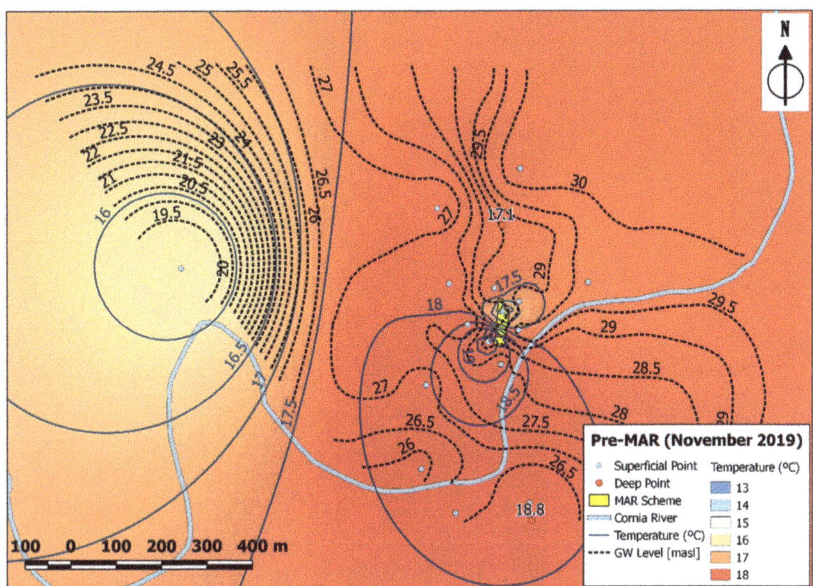

Figure 3. Groundwater temperature distribution in the aquifer before MAR operations started. Data taken from 25 November 2019 to 27 November 2019.

The initial temperature conditions in the aquifer at the beginning of the rainfall season (just after the end of the dry season), before the managed aquifer recharge operations started in 2019, can be seen in Figure 3. The local groundwater temperatures ranged between 15.7 and 19.6 °C in November 2019, with air temperature varying from 8 to 20 °C, and surface water temperature at about 15 °C in those days. A fairly homogeneous distribution of temperatures, higher than about 17 °C, is noticeable in the MAR scheme area. Two deeper points, REW_6 and REW_142, show temperatures of 17.1 and 18.8 °C, respectively (a map of temperature distribution only is available as Supplementary Materials, Figure S1). These relatively high groundwater temperatures highlight the presence of the above-mentioned geothermal flow.

2.2. The LIFE REWAT Managed Aquifer Recharge Scheme

The LIFE REWAT Managed Aquifer Recharge scheme is a two-stage infiltration basin using harvested rainwater from the Cornia River during high-flow periods. The scheme consists of diversion infrastructure and then two basins: a settling pond and the infiltration basin (Figure 4). Surface water is firstly diverted from the Cornia river into the decantation pond, where the suspended solids are deposited. Afterwards, the water enters into the infiltration pond. The infiltration pond was constructed in a topographic low, where the soil (sandy/silty gravels) provides a full hydraulic connection with the phreatic aquifer.

The MAR scheme is operated using a hi-tech high-frequency automated and remotely controlled system, and quasi real-time monitoring of water quantity and quality is run. This system is supported by the data gathered from different sensors installed in the area, recording different parameters into a database with a frequency of fifteen minutes.

2.3. Groundwater Head and Temperature Monitoring

For this study, groundwater head and temperature were monitored at selected points in the shallow aquifer (Table 1 and Figure 1; shallow points are named "Superficial"). These points, located upstream and downstream of the MAR scheme, were monitored before and during MAR operations, covering two weeks of full operations of the MAR scheme.

Deeper screened points (i.e., points at depths higher than 20 m; "Deep" points in Table 1) were also monitored (Figure 1), but their data are not used in the interpolation process, and only plotted against the temperature distribution in the shallow aquifer.

Figure 4. LIFE REWAT Managed Aquifer Recharge scheme.

The fieldwork measurements were carried out with two instruments, a portable water level meter (dipper) [30], and a thermo-dipper [31]. The dipper had precision of 1 cm, while the temperature sensor had accuracy of ± 0.1 °C ranging from -10 to $+50$ °C.

The study was carried out in winter, taking advantage of the existing difference in temperature between the surface water (cold, between 8 and 13 °C, and in quasi-equilibrium with the air temperature) and the groundwater. This way the colder surface water infiltrating in the basin could mix with/replace the warmer groundwater in the aquifer during the recharge operations. The experiment started on 9 February 2020. On that date at 15:00 (CET) the MAR scheme was set off for 52 h, being in full operation since 10 December

2019. On 11 February at 19:00 (CET) the scheme was turned on again. The temperature monitoring took place on three campaigns: C1 on 10 February 2020, and 11 February 2020; C2 on 18 February 2020; C3 on 25 February 2020. The experiment ended because of a large flooding event of the Cornia River occurring on 3 March 2020, when the managed recharge was temporarily suspended following operational protocols. During the experiment, we approximately recharged the aquifer at the rate of 5800 m^3/day.

Table 1. List of the piezometers and wells used in the experiment.

Point	Piezometer Type	Monitored Depth [m]	Point Depth [m]	Point	Type	Monitored Depth [m]	Point Depth [m]
REW_10	Superficial	2.70	2.80	REW_39	Superficial	15.00	15.50
REW_11	Superficial	5.00	6.14	REW_3	Superficial	10.00	12.00
REW_12	Superficial	8.00	11.84	REW_5	Superficial	5.90	6.00
REW_13	Superficial	6.00	6.50	REW_6	Deep	10.00	30.00
REW_14	Superficial	6.15	6.23	REW_119	Superficial	10.00	12.00
REW_15	Superficial	6.10	6.25	REW_142	Deep	40.00	43.00
REW_16	Superficial	4.90	5.00	REW_156	Superficial	3.90	4.00
REW_17	Superficial	6.00	7.05	REW_157	Superficial	5.50	5.60
REW_18	Superficial	6.80	6.90	REW_158	Superficial	5.20	5.25
REW_19	Superficial	8.50	8.88	REW_301	Superficial	2.00	3.76
REW_20	Superficial	14.60	14.70	REW_302	Superficial	1.60	2.88
REW_23	Superficial	13.00	14.00	REW_304	Superficial	1.50	3.71
REW_24	Superficial	8.00	8.16	REW_305	Superficial	4.00	4.94
REW_25	Superficial	7.00	7.57	REW_306	Superficial	4.10	4.15
REW_30	Superficial	10.00	12.00	REW_444	Deep	20.00	30.00
REW_36	Deep	21.00	30.00	-	-	-	-

Groundwater heads and temperatures were measured at 27 points in the phreatic aquifer and at 4 points at depth larger than 20 m from soil surface (Table 1). In order to avoid measuring the temperature of the groundwater superficially, hence subjectedto short-time changes in air temperature, the temperature measurements were taken from depths under 5 m of the water level if the depth of the piezometers allowed it. The measured values of heads and temperatures were finally spatially interpolated utilizing the Inverse Distance Weight interpolation feature of QGis 2.18.28 [32] and then some isolines were slightly modified to take the influence of the River Cornia into account. Additionally, water levels and temperature variations from different points were recorded automatically through a series of sensors in situ. All these values were recorded in the SCADA system of the MAR scheme with a frequency of fifteen minutes.

The meteo-climatic and hydrologic conditions were monitored during the experiment period and are summarized in Figure 5. The air temperature ranged between a minimum value of 0.5 °C and a maximum value of 19.4 °C, with an average value of 10.4 °C for the whole period [33]. Similarly, the water temperature from the Cornia River presented a mean value of 12.7 °C. The experiment was run with the river level remaining in baseflow conditions, at a constant level of 0.51 m, varying 1 or 2 cm during the day, at the Ponte per Montioni monitoring station [34]. During the study period, 3 days of rainfall were recorded, where only a total of 1.8 mm of rainfall was recorded on 19 February 2020 and 0.2 mm on 20 February 2020 and 24 February 2020, respectively (recorded at the rain gauge station of Suvereto) [35]. The amount of rainfall is therefore considered negligible in term of aquifer recharge affecting the experiment.

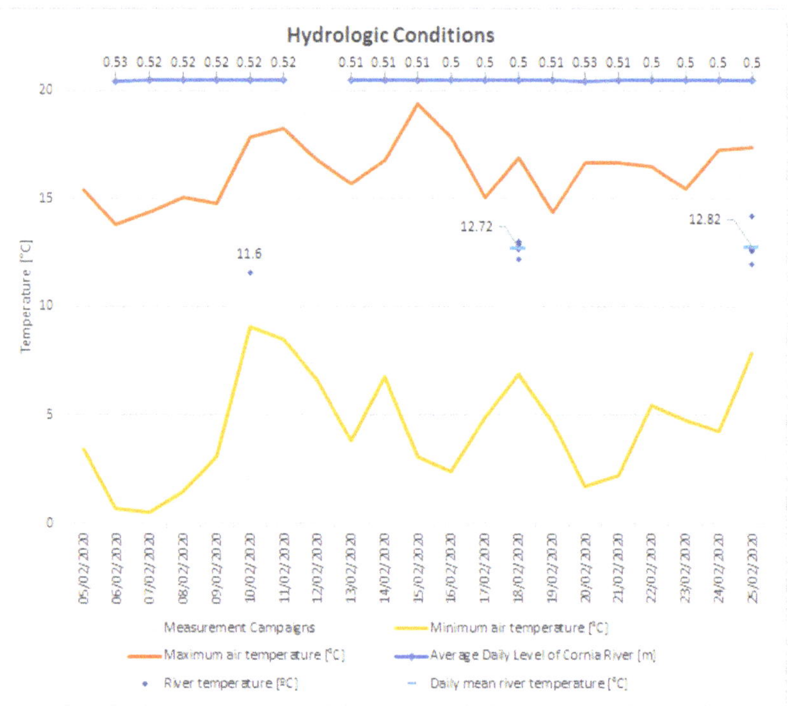

Figure 5. Hydrologic conditions at the MAR site during the experiment.

3. Results and Discussion

In Figure 6, the discharge curve of the infiltration basin together with the piezometric variations at the points REW_10, located in the infiltration basin, and REW_17 are presented. The 52 h interruption of the recharge operations, from 9 until 11 February 2020, was reflected in the water level of the infiltration basin and in the aquifer. Another short interruption of the recharge operations is also observable from 19 until 20 February 2020 as a result of the automatic operation of the scheme. The changes in the basin water level are reflected in a relatively short time in nearby points (e.g., the point REW_17, located around 150 m downstream of the infiltration basin). This behavior cannot be explained solely by the Darcy equation of flow in porous media, but on the analysis of the speed of the pressure wave (celerity) [36–38]. Thus, the hydraulic head changes in the groundwater may not accurately represent the actual movement of the recharged water volume itself. Therefore, complementary information, such as those provided by heat carried by groundwater, and analyses are required for the determination of the development of the recharged plume.

The temperature variations of the surface water in the infiltration basin, and of the groundwater at the point REW_10, screened at 2.7 m depth under the infiltration basin, can be seen in Figure 7. The change in temperature reflected on the groundwater point is in direct relation with the changes in the surface-water temperature in the infiltration basin. The temperature differences between these two points are relatively small. This relation suggests a displacement of the native groundwater by the infiltrated one or a mix of these two endmembers, with a dominant surface-water component.

The existing temperature in the aquifer after 2 months of MAR operations, and shortly before the described experiment started, is shown in Figure 8 (the map of temperature distribution only is available as Supplementary Materials, Figure S2). Compared to November 2019 (Figure 3), a cold area centered in the recharge basin (REW_10 at 9.6 °C) has developed following two main axes: one towards West and one approximately South.

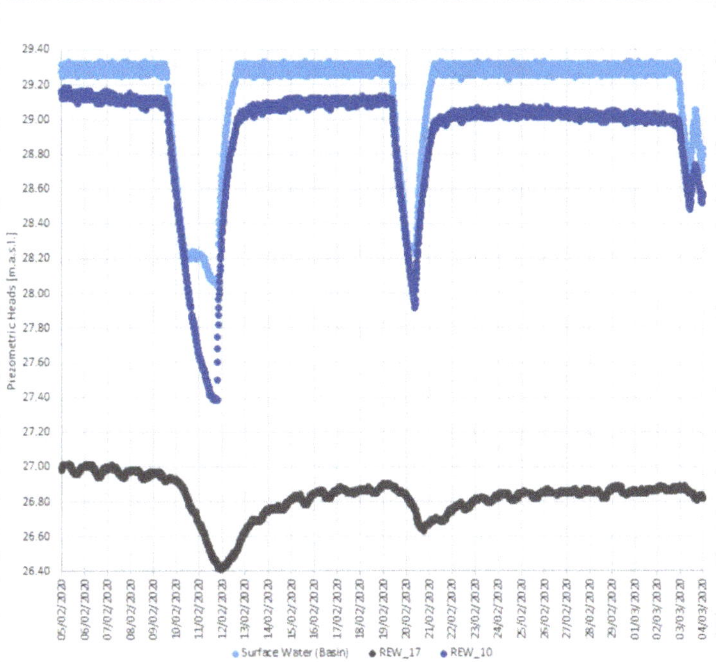

Figure 6. Recorded piezometric heads from in-situ sensors (m above mean sea level).

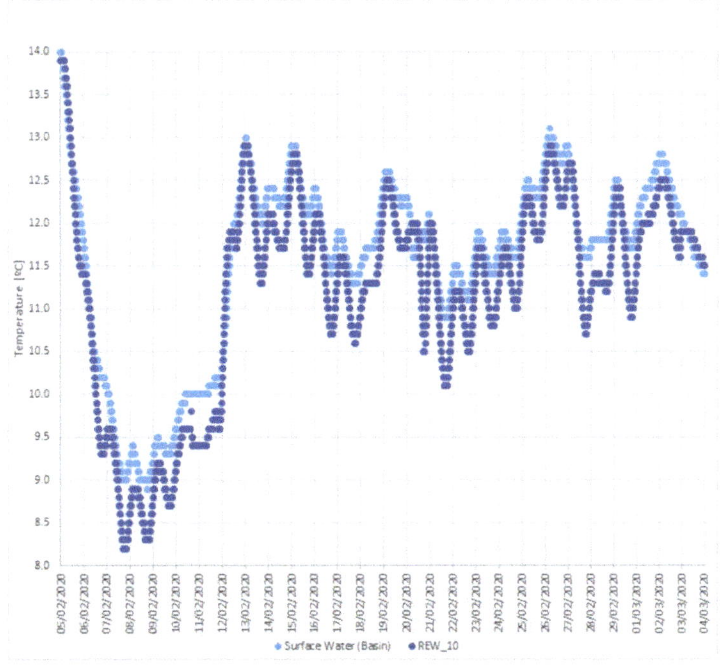

Figure 7. Recorded surface and groundwater temperature variations in the infiltration basin.

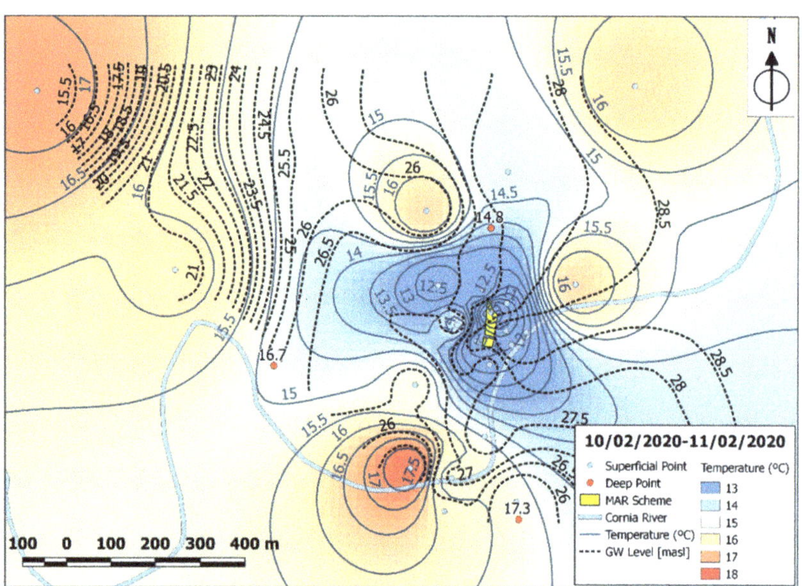

Figure 8. Groundwater temperature distribution with MAR operations halted for the experiment.

The MAR operations slightly modify the regional groundwater flow by superimposing two local, additional mainly East to West and North to South flow to the dominant in the area river recharge. A high thermal gradient is detected within the first 100 m around the infiltration basin. In our experiment, the groundwater flow perturbed the geothermal gradient by infiltration of relatively cool water in a recharge area contrasting with an upward flow of relatively warm water. Recharge is clearly affecting the aquifer temperature in the infiltration basin area.

Comparing Figure 8 with Figure 3, the points located on the North–East, upstream of the infiltration basin, hold a steady temperature above 16 °C, as seen before the recharge activities. This shows a very slow change in their temperature with time, in contrast with the areas directly impacted by the MAR activities. At the shallow point, REW_158, the temperature stays high at 17.8 °C. The same applies to the deeper points, REW_444, West of the basin, at 16.7 °C, and REW_142, at 17.3 °C, demonstrating the relevance of the geothermal flow in this section. The colder plume depicted South of the river is a result of the interpolation process, and no data are available to confirm these results.

Once the MAR operations restarted, on 11 February at 19.00 (CET) a cold temperature plume further developed following the above-mentioned directions. When observing the variation with time of the values of groundwater temperatures in Figure 9 and in Figure 10 (maps of temperature distribution only are available as Supplementary Materials, Figures S3 and S4), it is worth noting that the temperatures of the points located upstream of the infiltration basin (REW_19, REW_5, REW_3, and REW_30) still maintain constant values. This shows a minor development of the recharge bulb upstream of the MAR scheme, and the relevance in the area of a forced convection heat transport in agreement with the modified groundwater flow direction.

The temperature signal seems undetectable at REW_23, about 700 m West of the recharge area, while REW_158 still maintains a temperature higher than 17 °C. In this regard, two hypotheses may be made: (i) the recharge flow did not reach these points during the experiment time, and/or (ii) the upward geothermal warm flow potentially has a larger influence. In the second case, recharged groundwater would be mixing with the geothermal flow, but the rate of recharged water during the experiment would be low compared to the geothermal flow, then being unable to change the aquifer thermal state.

Taking the area enclosed within the 14 °C isotherm in the northern side of the Cornia River as a reference, this area expands with time. Starting with 110,000 m^2 on 11th February, the area grew up to 138,000 m^2 after 7 days, and up to 174,000 m^2 after 14 days.

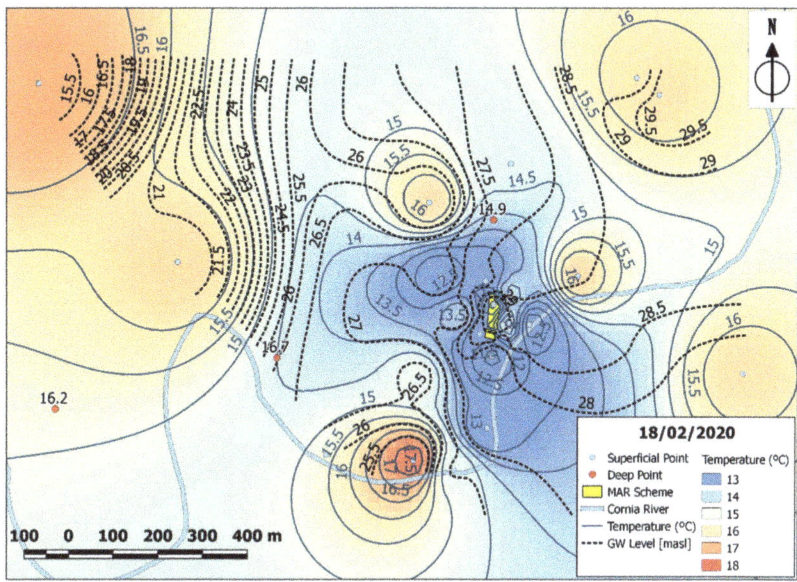

Figure 9. Groundwater temperature distribution after 7 days of the restart of MAR operations.

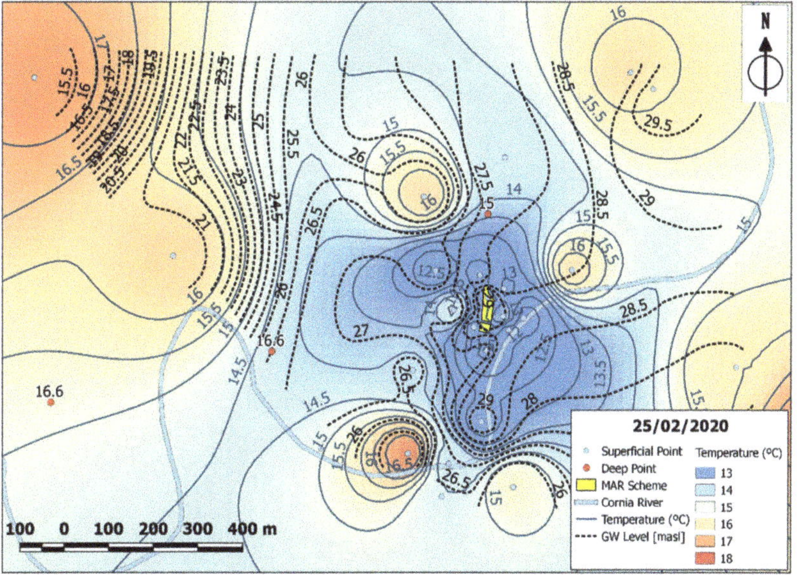

Figure 10. Groundwater temperature distribution after 14 days of the restart of MAR operations.

Because of the imposed head gradient in the MAR area, forced convection [20] seems to be the dominant heat transport mechanism in our experiment, while minor relevance seems to have conduction and transport with thermal dispersivity.

4. Conclusions

The experiment described here shows the use of heat carried by groundwater as a tracer in order to detect the development of the recharge plume in a Managed Aquifer Recharge scheme. As our experiment demonstrated, heat as a tracer is especially suited for delineating small-scale flow paths monitoring temperature in the aquifers [39]. Temperature, besides hydraulic head and groundwater chemistry data, is a readily available parameter that, in particular meteo-climatic conditions, may provide cost-effective observations to conduct hydrological investigations. Results show that in the experiment the cold-infiltrated surface water moves through the aquifer, allowing us to identify the development and extension in the two-dimensional space of the recharge plume resulting from the LIFE REWAT MAR infiltration basin operations. The results highlight two main components of convective heat transport, one towards the West and one to the South, forced by the hydraulic gradient set by the recharge operations. The upstream groundwater flow seems to limit the cold water movement on the eastern side of the MAR scheme. The recharge operations seem not to affect the deeper layers of the aquifer. Further analyses are needed to evaluate the mixing between the groundwater of geothermal origins and the recharged one.

Further works will include the assessment of heat transport in 2D along a cross-section monitoring temperature (with other parameters) at different depths of the aquifer. The joint use of groundwater head and temperature data in 3D groundwater modeling applications may support the parameterization of the aquifer system under investigation and the set-up of geochemical reactive transport models for the understanding of complex processes occurring during recharge. Finally, we suggest that along other parameters to be analyzed during the planning, design and investigation phase of Managed Aquifer Recharge schemes [40,41], groundwater temperature distribution is duly considered in order to accurately estimate groundwater flow direction and velocities prior to the modified state and following the beginning of MAR operations.

Supplementary Materials: The following supporting information can be downloaded at: https://www.mdpi.com/article/10.3390/hydrology9010014/s1, Figure S1: Groundwater temperature before MAR operations started. From 25 November 2019 to 27 November 2019. Figure S2: Groundwater temperature distributions with MAR operations halted for the experiment (10–11 February 2020). Figure S3: Groundwater temperature distribution after 7 days since restart of MAR operations (18 February 2020). Figure S4: Groundwater temperature distribution after 14 days since the restart of MAR operations (25 February 2020).

Author Contributions: Conceptualization, E.C. and R.R.; methodology, E.C. and R.R.; formal analysis, E.C.; investigation, E.C. and M.A.; resources, R.R.; data curation, E.C.; writing—original draft preparation, E.C. and R.R.; writing—review and editing, E.C. and R.R.; visualization, E.C.; supervision, R.R.; project administration, R.R.; funding acquisition, R.R. All authors have read and agreed to the published version of the manuscript.

Funding: This paper is presented within the framework of the MARSoluT ITN project (www.marsolut-itn.eu) (accessed on 10 December 2021). and the EU LIFE REWAT project (www.liferewat.eu) (accessed on 10 December 2021). The MARSoluT project receives funding from the European Union's Horizon 2020 research and innovation programme under the Marie Skłodowska-Curie grant agreement no. 814066. The LIFE REWAT project received funding from the European Union's Life Programme LIFE 14 ENV/IT/001290.

Acknowledgments: The authors wish to thank ASA spa (Mirko Brilli, Claudio Benucci and Patrizio Lainà) and Consorzio di Bonifica 5 Toscana Costa (Alessandro Fabbrizzi, Roberto Benvenuto and Giancarlo Vallesi) for technical support at the MAR scheme. E.C. performed the research as ESR6 PhD student in the MARSOLUT project. M.A (now at Studio Geologico Rafanelli) participated to this research as a post-graduate scholarship.

Conflicts of Interest: The authors declare no conflict of interest.

References

1. Arnell, N.W. Climate change and global water resources. *Glob. Environ. Chang.* **1999**, *9*, S31–S49. [CrossRef]
2. European Environment Agency. Water Use and Environmental Pressures. 2020. Available online: https://www.eea.europa.eu/th emes/water/european-waters/water-use-and-environmental-pressures/water-use-and-environmental-pressures (accessed on 30 December 2020).
3. Leduc, C.; Pulido-Bosch, A.; Remini, B. Anthropization of groundwater resources in the Mediterranean region: Processes and challenges. *Hydrogeol. J.* **2017**, *25*, 1529–1547. [CrossRef]
4. Dillon, P. Future management of aquifer recharge. *Hydrogeol. J.* **2005**, *13*, 313–316. [CrossRef]
5. Dillon, P.; Stuyfzand, P.; Grischek, T.; Lluria, M.; Pyne, R.D.G.; Jain, R.C.; Bear, J.; Schwarz, J.; Wang, W.; Fernandez, E.; et al. Sixty years of global progress in managed aquifer recharge. *Hydrogeol. J.* **2019**, *27*, 1–30. [CrossRef]
6. Wintgens, T.; Hochstrat, R.; Kazner, C.; Jeffrey, P.; Jefferson, B.; Melin, T. Managed Aquifer Recharge as a Component of Sustainable Water Strategies-A Brief Guidance for EU Policies. In *Water Reclamation Technologies for Safe Managed Aquifer Recharge*; IWA Publishing: London, UK, 2012; pp. 411–429.
7. Fernandez Escalante, E.; Henao Casas, J.D.; Vidal Medeiros, A.M.; San Sebastián Sauto, J. Regulations and guidelines on water quality requirements for Managed Aquifer Recharge. International comparison. *Acque Sotter. Ital. J. Groundw.* **2020**, *9*, 7–22. [CrossRef]
8. Prathapar, S.; Dhar, S.; Rao, G.T.; Maheshwari, B. Performance and impacts of managed aquifer recharge interventions for agricultural water security: A framework for evaluation. *Agric. Water Manag.* **2015**, *159*, 165–175. [CrossRef]
9. Maliva, R.G. Managed aquifer recharge: State-of-the-art and opportunities. *Water Sci. Technol. Water Supply* **2015**, *15*, 578–588. [CrossRef]
10. Bundschuh, J.; Sracek, O. *Introduction to Groundwater Geochemistry and Fundamentals of Hydrogeochemical Modelling*; Taylor & Francis Group: Abingdon, UK, 2012.
11. Ganot, Y.; Holtzman, R.; Weisbrod, N.; Nitzan, I.; Katz, Y.; Kurtzman, D. Monitoring and modeling infiltration-recharge dynamics of managed aquifer recharge with desalinated seawater. *Hydrol. Earth Syst. Sci.* **2017**, *21*, 4479–4493. [CrossRef]
12. Rossetto, R.; Barbagli, A.; De Filippis, G.; Marchina, C.; Vienken, T.; Mazzanti, G. Importance of the Induced Recharge Term in Riverbank Filtration: Hydrodynamics, Hydrochemical, and Numerical Modelling Investigations. *Hydrology* **2020**, *7*, 96. [CrossRef]
13. Rossetto, R.; Barbagli, A.; Borsi, I.; Mazzanti, G.; Vienken, T.; Bonari, E. Site investigation and design of the monitoring system at the Sant'Alessio Induced RiverBank Filtration plant (Lucca, Italy). *Rend. Online Soc. Geol. Ital.* **2015**, *35*, 248–251. [CrossRef]
14. Mawer, C.; Kitanidis, P.; Pidlisecky, A.; Knight, R. Electrical resistivity for characterization and infiltration monitoring beneath a managed aquifer recharge pond. *Vadose Zone J.* **2013**, *12*, vzj2011.0203. [CrossRef]
15. Nenna, V.; Pidlisecky, A.; Knight, R. Monitoring managed aquifer recharge with electrical resistivity probes. *Interpretation* **2014**, *2*, T155–T166. [CrossRef]
16. Haaken, K.; Furman, A.; Weisbrod, N.; Kemna, A. Time-Lapse Electrical Imaging of Water Infiltration in the Context of Soil Aquifer Treatment. *Vadose Zone J.* **2016**, *15*, 1–12. [CrossRef]
17. García-Menéndez, O.; Ballesteros, B.J.; Renau-Pruñonosa, A.; Morell, I.; Mochales, T.; Ibarra, P.I.; Rubio, F.M. Using electrical resistivity tomography to assess the effectiveness of managed aquifer recharge in a salinized coastal aquifer. *Environ. Monit. Assess.* **2018**, *190*, 100. [CrossRef]
18. Davis, K.; Li, Y.; Batzle, M. Time-lapse gravity monitoring: A systematic 4D approach with application to aquifer storage and recovery. *Geophysics* **2008**, *73*, WA61–WA69. [CrossRef]
19. Chapman, D.S.; Sahm, E.; Gettings, P. Monitoring aquifer recharge using repeated high-precision gravity measurements: A pilot study in South Weber, Utah. *Geophysics* **2008**, *73*, WA83–WA93. [CrossRef]
20. Anderson, M.P. Heat as a ground water tracer. *Groundwater* **2005**, *43*, 951–968. [CrossRef]
21. Banks, D. From Fourier to Darcy, from Carslaw to Theis: The analogies between the subsurface behaviour of water and heat. *Acque Sotter. Ital. J. Groundw.* **2012**, *1*, 3. [CrossRef]
22. Becker, M.W.; Bauer, B.; Hutchinson, A. Measuring artificial recharge with fiber optic distributed temperature sensing. *Groundwater* **2013**, *51*, 670–678. [CrossRef]
23. Mawer, C.; Parsekian, A.; Pidlisecky, A.; Knight, R. Characterizing heterogeneity in infiltration rates during managed aquifer recharge. *Groundwater* **2016**, *54*, 818–829. [CrossRef]
24. Racz, A.J.; Fisher, A.T.; Schmidt, C.M.; Lockwood, B.S.; Huertos, M.L. Spatial and temporal infiltration dynamics during managed aquifer recharge. *Groundwater* **2012**, *50*, 562–570. [CrossRef]
25. des Tombe, B.F.; Bakker, M.; Schaars, F.; van der Made, K.J. Estimating travel time in bank filtration systems from a numerical model based on DTS measurements. *Groundwater* **2018**, *56*, 288–299. [CrossRef]
26. Kurylyk, B.L.; Irvine, D.J.; Carey, S.K.; Briggs, M.A.; Werkema, D.D.; Bonham, M. Heat as a groundwater tracer in shallow and deep heterogeneous media: Analytical solution, spreadsheet tool, and field applications. *Hydrol. Processes* **2017**, *31*, 2648–2661. [CrossRef]
27. Rossetto, R.; De Filippis, G.; Borsi, I.; Foglia, L.; Cannata, M.; Criollo, R.; Vázquez-Suñé, E. Integrating free and open source tools and distributed modelling codes in GIS environment for data-based groundwater management. *Environ. Model. Softw.* **2018**, *107*, 210–230. [CrossRef]

28. Barazzuoli, P.; Bouzelboudjen, M.; Cucini, S.; Király, L.; Menicori, P.; Salleolini, M. Olocenic alluvial aquifer of the River Cornia coastal plain (southern Tuscany, Italy): Database design for groundwater management. *Environ. Geol.* **1999**, *39*, 123–143. [CrossRef]

29. Celati, R.; Grassi, S.; D'Amore, F.; Marcolini, L. The low temperature hydrothermal system of Campiglia, Tuscany (Italy): A geochemical approach. *Geothermics* **1991**, *20*, 67–81. [CrossRef]

30. PASI. Portable Groundwater Level Meter. Freatimetro. 2021. Available online: https://www.pndshop.it/dt/ct20183/art201821/freatimetro-bfk (accessed on 10 December 2021).

31. PASI. Portable Groundwater Level and Thermo Meter. Termofreatimetro. 2021. Available online: https://www.pndshop.it/dt/ct20183/art201823/freatimetro-wmf-02 (accessed on 10 December 2021).

32. QGIS Development Team. QGIS Geographic Information System. Open Source Geospatial Foundation Project. 2021. Available online: http://qgis.osgeo.org (accessed on 11 December 2021).

33. Servizio Idrologico e Geologico della Regione Toscana. Settore Idrologico e Geologico Regionale. DATI Termometria. 2021. Available online: https://www.sir.toscana.it/termometria-pub (accessed on 10 December 2021).

34. Servizio Idrologico e Geologico della Regione Toscana. Settore Idrologico e Geologico Regionale. DATI Idrometria. 2021. Available online: https://www.sir.toscana.it/idrometria-pub (accessed on 10 December 2021).

35. Servizio Idrologico e Geologico della Regione Toscana. Settore Idrologico e Geologico Regionale. DATI Pluviometria. 2021. Available online: https://www.sir.toscana.it/pluviometria-pub (accessed on 10 December 2021).

36. Chesnaux, R. Avoiding confusion between pressure front pulse displacement and groundwater displacement: Illustration with the pumping test in a confined aquifer. *Hydrol. Processes* **2018**, *32*, 3689–3694. [CrossRef]

37. Verseveld, W.J.V.; Barnard, H.R.; Graham, C.B.; McDonnell, J.J.; Brooks, J.R.; Weiler, M. A sprinkling experiment to quantify celerity–velocity differences at the hillslope scale. *Hydrol. Earth Syst. Sci.* **2017**, *21*, 5891–5910. [CrossRef] [PubMed]

38. Worthington, S.R.; Foley, A.E. Deriving celerity from monitoring data in carbonate aquifers. *J. Hydrol.* **2021**, *598*, 126451. [CrossRef]

39. Giambastiani, B.M.S.; Colombani, N.; Mastrocicco, M. Limitation of using heat as a groundwater tracer to define aquifer properties: Experiment in a large tank model. *Environ. Earth Sci.* **2012**, *70*, 719–728. [CrossRef]

40. Barbagli, A.; Jensen, B.N.; Raza, M.; Schueth, C.; Rossetto, R. Assessment of soil buffer capacity on nutrients and pharmaceuticals in nature-based solution applications. *Environ. Sci. Pollut. Res.* **2019**, *26*, 759–774. [CrossRef]

41. NRMMC; EPHC; NHMRC. *Australian Guidelines for Water Recycling, Managing Health and Environmental Risks, Vol 2C: Managed Aquifer Recharge*; Natural Resource Management Ministerial Council, Environment Protection and Heritage Council National Health and Medical Research Council: Australia, 2009; 237 p. Available online: http://www.environment.gov.au/water/publications/quality/water-recycling-guidelines-mar-24.html (accessed on 8 January 2022).

Technical Note

Impact of Stream-Groundwater Interactions on Peak Streamflow in the Floods

Jaewon Joo [1] and Yong Tian [2,*]

1   Department of Environmental Planning, Graduate School of Environmental Studies, Seoul National University, Seoul 08826, Korea; jjw3741@gmail.com

2   Guangdong Provincial Key Laboratory of Soil and Groundwater Pollution Control, School of Environmental Science and Engineering, Southern University of Science and Technology, Shenzhen 518055, China

*   Correspondence: tiany@sustech.edu.cn

Abstract: Floods are the one of the most significant natural disasters, with a damaging effect on human life and properties. Recent global warming and climate change exacerbate the flooding by increasing the frequency and intensity of severe floods. This study explores the role of groundwater during the floods at the Miho catchment in South Korea. The Hydrological-Ecological Integrated watershed-scale Flow model (HEIFLOW) model is used for the flood simulations to investigate the impact of groundwater and streamflow interactions during floods. The HEIFLOW model is assessed by the Nash–Sutcliffe model efficiency coefficient (NSE) and Root Mean Square Error (RMSE) for the surface water and groundwater domains, respectively. The model evaluation shows the acceptable model performance (0.64 NSE and 0.25 m–2.06 m RMSE) with the hourly time steps. The HEIFLOW shows potential as one of the methods for the flood risk management in South Korea. The major findings of this study indicate that the stream runoff at the Miho catchment is highly affected by the groundwater flows during the dry and flood seasons. Thus, the interactions between surface water and groundwater domains should be fully considered to mitigate the water hazards at the catchment scale.

Keywords: flood; surface and groundwater interactions; HEIFLOW

Citation: Joo, J.; Tian, Y. Impact of Stream-Groundwater Interactions on Peak Streamflow in the Floods. *Hydrology* **2021**, *8*, 141. https://doi.org/10.3390/hydrology8030141

Academic Editors: Il-Moon Chung, Sun Woo Chang, Yeonsang Hwang and Yeonjoo Kim

Received: 26 July 2021
Accepted: 6 September 2021
Published: 17 September 2021

Publisher's Note: MDPI stays neutral with regard to jurisdictional claims in published maps and institutional affiliations.

1. Introduction

Flood is one of the most significant natural disasters in the world that cause about USD 40 billion losses in human life and properties every year [1]. Recent global warming and climate change amplify the flooding by increasing frequency and intensity of severe floods in the near future [2]. Flood risk mitigation is a major challenge for hydrological scientists and civil engineers.

The traditional method of flood risk mitigation aims to at reduce the flood risks by land surface hydraulic structures such as dams, river embankments, and reservoirs [3]. Those hydraulic structures only focus on the surface water domain in the wet season without considering the impacts of groundwater and groundwater flooding. The groundwater domain in the surface floods are generally assumed to be fully saturated and most of peak flow is caused by primarily precipitation [4]. However, the variabilities in groundwater level can cause various flood situations. For example, the surface flood generally infiltrates into the aquifer. The infiltration of surface water can vary according to the conditions of soil moisture and groundwater levels. The initially wet condition of soil moisture contributes to the fast groundwater level rise. These conditions drive the steep and rapid hydrograph during floods. Thus, it is difficult to forecast the flood situation considering the complex process of surface and groundwater interactions [5].

Many hydrological models have been developed for river management and flood management, such as the Hydrologic Modeling System (HEC-1 and HMS), developed by the US Army Corp-Hydrologic Engineering Center [6], and the Revitalised Flood

Hydrograph (ReFH) rainfall-runoff model used for simulation of design flood events in the UK [3,7]. HEC-HMS is one of the most utilized hydrologic modeling tools in many countries (USA, Europe, and Asia) in order to simulate the influences of climate change [8] and land use on stream flow [9–11]. In addition, a number of studies have focused on the applications of data scarce catchments [3]. The conceptual hydrological model is generally used in data scarce catchments, and it does not properly consider groundwater flow. Recently, the hydrological data records and qualities have been improved from many efforts (e.g., Hydrological Survey Center in Korea).

In order to simulate reliable prediction of flooding, it is necessary to consider the interactions between surface water and groundwater domains. Nowadays, there is an increasing need for an integrated surface water and groundwater model [12]. However, the integrated surface water and groundwater model has been rarely used for flood assessment, and its significance has not been widely recognized. Understanding the interactions between surface water and groundwater is important for flood simulation, and provides useful knowledge about the complex flood processes [13]. This study explores the role of groundwater and streamflow interactions in the flood runoff at the Miho catchment in South Korea. The Hydrological-Ecological Integrated watershed-scale Flow model (HEIFLOW) model is used for the hourly flood simulations to investigate the impact of groundwater discharge on the peak stream runoff at the flood events.

2. Study Area and Data

The Miho catchment, which is located in the northern part of the Geum River Basin (GRB), is the largest catchment in the GRB in South Korea (Figure 1). The strategic water management plan is continuously required in the Miho catchment because the outflows from the Miho catchment highly affect the water quality and quantity of the downstream of GRB. The catchment area is approximately 1800 km^2 and the elevation ranges from 7 m to 631 m. The average precipitation indicates that about 70% of annual precipitation (1239 mm) is concentrated in the summer wet season [3].

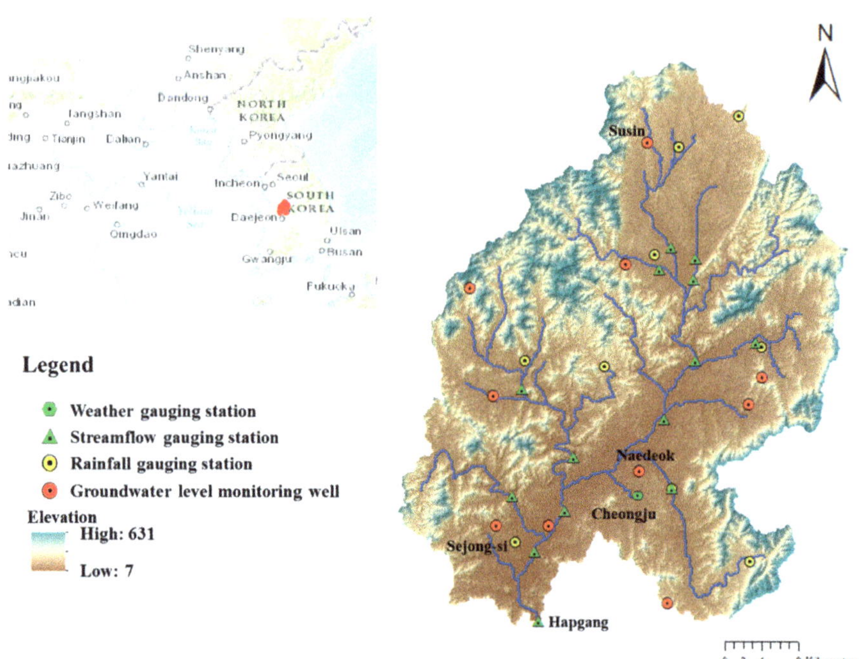

Figure 1. Digital Elevation Map of the Miho catchment with hydrological stations.

The data set for this study requires the geological, meteorological, and hydrological data sets for developing the HEIFLOW model. The geological data set in the Miho catchment employs a 30 m spatial resolution digital elevation model (DEM), a land-use map, soil maps, hydrogeological map, and bore hole information. Those geological data sets are obtained from the ASTER DEM (http://asterweb.jpl.nasa.gov accessed in 15 July 2021), Water Resources Management Information System (WAMIS, http://www.wamis.go.kr accessed on 26 July 2021) and Groundwater information Service (GIMS, http://www.gims.go.kr accessed on 26 July 2021). The hourly meteorological data from 2013 to 2014 are obtained by the WAMIS and Korea Meteorological Administration (KMA). The precipitation data sets are prepared from the eight rainfall gauging stations, the locations of which are shown in Figure 1. The hourly weather information such as the temperature, air pressure, relative humidity, wind speed and sunshine hours is employed from the Cheongju weather station. The Hapgang water level gauging station, which is located in the outlet of the Miho catchment, is selected to obtain the hourly streamflow observations (WAMIS). The hourly groundwater level data in the study catchment are provided by the 10 groundwater level monitoring wells in Figure 1 (GIMS).

3. Methods

This study employs Hydrological-Ecological Integrated watershed-scale Flow (HEI-FLOW) to describe the impacts of the groundwater on flood events. HEIFLOW is a three-dimensional distributed eco-hydrological coupling model, whose forerunner is Groundwater and Surface-water FLOW (GSFLOW). The GSFLOW generally simulates the hydrologic process, which integrates the Precipitation-Runoff Modeling System (PRMS) [14] with the Modular Groundwater Flow Model (MODFLOW-2005) [15] in the basin scale. However, it has limitations for the time step of simulation, ecological processes, land use changes, and dynamic land use. The modified version of GSFLOW was developed by Tian et al. [16] in order to improve the limitations of GSFLOW.

The model construction of HEIFLOW requires many processes such as watershed delineation, processing the input data sets, model parameterizations, calibration, and analysis of model results. Thus, the visual hydrological ecological integrated watershed-scale flow (VHF) [16] is used to construct the complex processes of HEIFLOW model for the Miho catchment. The model domain boundary and stream networks of Miho catchment are delineated with the uniform grids in both surface and groundwater domains to reduce the computation errors [17].

The surface water model domain in the Miho catchment is delineated into 7220 grids, which have a width and height of 500 m. These grids are defined as Hydrologic Response Unit (HRU) of PRMS and MODFLOW grids. The HRUs contain the input data of surface model domain of HEIFLOW such as elevation, basin area, aspect, latitude, longitude, land cover type, and soil type. The input data sets of HEIFLOW are required to estimate the initial model parameter values for the model calibration. The metrological input data for the Miho catchment were employed from the Cheongju weather station of KMA. The hourly rainfall data from the eight rain gauge stations were interpolated by the inverse distance weighting (IDW) method.

The initial parameter values of HRUs are estimated from the DEM, land use, soil type and vegetation data sets. The major surface model parameters are considered as the plant canopy density (covden_win and covden_sum), the maximum storage of the plant canopy for precipitation (snow_intcp, srain_intcp, and wrain_intcp), and the water contents of soil zone (soil_moist_init, soil moist_max, soil_rechr_init, and soil_rechr_max). The initial surface model parameter sets are further calibrated. The groundwater domain in the Miho catchment is divided into three layers for the groundwater modeling. The groundwater model domains of this study are represented by three layers (i.e., layer 1, layer 2 and layer 3 from top to bottom). All the layer types are set as convertible. A convertible layer means that it can be either confined or unconfined, depending on the elevation of the computed water table. The major parameter sets of groundwater domain contain the horizontal hydraulic conductivity (HK), vertical hydraulic conductivity (VK), specific storage (SS), and specific yield (SY). Both SS and SY are applied for the three layers since they are convertible. The parameter zone of groundwater domain is divided into 41 parameter zones by the information of hydrogeological map and bore hole in the Miho catchment, and the initial groundwater model parameters are adjusted by the daily GSFLOW model in the Miho catchment in previous research [17]. The input data sets for the surface model are generated by the VHF. In the HEIFLOW model, stream network is generally divided into the reaches and segments. The stream network of Miho catchment model contains the 123 segments with the 1269 reaches. The adjusted model parameter sets of HEIFLOW are represented in Table 1.

Table 1. Calibrated major parameter ranges for the HEIFLOW model in the Miho catchment [17].

Zone	Parameters	Minimum	Maximum	Unit
	covden_sum	0.1	0.9	dimensionless
	covden_win	0	0.1	dimensionless
Surface	srain_intcp	0	0.05	inches
	wrain_intcp	0.1	3	inches
	snow_intcp	0.1	3	inches
	soil_moist_max	5	18	inches
Soil	soil_moist_init	0.5	9	inches
	soil_rechr_max	3	9	inches
	soil_rechr_init	0.5	4.5	inches
	HK (layer 1)	0.5	10	meters per day
	HK (layer 2)	0.1	2	meters per day
	HK (layer 3)	0.02	0.4	meters per day
	VK (layer 1)	0.0083	0.33	meters per day
Groundwater	VK (layer 2)	0.00014	0.0056	meters per day
	VK (layer 3)	2.3×10^{-5}	0.0009	meters per day
	SY (layer 1–3)	0.04	0.11	dimensionless
	SS (layer 1–3)	1.0×10^{-5}	4.0×10^{-5}	meters^{-1}

4. Results and Discussions

The HEIFLOW model in the Miho catchment was calibrated by the daily GSFLOW modeling research from Joo et al. [17]. The HEIFLOW model was employed to verify the hydrological processes in the Miho catchment with hourly time step in 2013. Figure 2 shows the model evaluation of the Miho catchment by the Nash–Sutcliffe model efficiency coefficient (NSE). The gray line indicates the observed stream runoff at the outlet of the Mho catchment at the Hapgang water level gauging station, and the black dashed line represents the HEIFLOW model simulation. The blue bar graph in Figure 2 indicates the hourly rainfall at the Sejong-si rain gauge station, which is the nearest rain gauge station from the outlet of Miho catchment. The model performance of the Miho catchment indicates appropriate simulation runoff with 0.64 NSE in hourly time step. The stream hydrograph of the HEIFLOW model is generally underestimated in the low flow regime and is overestimated in the peak flow regime.

Figure 3 illustrates the interactions between the stream flow and groundwater. The gray line indicates the simulated stream runoff, and the dark dashed line represents the total groundwater discharge out (GW_out) to the streams. The temporal variability of GW_out shows the similar pattern with the stream runoff. GW_out in the dry season (see Figure 3) is generally larger than stream runoff because the streamflow is lost through evaporation and recharge to groundwater. The results in Figure 3 indicate that most of baseflow for the downstream of Miho catchment is sourced from the groundwater. These also represent the peak flow during flooding is highly influenced by the groundwater flow to the stream. Figure 4 compares the simulated and observed groundwater levels at daily time scale for two groundwater monitoring wells. The RMSE between the daily observed groundwater levels and corresponding simulated levels for the Susin and Naedeok monitoring wells are equal to 1.12 m and 0.25 m, respectively, while the RMSE between the daily observed and simulated groundwater levels for all the 10 monitoring wells ranges from 0.25 m to 2.06 m. As shown in Figure 4, the groundwater levels at the two wells are greatly influenced by stream-aquifer interactions. The model can capture the fluctuation pattern of groundwater level at daily time scale.

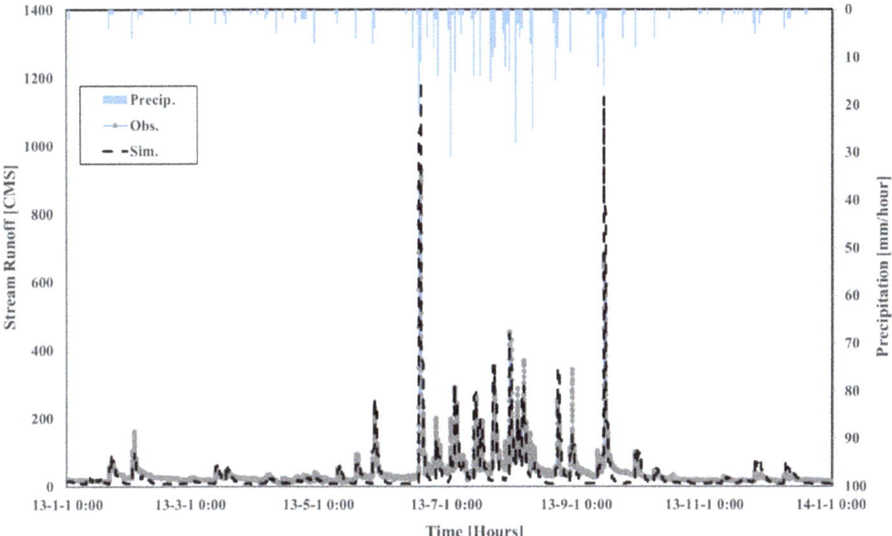

Figure 2. Comparison of the simulated and measured streamflow at the Hapgang stream gauging station in 2013 (NSE = 0.64).

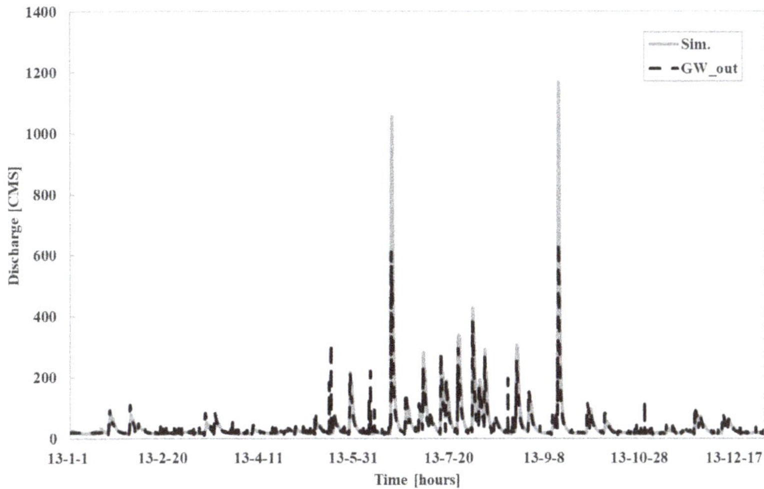

Figure 3. Comparison of the simulated stream runoff and the groundwater discharge to the stream at the Miho catchment.

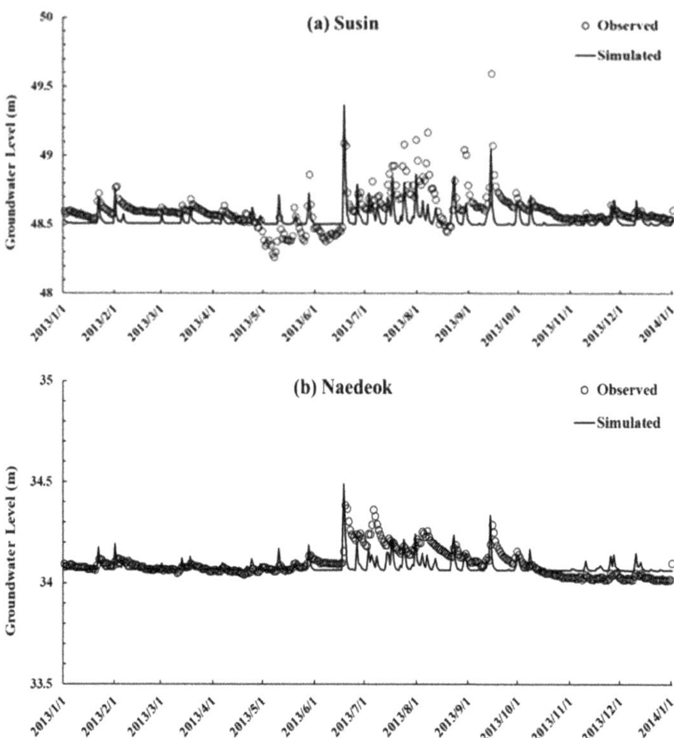

Figure 4. Comparison of the simulated and observed groundwater level at two monitoring wells:
(**a**) Susin and (**b**) Naedeok.

Figure 5 indicates the surface and groundwater interactions during the flood events. This study selects the four flood events in 2013. The details of flood events in this study are described in Table 2. The cumulated rainfall events are shown in Table 2 and the hourly rainfalls are illustrated as the blue bar graphs in Figure 5. Entire flood events in Figure 5a–d show that the groundwater discharges to the stream flow are highly related to the rainfall. In addition, the rising and falling limbs in stream hydrograph are generally affected from the variability of groundwater discharge to the stream. The responses of the GW_out and stream runoff indicate that stream runoff is faster response than the GW_out from the rainfall. The flood events in Figure 5 have the multiple rainfall events except Figure 5c. The GW_out in Figure 5 shows that GW_out patterns affect the peak discharge in the stream. For example, GW_out in Figure 5a,c showa similar patterns, and these two events have the multiple peaks in stream discharge. Both events also show that GW_out is dramatically reduced after the first peak of GW_out. The variabilities of GW_out are causing to mitigate the peak stream discharge and make multiple peaks in the streamflow hydrograph. However, the event 4 in Figure 5d indicates the different fluctuation of GW_out to the event 1 and 2. The fluctuation patterns of GW_out in event 1 and event 4 show that the peak stream discharge of event 4 is approximately 100 m3/s larger than event 1 although the rainfall is smaller than event 1.

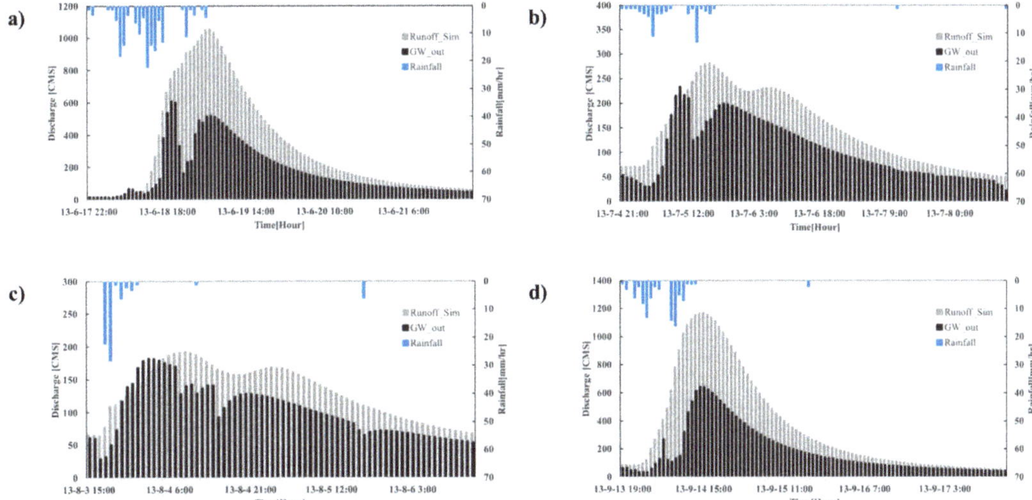

Figure 5. (**a–d**) The interactions between the groundwater and stream flow during flood events.

Table 2. Flood events of the Miho catchment at the Hapgang gauging station.

Events No.	Period of Flood Events	Rainfall
Event _1	2013-06-17 22:00 to 2013-06-21 23:00	157 mm
Event _2	2013-07-04 21:00 to 2013-07-08 13:00	59 mm
Event _3	2013-08-03 15:00 to 2013-08-06 14:00	70 mm
Event _4	2013-09-13 19:00 to 2013-09-17 18:00	91 mm

5. Conclusions

This study tested the HEIFLOW model with the hourly time step at the Miho catchment in South Korea. The integrated surface and groundwater model for the flood event is successfully verified with the 0.64 NSE. The major conclusions of this study are as follows.

First, the HEIFLOW enables complex interactions to be simulated between the groundwater domain and stream. The model verification results indicate acceptable simulation in entire flood events. Second, the hourly flood simulation employing the HEIFLOW shows potential as one of the methods for the flood risk management in South Korea. These also have the advantage of understanding the interactions between surface and groundwater domains. Finally, the results indicate that the hydrological response at the Miho catchment is highly affected by the groundwater conditions. The interactions between surface water and groundwater domains should be fully considered to mitigate the water hazards at the catchment scale.

This study is the first application of HEIFLOW model in South Korea. Thus, further study will test the HEIFLOW model in other catchments to generalize our suggestion. It is also required to test this flood simulation into the recent historical severe flood event and regions in South Korea.

Author Contributions: Conceptualization, J.J. and Y.T.; Methodology, J.J. and Y.T.; Writing—Original draft, J.J.; Writing—Review and editing, Y.T. All authors have read and agreed to the published version of the manuscript.

Funding: This research was funded by the National Natural Science Foundation of China (No. 41890852 and No. 42071244).

Acknowledgments: The data used in this study, if not collected by the authors or acknowledged in the text, were provided by the Water Management Information System (WAMIS), Korea Meteorological Administration (KMA) and Groundwater Information Service (GIMS).

Conflicts of Interest: The authors declare no conflict of interest.

References

1. Gaume, E.; Bain, V.; Bernardara, P.; Newinger, O.; Barbuc, M.; Bateman, A.; Blaškovičová, L.; Bloschl, G.; Borga, M.; Dumitrescu, A.; et al. A compilation of data on European flash floods. *J. Hydrol.* **2009**, *367*, 70–78. [CrossRef]
2. Yu, X.; Moraetis, D.; Nikolaidis, N.P.; Li, B.; Duffy, C.; Liu, B. A coupled surface-subsurface hydrologic model to assess groundwater flood risk spatially and temporally. *Environ. Model. Softw.* **2019**, *114*, 129–139. [CrossRef]
3. Joo, J.; Kjeldsen, T.; Kim, H.-J.; Lee, H. A comparison of two event-based flood models (ReFH-rainfall runoff model and HEC-HMS) at two Korean catchments, Bukil and Jeungpyeong. *KSCE J. Civ. Eng.* **2014**, *18*, 330–343. [CrossRef]
4. Acreman, M.; Holden, J. How Wetlands Affect Floods. *Wetlands* **2013**, *33*, 773–786. [CrossRef]
5. Kreibich, H.; Thieken, A.; Grunenberg, H.; Ullrich, K.; Sommer, T. Extent, perception and mitigation of damage due to high groundwater levels in the city of Dresden, Germany. *Nat. Hazards Earth Syst. Sci.* **2009**, *9*, 1247–1258. [CrossRef]
6. Feldman, A.D. *Hydrologic Modeling System HEC-HMS: Technical Reference Manual*; US Army Corps of Engineers Hydrologic Engineering Center: Washington, DC, USA, 2000.
7. Kjeldsen, T.R. *The Revitalised FSR/FEH Rainfall-Runoff Method*; Center for Ecology & Hydrology: Wallingford, Oxfordshire, UK, 2007.
8. Markus, M.; Angel, J.R.; Yang, L.; Hejazi, M.I. Changing estimates of design precipitation in Northeastern Illinois: Comparison between different sources and sensitivity analysis. *J. Hydrol.* **2007**, *347*, 211–222. [CrossRef]
9. McColl, C.; Aggett, G. Land-use forecasting and hydrologic model integration for improved land-use decision support. *J. Environ. Manag.* **2007**, *84*, 494–512. [CrossRef] [PubMed]
10. Ali, M.; Khan, S.J.; Aslam, I.; Khan, Z. Simulation of the impacts of land-use change on surface runoff of Lai Nullah Basin in Islamabad, Pakistan. *Landsc. Urban Plan.* **2011**, *102*, 271–279. [CrossRef]
11. Du, J.; Qian, L.; Rui, H.; Zuo, T.; Zheng, D.; Xu, Y.; Xu, C.-Y. Assessing the effects of urbanization on annual runoff and flood events using an integrated hydrological mod-eling system for Qinhuai River basin, China. *J. Hydrol.* **2012**, *464*, 127–139. [CrossRef]
12. Padilla, F.; Hernández, J.-H.; Juncosa, R.; R.-Vellando, P. Modelling integrated extreme hydrology. *Int. J. Saf. Secur. Eng.* **2016**, *6*, 685–696.
13. Bernard-Jannina, L.; Britob, B.; Sun, X.; Jauchb, E.; Nevesb, R.; Sauvagea, S.; Sánchez-Péreza, J.-M. Spatially distributed modelling of surface water-groundwater exchanges during overbank flood events–a case study at the Garonne River. *Adv. Water Resour.* **2016**, *94*, 146–159. [CrossRef]
14. Leavesley, G.; Lichty, R.; Troutman, B.; Saindon, L. Precipitation-runoff modeling system; user's manual. *Water-Resour. Investig. Rep.* **1983**, *83*, 4238.
15. Harbaugh, A.W. *MODFLOW-2005, the US Geological Survey Modular Ground-Water Model: The Ground-Water Flow Process*; US Department of the Interior: Washington, DC, USA; US Geological Survey: Reston, VA, USA, 2005.
16. Tian, Y.; Zheng, Y.; Han, F.; Zheng, C.; Li, X. A comprehensive graphical modeling platform designed for integrated hydrological simulation. *Environ. Model. Softw.* **2018**, *108*, 154–173. [CrossRef]
17. Joo, J.; Tian, Y.; Zheng, C.; Zheng, Y.; Sun, Z.; Zhang, A.; Chang, H. An integrated modeling approach to study the surface water-groundwater interactions and influence of tem-poral damping effects on the hydrological cycle in the Miho catchment in South Korea. *Water* **2018**, *10*, 1529. [CrossRef]

Article

Hydrological Connectivity in a Permafrost Tundra Landscape near Vorkuta, North-European Arctic Russia

Nikita Tananaev [1,*], Vladislav Isaev [2], Dmitry Sergeev [3], Pavel Kotov [2] and Oleg Komarov [2]

1. P.I. Melnikov Permafrost Institute, Siberian Branch, Russian Academy of Sciences, 677010 Yakutsk, Russia
2. Department of Geology, Lomonosov Moscow State University, 119991 Moscow, Russia; tpomed@yandex.ru (V.I.); kotovpi@mail.ru (P.K.); tpomed@rambler.ru (O.K.)
3. Sergeev Institute of Environmental Geoscience, 101000 Moscow, Russia; sergueevdo@mail.ru
* Correspondence: TananaevNI@mpi.ysn.ru

Abstract: Hydrochemical and geophysical data collected during a hydrological survey in September 2017, reveal patterns of small-scale hydrological connectivity in a small water track catchment in the north-European Arctic. The stable isotopic composition of water in different compartments was used as a tracer of hydrological processes and connectivity at the water track catchment scale. Elevated tundra patches underlain by sandy loams were disconnected from the stream and stored precipitation water from previous months in saturated soil horizons with low hydraulic conductivity. At the catchment surface and in the water track thalweg, some circular hollows, from 0.2 to 0.4 m in diameter, acted as evaporative basins with low deuterium excess (d-excess) values, from 2‰ to 4‰. Observed evaporative loss suggests that these hollows were disconnected from the surface and shallow subsurface runoff. Other hollows were connected to shallow subsurface runoff, yielding d-excess values between 12‰ and 14‰, close to summer precipitation. 'Connected' hollows yielded a 50% higher dissolved organic carbon (DOC) content, 17.5 ± 5.3 mg/L, than the 'disconnected' hollows, 11.8 ± 1.7 mg/L. Permafrost distribution across the landscape is continuous but highly variable. Open taliks exist under fens and hummocky depressions, as revealed by electric resistivity tomography surveys. Isotopic evidence supports upward subpermafrost groundwater migration through open taliks under water tracks and fens/bogs/depressions and its supply to streams via shallow subsurface compartment. Temporal variability of isotopic composition and DOC in water track and a major river system, the Vorkuta River, evidence the widespread occurrence of the described processes in the large river basin. Water tracks effectively drain the tundra terrain and maintain xeric vegetation over the elevated intertrack tundra patches.

Keywords: permafrost hydrology; Russian Arctic; water tracks; hydrological connectivity; stable water isotopes; dissolved organic carbon; electrical resistivity tomography; taliks

Citation: Tananaev, N.; Isaev, V.; Sergeev, D.; Kotov, P.; Komarov, O. Hydrological Connectivity in a Permafrost Tundra Landscape near Vorkuta, North-European Arctic Russia. *Hydrology* **2021**, *8*, 106. https://doi.org/10.3390/hydrology8030106

Academic Editors: Il-Moon Chung, Sun Woo Chang, Yeonsang Hwang and Yeonjoo Kim

Received: 19 June 2021
Accepted: 17 July 2021
Published: 22 July 2021

Publisher's Note: MDPI stays neutral with regard to jurisdictional claims in published maps and institutional affiliations.

1. Introduction

Hydrologic connectivity is a complex concept referring to water transfer in the landscape, or between landscapes, or, more generally, within or between the water cycle units, and its (dis)continuity along the major water transport pathways acting on the watershed [1–3]. It includes both lateral water transfer along slopes, including channelized runoff, and vertical water transfer between surface and subsurface compartments, or between different groundwater aquifers at different depths [4]. Connectivity exists between larger domains, e.g., surface runoff and groundwater flow, landscape elements and fluxes—*structural connectivity*, and between processes—*functional connectivity* [4,5].

Permafrost significantly alters the water cycling through the affected landscapes compared to that in temperate catchments [6–8]. In continuous permafrost, water transport is mostly confined to the active layer, which rarely exceeds 3 m depth, and to vertical and lateral talik zones. Water migration in soils is partly driven by processes related to phase transition in soils [9,10]. The hydrological system structure is simplified, and the existing connections between compartments are exposed to observation [11,12]. Open talik zones are frequently detected under the largest lakes using geophysical methods [13–15] and may serve as pathways for both upward and downward water migration, connecting surface water to intra- and subpermafrost aquifers [16,17]. In discontinuous permafrost, with a deeper active layer (down to 5 m), persistence of residual thaw layers, permafrost fragmentation, and abundance of talik zones, the potential for water exchange between the compartments is significantly higher [18–20].

Stable water isotopes, ^{2}H (deuterium) and ^{18}O, are widely used to track water sources and hydrologic connectivity across the compartments and ecosystem classes, including boreal and permafrost-affected catchments [21–24]. The isotopic evaporation signal allows the tracing of connectivity between wetlands and perennial streams and the modeling of surficial wetland runoff contribution to streams during summer [25]. The isotope mass balance method reveals the specifics of the permafrost thaw cycle in continuous permafrost [26,27] and is successfully used in studies of both contemporary lakes and lacustrine paleoenvironments [28].

In discontinuous permafrost, taliks of different kinds, i.e., vertical open and closed taliks, connected to lateral intra-permafrost taliks and residual thaw layers, are responsible for conveying water from the slopes toward the streams [29,30]. In the Northern Yenisey region, isotopically heavier water, originating from late summer precipitation and thermokarst lakes subject to evaporation, was found to contribute significantly to the winter runoff through the residual thaw layer, an interface between the seasonally freezing layer and the top of the permafrost [31]. In the Ob River basin, a strong evaporation signal persists in most river samples in late autumn and around the spring freshet peak dates, evidencing subsurface connections between lakes and rivers of the region [32]. Lake-to-river connectivity is also maintained through sub-lacustrine taliks, both open and closed, developing even under shallow thermokarst lakes [33]. Geophysical techniques, notably electrical resistivity tomography, are useful in describing the complex frozen ground configuration in discontinuous permafrost [34,35].

The climate change presently occurring in the Arctic, followed by the deepening of the active layer, may lead to the rebuilding of existing connectivity patterns through changes in the saturated zone boundary, to an increase in non-frozen soil volume where water migration is possible, and to an increase in groundwater discharge on hillslopes [36], affecting future dissolved organic carbon (DOC) and other constituent fluxes [37]. Trends in regional climate and hydrology may also imply changes in fluvial activity [38], though the latter is showing only minor signs in the first-order fluvial network of the region. The potential effects of climate change and permafrost degradation on hydrological connectivity and water and material fluxes, including biogeochemical cycling, are still poorly understood. Recent reviews acknowledge important knowledge gaps in the subsurface hydrology of permafrost regions, including existing water and carbon transport pathways, their relation to frozen grounds, and alterations in subsurface routing resulting from permafrost degradation [39,40]. Ultimately, permafrost degradation is expected to alter hydrological connectivity in affected catchments, resulting in water flow redistribution between the surface and compartments and changes in seasonal water discharge [20,39].

This study was conceived to address these gaps and to better understand hydrological connectivity and water and DOC transport in a discontinuous permafrost environment at a small scale. We present new data on water stable isotope composition and DOC concentrations from several subarctic streams and water bodies in minor tundra water track catchments near Vorkuta, north-European Russia. These data are used to trace water origin in these water objects under late summer conditions, close to the maximum thaw period and to evaluate microscale hydrologic connectivity in the landscape. Geophysical survey data are used to support the discussion on surface water interaction with groundwater. The study region, with its mild and humid subarctic climate, may serve as a model region for other permafrost regions in transition under observed climate change.

2. Study Area

Fieldwork was performed in September 2017 in north-European Russia, on the margin of the Bolshaya Zemlya tundra region, about 30 km to the south-west from Vorkuta, Komi Republic, the closest large city (Figure 1a). The major sampling effort was concentrated around Khanovey key study site (N 67°17.193′, E 63°39.252′; Figure 1b), an abandoned settlement for railroad workers on the right bank of the Vorkuta River, where the seasonal permafrost research station is maintained by the Department of Geocryology, Moscow State University [41]. The studied location occupies a typical periglacial landscape at the southern margin of the Bolshaya Zemlya tundra, a hilly terrain with gently rolling slopes dissected by hummocky depressions and first-order stream valleys. Permian bedrock, exposed locally in the Vorkuta River bluffs, is overlaid with Quaternary deposits with thickness from 10 to 15 m, mostly loams and loamy clays, with variable ice content and cryostructure [41]. Surficial loamy clays are highly thixotropic, and easily lose structural integrity on stress. Sandy deposits were described in shallow excavations in the adjacent areas but were never encountered at the watershed in question. A topsoil organic layer is omnipresent and has a thickness of 0.3–0.5 m.

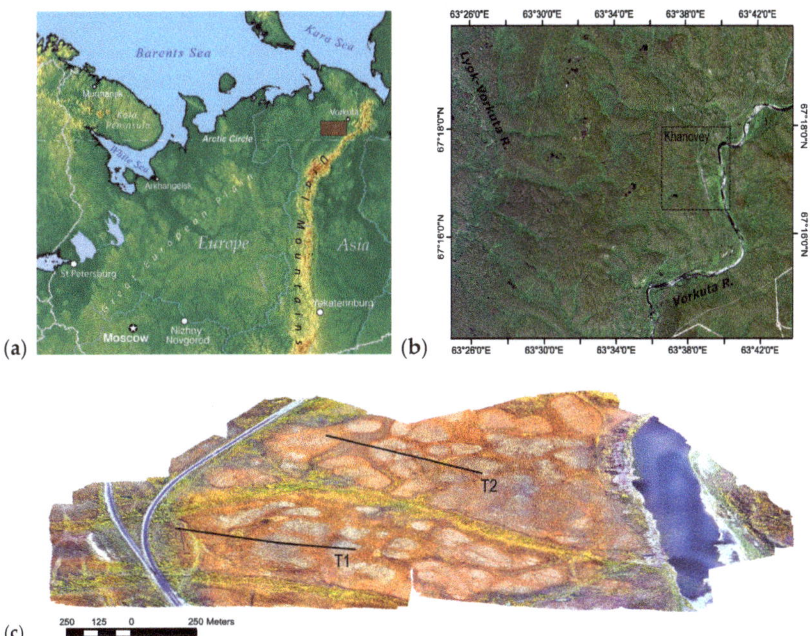

Figure 1. Geographical location of the study site, (**a**) in north-European Russia, (**b**) at the Vorkuta River valley slope; (**c**) reference orthophoto image of the studied water track section, north is up, T1 and T2 denote the electric resistivity tomography profiles.

The meso-scale topography is dominated by *uvals*, a local name for smoothed hilly chains. These hilly chains have a submeridional orientation and elevation between 170 and 200 m a.s.l. and are divided by the valleys of the Usa River and its major right tributaries, the Vorkuta River and Seida River. The *uvals'* surface is an undulating plain, hosting numerous lakes, peatlands, and a network of overwetted depressions, with mires presumably of thermokarst origin. The hillslopes descending toward the major rivers are transformed by the joint action of fluvial processes, thermal erosion, and linear thermokarst, with the widespread occurrence of water tracks. This network evolves from chaotic in the interfluve zone to a linearly shaped fluvial network hosting intermittent streams (Figure 2), then again becoming poorly organized toward the foot slope with fen-like features. Overwide valleys surrounding the remaining elevated tundra patches resemble those of organic-rich and wide water track classes described by Trochim et al. [42].

Figure 2. Typical midslope landscape of the study region, with minor (first-order) water tracks on the background, a second-order water track feature crossing the image left to right, all easily detectable by its contrasting foliage color, elevated tundra patches, and intertracks with *Betula nana* L. and lichen patches.

Interfluves are scarcely vegetated because of the snow cover removal by heavy winds, active in the region during winter, and subsequently lower ground temperatures, hence only lichens are omnipresent at these surfaces, associated with *Arctous alpinus* (L.) Nied. and crowberry (*Empetrum nigrum* L.). Locally, moss-dominated mires with *Sphagnum* spp. occur in topographical depressions in the interfluve belt. Gentle slopes are covered by creeping willow (*Salix arctica* Pall.); dwarf birch (*Betula nana* L.); and northern Labrador tea (*Rhododendron tomentosum* Harmaja); and small deciduous shrubs and plants: blueberry (*Vaccinium cyanococcus* Rydb.), blackberry (*Vaccinium myrtillus* L.), lingonberry (*Vaccinium vitis-idaea* L.), bearberry (*Arctostaphylos uva-ursi* Spreng.), and crowberry. Water track valleys are willow dominated, mainly *Salix phylicifolia* L., associated with *Equisetum arvense* L. and *Carex* spp.

The studied territory occupies the interfluve area between the Vorkuta River and its right tributary, the Lyok-Vorkuta River, and the valley slope inclined toward the east, to the Vorkuta River valley (Figure 1b). This slope is dissected by three first-order water track valleys, one of which was studied in detail in its middle and lower reach, downstream from the railroad crossing (Figure 1c). The water track valley is oriented west to east. Its headwaters are connected, via a network of wet depressions and mires, to the headwaters of all major neighboring water tracks, so that no interfluve exists between the water track systems draining in different directions. For this reason, the basin area of the studied water track, is estimated, with significant uncertainty, to be around 0.901 ± 0.055 km^2. This uncertainty is not related to the digital elevation model (DEM) resolution but reflects the fluvial network structure, as seen in Figure 1c. No clear line separating the two neighboring water track catchments can be easily drawn, because the flow direction can hardly be determined in the interconnected polygonal network in intertrack spaces.

The regional climate is subarctic, summer is short and cool, and winter is long and cold, lasting over eight months from October to May (Table 1). At the same time, the period without negative daily temperatures is only 70 days in an average year. Mean annual daily temperature observed at Vorkuta meteorological station (N 67°29.52′, E 63°58.53′) is −5.6 °C. Precipitation is approximately 530 mm, of which from 50% to 70% falls as snow, which can occur at any month of the year.

Table 1. Mean monthly air temperature (T, °C) and precipitation (P, mm), observed at Vorkuta meteorological station (1927–2019).

	Jan	Feb	Mar	Apr	May	Jun	Jul	Aug	Sep	Oct	Nov	Dec
T	−19.9	−19.7	−15.4	−9.3	−2.2	7.1	12.7	9.7	4.2	−4.2	−12.9	−17.0
P	22	18	20	24	32	48	59	59	54	40	30	25

Permafrost is continuous, with thickness varying from 50 to 100 m, and mean annual ground temperatures between −0.5 and −1.0 °C, at zero annual amplitude depth, around 12 m. A residual thaw layer occurs annually between the base of the seasonally freezing layer, ca. 2 to 3 m, and the top of permafrost at depths of 3.5 to 4.0 m. The prevailing permafrost cryostructure is massive; within the active layer, distinct traces of melted segregation ice lenses can be found–small lenticular unconformities parallel to the ground surface, which were filled by ice during winter and melted later in summer. Important cryogenic processes include thermokarst and fluvial thermal erosion. Frost boils are common cryogenic features, mostly occurring in a narrow belt surrounding the water track valleys.

3. Materials and Methods

Field observations were performed from the 5–19 September 2017. Water samples for the analyses of stable water isotope, ^{2}H and ^{18}O, were collected regularly, once in 2–3 days, from the Vorkuta River and the stream at the water track thalweg, draining into the river near the base camp at the Khanovey station, near its mouth (Figure 1c). Multiple samples were taken along the water track thalweg in its lower reach, downstream from the railroad crossing. Several samples were taken from natural hollows, circular depressions from 0.2 to 0.4 m in diameter, occurring on the ground surface in the water track valley and on slopes. Several soil pits, 40 to 90 cm deep, were dug at various locations in the water track valley and in the open tundra to sample shallow subsurface groundwater. Rainwater was sampled in Vorkuta, from an intense rain shower occurring on 11 September 2017. Subpermafrost groundwater was sampled from an artesian well, No.39-B, near the Khanovey railway station, feeding from a regional aquifer at a depth of around 80 m. Water samples were collected in 15 mL conical centrifuge tubes, sealed with Parafilm©, and stored at 4 °C before they were transported to the lab.

Stable water isotope samples ($n = 35$) were analyzed by multiflow-isotope ratio mass spectrometry (MF-IRMS, with instruments from Elementar, Germany) at SHIVA Isotopic Platform, EcoLab, Toulouse, France, in December 2017. The internal standard was Vienna Standard Mean Ocean Water (VSMOW2), and resulting values were expressed in δ notation relative to this standard [43]. The analytical precision of the method is ±0.1‰ for δ^{18}O, and ±1.0‰ for δ^2H. Each water sample was measured in duplicate and averaged, so each δ^{18}O/δ^2H value presented in the paper is a mean value. Deuterium excess (d-excess, or d_{ex}, ‰) was calculated as $d_{ex} = \delta^2$H $- 8 \cdot \delta^{18}$O.

Dissolved organic carbon (DOC) samples ($n = 33$) were collected in 20 mL LDPE bottles, pre-washed with weak sulfuric acid and rinsed with MilliQ water. All samples were acidified in the field directly after collection with two drops of 30% H_2SO_4 to suppress biological activity and stored at 4 °C until transported to the lab. The analyses were carried out at VNIRO (All-Russian Research Institute of Fisheries and Oceanography, Moscow), on a Shimadzu TOC-Vcph analyzer (Shimadzu, Japan), and are precise to ±0.1 mg/L.

Electric resistivity tomography (ERT) surveys were performed with 'SKALA-64' ERT station (NEMPHIS, Russia), using a combined three-electrode protocol (AMN-MNB) with a remote electrode installed at a distance of 600–800 m from the profile, which was considered as infinity. The ERT surveys were done at currents between 35 and 70 mA, and survey data were treated with Res2dInv (Geotomo Software, Malaysia, https://www.geotomosoft. com/, accessed on 13 July 2021) and X2ipi (Aleksey Bobachev, Moscow State University, http://x2ipi.ru/en, accessed on 13 July 2021) software.

Three shallow boreholes around the study area were instrumented with ground temperature sensors and dataloggers: one borehole was equipped with a Hobo 2-sensor datalogger at 0.5 and 2.0 m depth, and two boreholes had Hobo 4-sensor dataloggers at 0.5, 1.0, 2.0, and 5.0 m depths. The ground temperature readings are accurate to $\pm 0.1\,^\circ$C. Ground temperatures from -0.2 to $-0.5\,^\circ$C at 5 m depth were observed in two of them, and around $+0.8\,^\circ$C in the third borehole, in the water track thalweg.

4. Results

4.1. Water Stable Isotopes

The closest stations of the IAEA Global Network for Isotopes in Precipitation (GNIP) [44] network, providing baseline data for the isotopic composition of meteoric waters, are located in Pechora and Salekhard, several hundreds of kilometers from the studied location (Table 2). Their data show the effect of different vapor sources and the transformation of isotopic composition of regional precipitation as air masses cross the Polar Ural Mountains. At the Pechora site, more than 800 km to the SW from Vorkuta (N $65^\circ07.30'$, E $57^\circ08.98'$), the local meteoric water line (LMWL) is close to the global one (GMWL). In Salekhard (N $66^\circ32.20'$, E $66^\circ37.94'$), ca. 300 km to the SE from Vorkuta on the eastern side of the Ural Mountains in the Ob' River estuary, the LMWL plots below the GMWL with an intercept $b = 1.83$ (± 1.79).

Table 2. Isotopic composition of the major water sources in the studied region and at closest GNIP stations.

Station	Sample Source	n [1]	Data Source	δ^{18}O, ‰SMOW	δ^2H, ‰SMOW	d_{ex}, ‰
Pechora	Rain (VIII)	7/3	[44]	-12.09 ± 1.76	-80.3 ± 25.0	11.1 ± 8.6
	Rain (IX)	7/3	[44]	-12.97 ± 2.10	-87.0 ± 20.0	8.7 ± 3.5
Salekhard	Rain (VIII)	5	[44]	-13.18 ± 0.68	-100.1 ± 6.0	5.4 ± 3.0
	Rain (IX)	5	[44]	-13.35 ± 1.71	-102.4 ± 11.1	4.5 ± 4.0
Khanovey	Rain (IX)	1	This work	-14.7	-105	12.3
	Bog	1	"	-12.72	-92.89	8.9
	Groundwater	1	"	-15.8	-110	16.1
	Hollows	6	"	-11.46 ± 0.84	-83.53 ± 2.62	8.14 ± 4.9
	Lake	1	"	-10.3	-82.9	-0.7
	River/stream	20	"	-12.61 ± 0.33	-86.48 ± 3.04	14.4 ± 3.3
	Soil pits	4	"	-13.33 ± 1.40	-94.91 ± 9.21	11.7 ± 4.4

[1] n is number of samples; where separated by a slash, first value refers to number of δ^{18}O samples, second value, to the number of δ^2H and d-excess values.

The local meteoric water line (LMWL), plotted using all data except the most enriched samples subject to evaporative loss, is close to the global one (GMWL) and plots slightly above GMWL (Figure 3). The LMWL equation is

$$\delta^2\text{H} = 7.65 \times \delta^{18}\text{O} + 9.8\ (\text{‰}),\qquad(1)$$

The single precipitation event during the field campaign occurred on 10 September 2017 and was sampled in Vorkuta city (N $67^\circ28.92'$, E $64^\circ01.48'$), about 30 km to the northeast from the Khanovey study site. It shows a more depleted isotopic signature compared to long-term September averages for Pechora (Table 2), and a relatively high d-excess value, evidencing significant kinetic fractionation and distant moisture sources.

The local evaporation line (Figure 3) is drawn through a rain sample point at the bottom left of the plot, and across the field of points corresponding to samples with a high degree of evaporative loss. An evaporation effect is clear in a thermokarst lake sample that has a negative d-excess value and isotopically enriched composition, as well as in several samples from microtopographical hollows, where the d-excess was low positive (Table 2). The slope of the local evaporation line is around 5.0, when connecting rainfall to the highly enriched thermokarst lake point, and around 6.2 when only hollows and soil pits are considered; the averaged evaporation line, shown on Figure 3, yields a slope of 5.7.

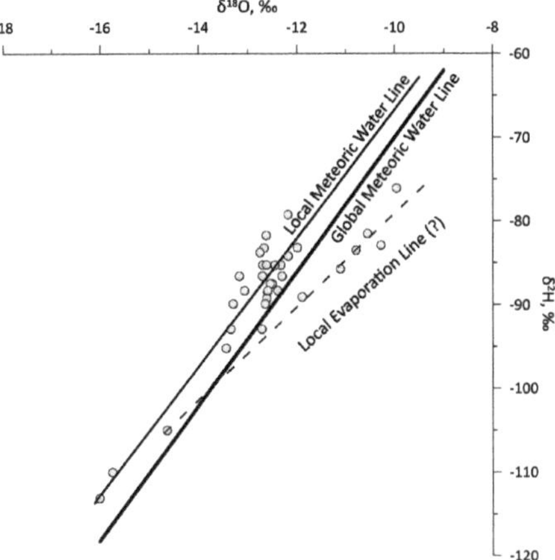

Figure 3. Local meteoric water line for water samples, collected in the Khanovey study area, in relation to the global MWL and the potential evaporation line.

The isotopic composition of sampled rivers and streams departs significantly from the GMWL toward higher *d*-excess and $\delta^{18}O$ (Table 2 and Figure 3) and plots uniformly above the other end member points on the *d*-excess–$\delta^{18}O$ diagram (Figure 4), evidencing various degrees of water fractionation in surface and shallow subsurface compartments [23]. It is presumably closer to the average isotopic composition of July and August rains, in accordance with the lower relative humidity of these summer months.

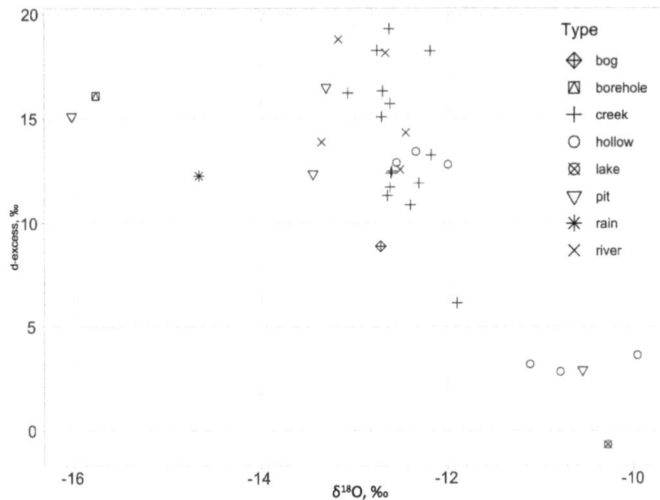

Figure 4. The *d*-excess–$\delta^{18}O$ diagram of sampled water bodies in the Khanovey study site.

Below the point cloud comprised mostly of surface waters, two separate end members are plotted in the opposite corners of the *d*-excess–$\delta^{18}O$ diagram (Figure 4). In the top left corner, a single sample of deep subpermafrost groundwater is plotted, depleted in heavy ^{18}O isotope, with high *d*-excess and isotopic signature consistent with that of a confined groundwater aquifer [45]. In the bottom right corner, samples that have undergone substantial evaporative loss are plotted, as discussed above, including the lake and several hollows (Figure 4).

4.2. Dissolved Organic Carbon

The DOC concentration measured at the Khanovey study site, is relatively low, averaging 10.4 ± 5 mg/L across the dataset. However, it is highly variable across water body types (Figure 5). In general, the highest DOC content is observed in standing water, i.e., bogs and hollows, whilst it is significantly lower in streams and rivers. The most variable DOC content was in hollows, while for other compartments it was more stable.

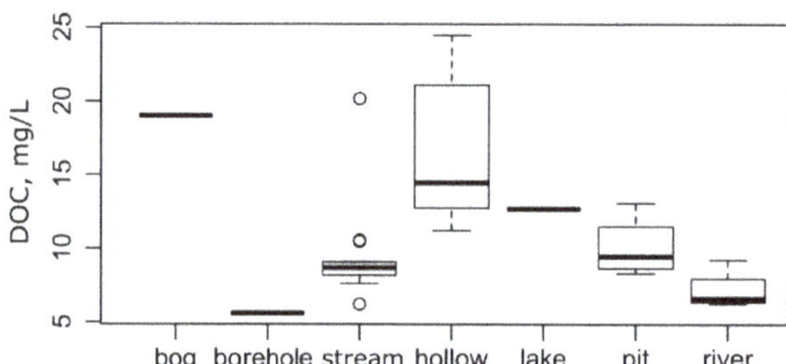

Figure 5. The dissolved organic carbon (DOC) concentrations across different water bodies in the Khanovey region.

4.3. Electric Resistivity Tomography (ERT)

The ERT surveys show a highly diverse distribution of high- and low-resistivity grounds in the studied sections (Figure 6), in most cases correlating closely with the hydrographic network's features.

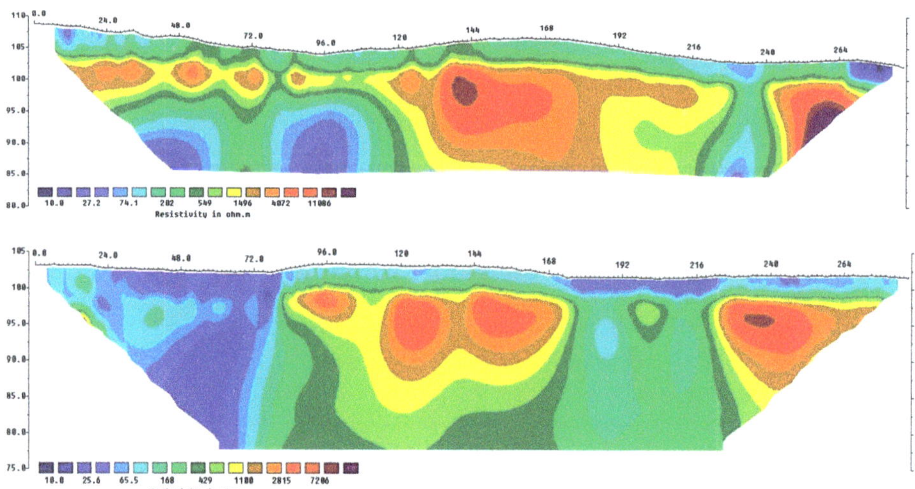

Figure 6. Electrical resistivity tomography profiles across the transects T1 (**top**) and T2 (**bottom**), Figure 1c for geographical reference. Vertical axis is altitude, in m a.s.l.; the lateral distance from the starting point, in m, is given along the profile.

Geophysical evidence suggests a relatively thick and steady permafrost, exceeding 30 m, persisting under raised non-dissected tundra patches and peat plateaus; a shallow and thin permafrost under a recent mire on the left side of T1 transect (see Figures 1c and 6); and open taliks under bogs, fens, or hummocky depressions. In all cases, the talik walls are subvertical, with thaw bulbs slightly expanding downward. This underscores the vulnerability of contemporary permafrost and also suggests significant groundwater circulation in the talik zones [29].

5. Discussion

5.1. Hydrological Connectivity at a Catchment Scale

The available data on water stable isotopes and their variability, presented in Table 2 and Figure 7, allow a generalized description of hydrological connectivity at a scale of a minor water track catchment.

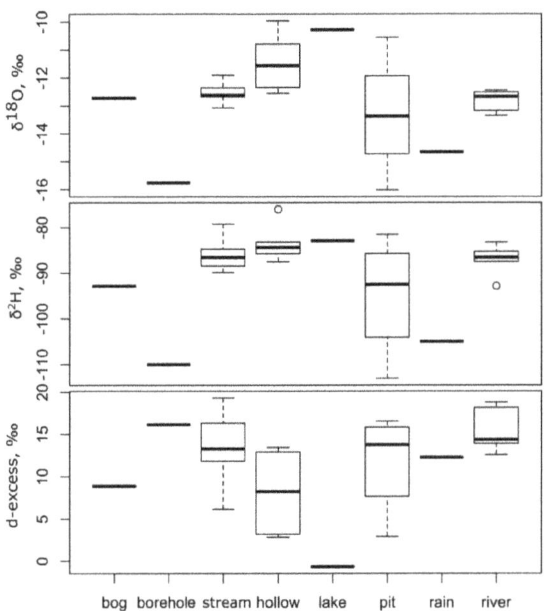

Figure 7. Boxplots of stable water isotope and *d*-excess values in the water track catchment in the Khanovey study site; median, 25%, and 75% probability of exceedance shown as bars, highest and lowest as whiskers, and outliers as separate points (see also Table 2).

Elevated tundra patches underlain by sandy loams were detached from the hydrological system at the time of observations. The upper part of their soil profile, above 0.6–0.9 m, was unsaturated. The saturated zone groundwater was sampled in a soil pit and had $\delta^{18}O = -10.55‰$, $\delta^2H = -81.52‰$, and $d_{ex} = 2.9‰$ (Figure 4), showing signs of evaporative loss and representing the isotopically heavier precipitation of preceding summer months. Other soil pits were opened in the water track and secondary drainage network thalwegs and contained water, intermediate between sampled September rainfall and stream/river water, with $\delta^{18}O = -13.3‰ \dots -13.4‰$, $\delta^2H = -89.9‰ \dots -95.2‰$, and $d_{ex} = 9.9‰ \dots 13.1‰$. One soil pit sample appeared close to the subpermafrost groundwater, potentially evidencing upward groundwater migration and discharge through the talik zones. This will be discussed in Section 5.4 in more detail.

On the tundra surface, multiple hollows were sampled, which were expected to show signs of evaporative loss. Biological fractionation from photosynthetic plant and algae activity and preferential evaporation of lighter oxygen isotope ^{16}O could potentially result in enrichment in heavy oxygen isotope, but this process is not expected to play an important role in late autumn, when the plants are in the end of their annual lifecycle. Surprisingly, only three out of six sampled hollows were evaporative basins, with mean $\delta^{18}O = -10.6‰ \pm 0.6‰$ and $d_{ex} = 3.23‰ \pm 0.39‰$. The other three hollows contained water that was isotopically similar to stream and river water, with $\delta^{18}O = -12.4‰ \pm 0.4‰$, $\delta^2H = -87.1‰ \pm 5.1‰$, and $d_{ex} = 11.7‰ \pm 3‰$, assuming their direct connection to shallow subsurface groundwater and the water track stream. This connection can be assured by the transmissivity feedback effect [46], occurring widely in the organic topsoil outside the elevated tundra patches on better drained soils. Heavily silted and clayey soils, observed locally in the catchment, are saturated and highly thixotropic. At rest, a stiffened thixotropic layer is expected to show barrier functions for water migration [47], barring water infiltration, leading to quick saturation of the overlying soil, and initiating rapid water drainage through the organic topsoil.

Multiple open-surface bogs in the drainage network depressions exist in the region, and one of them was sampled for water stable isotopes. This water object shows an isotopic signature intermediate between the hollows and the closest water track stream, which is consistent with the local buffer role of such bogs on the microscale slopes [48], transforming pluvial runoff and isotopic signal and conveying it to streamflow.

5.2. Temporal Evolution of Water Isotopic Composition

Paired samples were taken in the Vorkuta River and at the water track mouth to follow the coevolution of their isotopic composition (Figure 8). In the major stream, the Vorkuta River, it was gradually shifting toward lighter $\delta^{18}O$ and δ^2H values, reflecting the autumn flow recession and increasing groundwater input and also a potential change in rainwater isotopic signature (Figure 8a,b). No similar variation in $\delta^{18}O$ values was observed in the water track stream, but at the same time, a progressive depletion in δ^2H was recorded (Figure 8c).

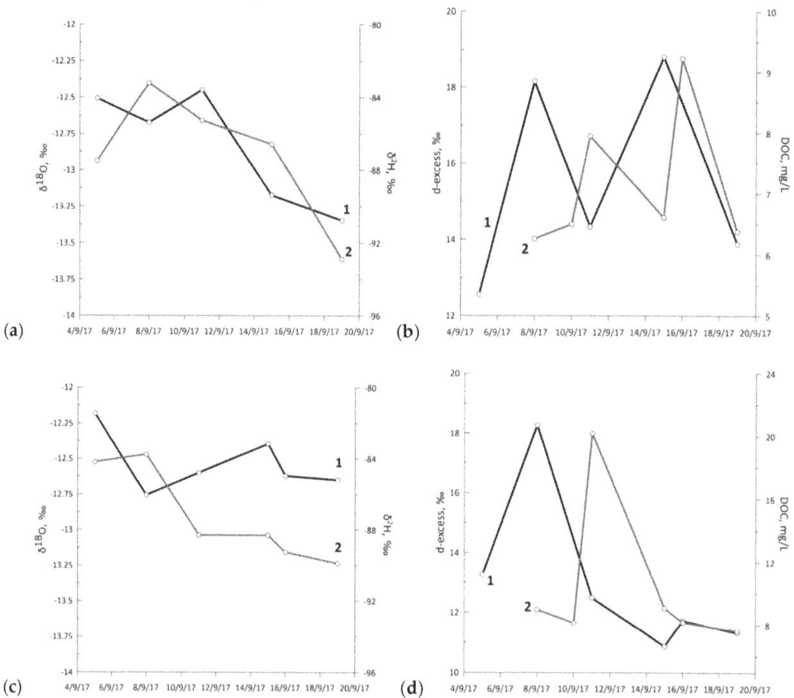

Figure 8. Temporal evolution of isotopic composition (**a**,**c**), *d*-excess and dissolved organic carbon (DOC) concentrations (**b**,**d**) in the Vorkuta River (**a**,**b**) and the water track (**c**,**d**). Numbers denote changes in variables shown (1) on the left axis and (2) on the right axis of each graph.

Deuterium excess values were highly variable with time in the Vorkuta River, jumping from 12‰ to 18‰ over the time span of several days (Figure 8b). Runoff inputs from sources adjacent to the observation point were responsible for the first peak, which coincided with the same *d*-excess peak in the water track stream (Figure 8d), while more distant sources could potentially contribute to the second peak. The Vorkuta River has a large basin, around 4000 km^2 at the sampling site, and smoother variations in water chemistry are expected. However, the water isotopic signal of the river could be reset by human activities upstream, i.e., the dam of the Heat Production Station No. 2 upstream of Vorkuta, where the flow is almost ceasing during low-flow periods. In this case, our samples reflect the changes in the isotopic composition of major Vorkuta River tributaries, including Ayach-Yaga River and Yun'-Yaga River, as well as numerous minor tributaries and water tracks. The low temporal resolution of the survey may add to the lack of smooth variations on the graph.

A DOC peaks follows the *d*-excess peak with a three-day lag in the water track stream (Figure 8d). The origin of these peaks is unclear, as is their causal relation. We may hypothesize that the *d*-excess peak at a small catchment scale could have been produced by rain events, when first, pre-event water yielding high *d*-excess values is pushed from the water storage within the catchment. The DOC-rich event waters from the subsurface compartment reach the fluvial network several days later, inducing hysteretic behavior of DOC, when its values are lower on the rising limb and higher on the falling limb of the stormflow hydrograph [49].

This effect is significantly better expressed and occurs twice in the larger Vorkuta River runoff most probably because its large catchment effectively integrates hydrological signals from numerous rain events in its different parts. It is important to note that the *d*-excess variability during the peak is comparable in the small water track and in the larger Vorkuta River, while the DOC increase is significantly higher in the smaller stream.

5.3. DOC Export from the Catchment

The DOC concentrations are relatively low and only slightly vary within each water object type, except the small water track stream (Figure 5). Its variability in pits and hollows follows the variability detected in the isotopic signature. The hollows subject to evaporative loss and disconnected from subsurface flow (Figure 4), also yielded lower DOC values, 11.8 ± 1.7 mg/L on average. The pits and hollows connected to fast subsurface flow had DOC concentrations about 50% higher than disconnected ones (17.5 ± 5.3 mg/L), showing the important role they play in the DOC lateral fluxes. Lower DOC content in disconnected hollows may reflect multiple processes, from photodegradation to aerobic microbial degradation, with complex interaction between them [50].

The water track stream discharge is estimated to vary between 2 and 4 L/s, this estimate is based on occasional hydrologic observations from previous years, performed around the same dates in early September and in comparable weather conditions. Using the average DOC concentration in the water track, 9.45 mg/L, and an average daily discharge of around 3 L/s, the daily DOC export from the water track catchment for an average day in late autumn equals 2.4 kg/km^2.

5.4. Subpermafrost Groundwater Input

An 80 m deep artesian well sampled during the course of the study intercepts subpermafrost groundwater with a distinct isotopic signature (Table 2, Figure 4). Isotopically similar water, highly depleted in ^{18}O (δ^{18}O = −15.8‰) and with *d*-excess above 15‰, was sampled from a soil pit near the borehole located in a water track thalweg, between 48 and 72 m from the start point of the ERT T2 profile (see Figures 1c and 6 for reference). This soil pit exposed sandy loams down to 0.9 m and, when opened, remained almost dry for 2 to 3 h, until the water level at the pit slowly settled at a 0.6 m depth from the surface. The water source in this soil pit, located in the topographical depression can be either from the surrounding elevated tundra patches or from the subsurface compartment, including intra- and subpermafrost groundwater. The ERT data suggests the existence of a thin relict permafrost layer, underlain by non-frozen ground and interrupted by subvertical talik zones. We suppose that water observed in this soil pit originates from the deep groundwater aquifer, close in isotopic composition to the groundwater from the artesian well (see Table 2).

The elevated tundra patches are shown to yield significantly isotopically heavier water, and we cannot propose any viable mixing mechanism or end member that could change the isotopic composition of this water in a way that brings it close to the discussed soil pit water. The occasional evaporation of the water sample during storage would lead to heavier isotopic composition of this sample, and not a depleted one, so this possibility can also be ruled out. From this, we can conclude that, even if the single sample is not an entirely convincing evidence, when combined with geophysical data, it is highly likely that there exists an important connection between shallow subsurface groundwater in water track valleys with deeper groundwater aquifers through upward water migration.

5.5. Water Tracks: Current and Future Development

Water track are widespread periglacial features of the Russian Arctic, and they dominate the slope topography in studied region of north-European Arctic Russia. Unlike Western Siberia and north-eastern Russia, the water track drainage network is highly developed, with deeply incised water track valleys overgrown by willows, and intertrack areas with a polygonally shaped secondary drainage network resembling thermokarst-related patterns of permafrost degradation.

Hydrological connectivity in the water track landscape controls surface and subsurface water fluxes and is, as such, an important ecohydrological factor of the vegetation community structure. Linear thermokarst features of the Bylot Island, Nunavut, were recently reported to control tundra landscapes by promoting the transition from wet to mesic tundra vegetation in areas adjacent to thermo-erosional gullies [51]. At the Khanovey study site, and generally in the north-European Arctic Russia, water tracks appear to play a comparable role, sustaining drier habitats on the tops of elevated tundra patches and moist habitats in the water track thalwegs.

Contemporary climate in the Khanovey region is capable of maintaining wet tundra habitats, as evidenced by the abundance of fens in the water track channels on the interfluves of major rivers of the studied region, at slope summits, shoulders, and partially on backslopes. At these positions, slope steepness is insufficient to rapidly convey water downslope even through water tracks, which leads to the persistent waterlogging of such locations. At footslope positions with steeper slopes, water tracks are increasingly efficient to drain the adjacent tundra. Surface subsidence accompanied by permafrost degradation (linear thermokarst) increases the local height difference between the elevated tundra patches and the water track thalwegs and enhances drainage of these patches. As a result, the intertrack elevated tundra hosts mostly xeric communities, with dwarf shrubs and ericaceous species, i.e., red bearberry, crowberry, Labrador tea, and dwarf birch. Water tracks lower the groundwater table and drain the surrounding tundra effectively enough to maintain xeric habitats.

The water track network in the study site and adjacent territories is still developing, both vertically and laterally. The studied water track valley (Figure 1c) is supposedly freshly incised, because its longitudinal profile is not in equilibrium and hosts several waterfalls, up to ca. 1.5 m high. The evolution of the drainage network at the intertrack surface continues. The dominant local slope is directed toward the water track valley, rather than toward the base level of the Vorkuta River. Because of this, we observe the development of new linear depressions on the left side of the water track valley, with depth between 0.15 and 0.30 m, associated with terrain highly disturbed by hummocks.

Climate change is expected to alter the functioning of the water track system of the region, but the direction of future change is unclear. Permafrost degradation is expected to promote the gradual lateral thawing of elevated tundra patches now frozen [52]. Visual inspection of satellite imagery shows that there is significant difference between water track networks on the south- and north-facing slopes; hence, increasing insolation and air temperature in the future climate are expected to produce a significant geomorphic response.

6. Conclusions

The hydrological snapshot of a small tundra catchment in North-European Russia provides several insights into the hydrological connectivity within its limits. Elevated tundra patches appear to be disconnected from the stream, while the slopes and the riparian zone contribute actively, though locally, to the stream runoff. Minor hollows are found to be either connected to the shallow subsurface runoff or disconnected from it. The difference in connectivity is traceable via d-excess; the disconnected hollows act as evaporative basins and have lower d-excess values. The connected hollows also serve as important DOC flux conveyors, with DOC concentrations up to 50% higher than in disconnected hollows.

Author Contributions: Conceptualization and methodology, N.T.; field investigation, N.T., V.I., D.S., P.K. and O.K.; writing—original draft preparation, N.T.; writing—review and editing, N.T., D.S. and P.K.; visualization, N.T. and O.K. All authors have read and agreed to the published version of the manuscript.

Funding: This study was conducted with support from RuNoCore: Russian–Norwegian research-based education in cold region engineering (#CPRU-2017/10015), funded by the Norwegian Center for International Cooperation in Education (SIU). The manuscript was prepared under the implementation of the state task and the research plan of the IEG RAS, Research Topic AAAA-A19-119021190077-6, and the Melnikov Permafrost Institute SB RAS, Research Topic AAAA-A20-120111690008-9.

Data Availability Statement: The analyzed dataset is available through Figshare, an open repository, and accessible by DOI:10.6084/m9.figshare.14502003.

Acknowledgments: The authors acknowledge the support from the Northern Railroad services in arranging the field surveys.

Conflicts of Interest: The authors declare no conflict of interest. The funders had no role in the design of the study; in the collection, analyses, or interpretation of data; in the writing of the manuscript; or in the decision to publish the results.

References

1. Bracken, L.J.; Croke, J. The concept of hydrological connectivity and its contribution to understanding runoff-dominated geomorphic systems. *Hydrol. Process.* **2007**, *21*, 1749–1763. [CrossRef]
2. Cammeraat, L.H. A review of two strongly contrasting geomorphological systems within the context of scale. *Earth Surf. Process. Landf.* **2002**, *27*, 1201–1222. [CrossRef]
3. Pringle, C. What is hydrologic connectivity and why is it ecologically important? *Hydrol. Process.* **2003**, *17*, 2685–2689. [CrossRef]
4. Bracken, L.J.; Wainwright, J.; Ali, G.A.; Tetzlaff, D.; Smith, M.W.; Reaney, S.M.; Roy, A.G. Concepts of hydrological connectivity: Research approaches, pathways and future agendas. *Earth Sci. Rev.* **2013**, *119*, 17–34. [CrossRef]
5. Guzmán, P.; Anibas, C.; Batelaan, O.; Huysmans, M.; Wyseure, G. Hydrological connectivity of alluvial Andean valleys: A groundwater/surface-water interaction case study in Ecuador. *Hydrogeol. J.* **2016**, *24*, 955–969. [CrossRef]
6. Woo, M.-K. *Permafrost Hydrology*; Springer: Berlin/Heidelberg, Germany, 2012; 575p.
7. Tananaev, N.; Teisserenc, R.; Debolsky, M. Permafrost hydrology research domain: Process-based adjustment. *Hydrology* **2020**, *7*, 6. [CrossRef]
8. Gao, H.; Wang, J.; Yang, Y.; Pan, X.; Ding, Y.; Duan, Z. Permafrost hydrology of the Qinghai-Tibet Plateau: A review of processes and modeling. *Front. Earth Sci.* **2021**, *8*, 576838. [CrossRef]
9. Boike, J.; Roth, K.; Overduin, P.P. Thermal and hydrologic dynamics of the active layer at a continuous permafrost site (Taymyr Peninsula, Siberia). *Water Resour. Res.* **1998**, *34*, 355–363. [CrossRef]
10. Shepelev, V.V. Suprapermafrost waters of the cryolithosphere and their classification. *Geogr. Nat. Resour.* **2009**, *30*, 151–155. [CrossRef]
11. Fotiev, S.M. Underground waters of cryogenic areas of Russia (classification). *Earth Cryosphere* **2013**, *17*, 41–59. (In Russian)
12. Gooseff, M.N.; Wlostowski, A.; McKnight, D.M.; Jaros, C. Hydrologic connectivity and implications for ecosystem processes—Lessons from naked watersheds. *Geomorphology* **2017**, *277*, 63–71. [CrossRef]
13. Briggs, M.A.; Walvoord, M.A.; McKenzie, J.M.; Voss, C.I.; Day-Lewis, F.D.; Lane, J.W. New permafrost is forming around shrinking Arctic lakes, but will it last? *Geophys. Res. Lett.* **2014**, *41*, 1585–1592. [CrossRef]
14. Ling, F.; Wu, Q.; Zhang, T.; Niu, F. Modelling open-talik formation and permafrost lateral thaw under a thermokarst lake, Beiluhe basin, Qinghai-Tibet Plateau. *Permafr. Periglac. Process.* **2012**, *23*, 312–321. [CrossRef]

15. You, Y.; Yu, Q.; Pan, X.; Wang, X.; Guo, L. Geophysical imaging of permafrost and talik configuration beneath a thermokarst lake. *Permafr. Periglac. Process.* **2017**, *28*, 470–476. [CrossRef]
16. Pavlova, N.; Lebedeva, L.; Efremov, V. Lake water and talik groundwater interaction in continuous permafrost, Central Yakutia. *E3S Web Conf.* **2019**, *98*, 7024. [CrossRef]
17. Wellman, T.P.; Voss, C.I.; Walvoord, M.A. Impacts of climate, lake size, and supra- and sub-permafrost groundwater flow on lake-talik evolution, Yukon Flats, Alaska (USA). *Hydrogeol. J.* **2013**, *21*, 281–298. [CrossRef]
18. Connon, R.F.; Quinton, W.L.; Craig, J.R.; Hayashi, M. Changing hydrologic connectivity due to permafrost thaw in the lower Liard River valley, NWT, Canada. *Hydrol. Process.* **2014**, *28*, 4163–4178. [CrossRef]
19. Walvoord, M.A.; Kurylyk, B.L. Hydrologic impacts of thawing permafrost—A review. *Vadoze Zone J.* **2016**, *15*, 1–20. [CrossRef]
20. Lamontagne-Hallé, P.; McKenzie, J.M.; Kurylyk, B.L.; Zipper, S.C. Changing groundwater discharge dynamics in permafrost regions. *Environ. Res. Lett.* **2018**, *13*, 084017. [CrossRef]
21. Hayashi, M.; Quinton, W.L.; Pietroniro, A.; Gibson, J.J. Hydrologic functions of wetlands in a discontinuous permafrost basin indicated by isotopic and chemical signatures. *J. Hydrol.* **2004**, *296*, 81–97. [CrossRef]
22. Tetzlaff, D.; Piovano, T.; Ala-Aho, P.; Smith, A.; Carey, S.K.; Marsh, P.; Wookey, P.A.; Street, L.E.; Soulsby, C. Using stable isotopes to estimate travel times in a data-sparse Arctic catchment: Challenges and possible solutions. *Hydrol. Process.* **2018**, *32*, 1936–1952. [CrossRef]
23. Throckmorton, H.M.; Newman, B.D.; Heikoop, J.M.; Perkins, G.B.; Feng, X.; Graham, D.E.; O'Malley, D.; Vesselinov, V.V.; Young, J.; Wullschleger, S.D.; et al. Active layer hydrology in an arctic tundra ecosystem: Quantifying water sources and cycling using water stable isotopes. *Hydrol. Process.* **2016**, *30*, 4972–4986. [CrossRef]
24. Wan, C.; Gibson, J.J.; Peters, D.L. Isotopic constraints on water balance of tundra lakes and watersheds affected by permafrost degradation, Mackenzie Delta region, Northwest Territories, Canada. *Sci. Total Environ.* **2020**, *731*, 139176. [CrossRef] [PubMed]
25. Brooks, J.R.; Mushet, D.M.; Vanderhoof, M.K.; Leibowitz, S.G.; Christensen, J.R.; Neff, B.P.; Rosenberry, D.O.; Rugh, W.D.; Alexander, L.C. Estimating wetland connectivity to streams in the prairie pothole region: An isotopic and remote sensing approach. *Water Resour. Res.* **2018**, *54*, 955–977. [CrossRef] [PubMed]
26. Gibson, J.J.; Yi, Y.; Birks, S.J. Isotopic tracing of hydrologic drivers including permafrost thaw status for lakes across Northeastern Alberta, Canada: A 16-year, 50- lake assessment. *J. Hydrol. Reg. Stud.* **2019**, *26*, 100643. [CrossRef]
27. Welp, L.R.; Randerson, J.T.; Finlay, J.C.; Davydov, S.P.; Zimova, G.M.; Davydova, A.I.; Zimov, S.A. A high-resolution time series of oxygen isotopes from the Kolyma River: Implications for the seasonal dynamics of discharge and basin-scale water use. *Geophys. Res. Lett.* **2005**, *32*, L14401. [CrossRef]
28. Gibson, J.J.; Birks, S.J.; Yi, Y. Stable isotope mass balance of lakes: A contemporary perspective. *Quat. Sci. Rev.* **2016**, *131*, 316–328. [CrossRef]
29. Devoie, E.G.; Craig, G.R.; Connor, R.F.; Quinton, W.L. Taliks: A tipping point in discontinuous permafrost degradation in peatlands. *Water Resour. Res.* **2019**, *55*, 9838–9857. [CrossRef]
30. Marchenko, S.S.; Gorbunov, A.P.; Romanovsky, V.E. Permafrost warming in the Tien Shan Mountains, Central Asia. *Glob. Planet. Chang.* **2007**, *56*, 311–327. [CrossRef]
31. Streletsky, D.A.; Tananaev, N.I.; Opel, T.; Shiklomanov, N.I.; Nyland, K.; Streletskaya, I.D.; Tokarev, I.; Shiklomanov, A.I. Permafrost hydrology in changing climatic conditions: Seasonal variability of stable isotope composition in rivers in discontinuous permafrost. *Environ. Res. Lett.* **2015**, *10*, 095003. [CrossRef]
32. Ala-aho, P.; Soulsby, C.; Pokrovsky, O.S.; Kirpotin, S.N.; Karlsson, J.; Serikova, S.; Vorobyev, S.N.; Manasypov, R.M.; Loiko, S.; Tetzlaff, D. Using stable isotopes to assess surface water source dynamics and hydrological connectivity in a high-latitude wetland and permafrost influenced landscape. *J. Hydrol.* **2018**, *556*, 279–293. [CrossRef]
33. Roy-Leveillee, P.; Burn, C.R. Near-shore talik development beneath shallow water in expanding thermokarst lakes, Old Crow Flats, Yukon. *J. Geophys. Res. Earth Surf.* **2017**, *122*, 1070–1089. [CrossRef]
34. Disher, B.S.; Connon, R.F.; Haynes, K.M.; Hopkinson, C.; Quinton, W.L. The hydrology of treed wetlands in thawing discontinuous permafrost regions. *Ecohydrology* **2021**, *14*, e2296. [CrossRef]
35. Sjöberg, Y.; Marklund, P.; Pettersson, R.; Lyon, S.W. Geophysical mapping of palsa peatland permafrost. *Cryosphere* **2015**, *9*, 465–478. [CrossRef]
36. Evans, S.G.; Ge, S. Contrasting hydrogeologic responses to warming in permafrost and seasonally frozen ground hillslopes. *Geophys. Res. Lett.* **2017**, *44*, 1803–1813. [CrossRef]
37. Walvoord, M.A.; Striegl, R.G. Increased groundwater to stream discharge from permafrost thawing in the Yukon River basin: Potential impacts on lateral export of carbon and nitrogen. *Geophys. Res. Lett.* **2007**, *34*, L12402. [CrossRef]
38. Goudie, A.S. Global warming and fluvial geomorphology. *Geomorphology* **2006**, *79*, 384–394. [CrossRef]
39. McKenzie, J.M.; Kurylyk, B.L.; Walvoord, M.A.; Bense, V.F.; Fortier, D.; Spence, C.; Grenier, C. Invited perspective: What lies beneath a changing Arctic? *Cryosphere* **2021**, *15*, 479–484. [CrossRef]
40. Neilson, B.T.; Cardenas, M.B.; O'Connor, M.T.; Rasmussen, M.T.; King, T.V.; Kling, G.W. Groundwater flow and exchange across the land surface explain carbon export patterns in continuous permafrost watersheds. *Geophys. Res. Lett.* **2018**, *45*, 7596–7605. [CrossRef]

41. Isaev, V.; Kotov, P.; Sergeev, D. Technogenic hazards of Russian North Railway. In *Transportation Soil Engineering in Cold Regions. Lecture Notes in Civil Engineering. Vol. 49*; Petriaev, A., Konon, A., Eds.; Springer: Singapore, 2020; Volume 1, pp. 311–320. [CrossRef]

42. Trochim, E.D.; Jorgenson, M.T.; Prakash, A.; Kane, D.L. Geomorphic and biophysical factors affecting water tracks in northern Alaska. *Earth Space Sci.* **2016**, *3*, 123–141. [CrossRef]

43. IAEA. *Reference Sheet for International Measurement Standards*; VSMOW2/SLAP2. Rev. 1; IAEA: Vienna, Austria, 2017; Available online: https://nucleus.iaea.org/sites/ReferenceMaterials/Shared%20Documents/ReferenceMaterials/StableIsotopes/VSMOW2/VSMOV2_SLAP2.pdf (accessed on 13 July 2021).

44. International Atomic Energy Agency/World Meteorological Organization (IAEA/WMO). Global Network of Isotopes in Precipitation (GNIP). Available online: https://nucleus.iaea.org/Pages/GNIPR.aspx (accessed on 18 April 2021).

45. De Wet, R.F.; West, A.G.; Harris, C. Seasonal variation in tap water δ^2H and δ^{18}O isotopes reveals two tap water worlds. *Sci. Rep.* **2020**, *10*, 13544. [CrossRef]

46. Kendall, K.A.; Shanley, J.B.; McDonnell, J.J. A hydrometric and geochemical approach to test the transmissivity feedback hypothesis during snowmelt. *J. Hydrol.* **1999**, *219*, 188–205. [CrossRef]

47. Ren, Y.; Yang, S.; Andersen, K.H.; Yang, Q.; Wang, Y. Thixotropy of soft clay: A review. *Eng. Geol.* **2021**, *287*, 106097. [CrossRef]

48. Veizaga, E.A.; Ocampo, C.J.; Rodríguez, L. Hydrological and hydrochemical behavior of a riparian zone in a high-order flatland stream. *Environ. Monit. Assess.* **2019**, *191*, 10. [CrossRef]

49. Brown, V.A.; McDonnell, J.J.; Burns, D.A.; Kendall, C. The role of event water, a rapid shallow flow component, and catchment size in summer stormflow. *J. Hydrol.* **1999**, *217*, 171–190. [CrossRef]

50. Cory, R.M.; Kling, G.W. Interactions between sunlight and microorganisms influence dissolved organic matter degradation along the aquatic continuum. *Limnol. Oceanogr. Lett.* **2018**, *3*, 102–116. [CrossRef]

51. Perreault, N.; Lévesque, E.; Fortier, D.; Lamarque, L.J. Thermo-erosion gullies boost the transition from wet to mesic tundra vegetation. *Biogeosciences* **2016**, *13*, 1237–1253. [CrossRef]

52. Douglas, T.A.; Hiemstra, C.A.; Anderson, J.E.; Barbato, R.A.; Bjella, K.L.; Deeb, E.J.; Gelvin, A.B.; Nelsen, P.E.; Newman, S.D.; Saari, S.P.; et al. Recent degradation of Interior Alaska permafrost mapped with ground surveys, geophysics, deep drilling, and repeat airborne LiDAR. *Cryosphere Discuss.* **2021**, 1–39. [CrossRef]

Article

Groundwater-Surface Water Interaction in the Nera River Basin (Central Italy): New Insights after the 2016 Seismic Sequence

Lucio Di Matteo *, Alessandro Capoccioni, Massimiliano Porreca and Cristina Pauselli

Department of Physics and Geology, University of Perugia, 06123 Perugia, Italy;
alessandro.capoccioni@studenti.unipg.it (A.C.); massimiliano.porreca@unipg.it (M.P.);
cristina.pauselli@unipg.it (C.P.)
* Correspondence: lucio.dimatteo@unipg.it

Abstract: The highest part of the Nera River basin (Central Italy) hosts significant water resources for drinking, hydroelectric, and aquaculture purposes. The river is fed by fractured large carbonate aquifers interconnected by Jurassic and Quaternary normal faults in an area characterized by high seismicity. The 30 October 2016, seismic sequence in Central Italy produced an abrupt increase in river discharge, which lasted for several months. The analysis of the recession curves well documented the processes occurring within the basal aquifer feeding the Nera River. In detail, a straight line has described the river discharge during the two years after the 2016 seismic sequence, indicating that a turbulent flow characterized the emptying process of the hydrogeological system. A permeability enhancement of the aquifer feeding the Nera River—due to cleaning of fractures and the co-seismic fracturing in the recharge area—coupled with an increase in groundwater flow velocity can explain this process. The most recent recession curves (2019 and 2020 periods) fit very well with the pre-seismic ones, indicating that after two years from the mainshock, the recession process recovered to the same pre-earthquake conditions (laminar flow). This behavior makes the hydrogeological system less vulnerable to prolonged droughts, the frequency and length of which are increasingly affecting the Apennine area of Central Italy.

Keywords: groundwater; Nera River; carbonate aquifer; recession curves; seismic sequence

Citation: Di Matteo, L.; Capoccioni, A.; Porreca, M.; Pauselli, C. Groundwater-Surface Water Interaction in the Nera River Basin (Central Italy): New Insights after the 2016 Seismic Sequence. *Hydrology* **2021**, *8*, 97. https://doi.org/10.3390/hydrology8030097

Academic Editors: Il-Moon Chung, Sun Woo Chang, Yeonsang Hwang and Yeonjoo Kim

Received: 10 June 2021
Accepted: 24 June 2021
Published: 27 June 2021

Publisher's Note: MDPI stays neutral with regard to jurisdictional claims in published maps and institutional affiliations.

1. Introduction

The interaction between groundwater (GW) and surface water (SW) influences river water quantity and quality. The understanding of the processes and dynamics of GW–SW interactions are fundamental for the accurate assessment, integrated management, and environmental protection of water resources [1–5]. Although the anthropic pressure on many river basins is increasing [6], climate change is negatively impacting the river discharge, especially in regions characterized by a reduction of snow and rainfall during the recharge periods [7].

The Nera River in Central Italy represents a hydrogeological system where SWs are mainly provided by GWs, thanks to a set of permanent linear springs, the water of which comes from large fractured and karstified basal carbonate aquifers [8–11]. Aquifers hosted in the upper part of the Nera River supply water to a multipurpose system (drinking water, hydropower energy production, and fish farming). The study area is located in a region affected by a decrease in rainfall during the recharge period, which occurs from autumn to early spring [12–17]. A recent review of rainfall trends published by Caporali et al. [18] reveals a more pronounced negative trend in winter periods in Central Italy than in Northern Italy. This general trend is coupled with the increase in length and frequency of drought periods in the last two decades [15,19,20]. Moreover, as reported by Diodato and Bellocchi [21], the number of snowy days declined in peninsular Italy from the end of the Little Ice Age (LIA) and, markedly, after the 1940s. Since snowmelt and rainfall affects the groundwater recharge, river and spring discharges increasingly suffer from the reduction of these two fundamental components.

Considering the high seismicity of the Apennine ridge of Central Italy, locally GW–SW interaction can change due to co-seismic effects produced by earthquakes. The 2016 seismic sequence in Central Italy deeply affected the GW circulation, the changes of which had implications for the management of water resources [22–26]. In general, the seismic sequence induced changes to aquifer permeability and pore water pressure, with consequent variations of hydraulic gradient, which in a few months or years, tend to recover to close to what they were before the earthquake, producing transient effects on river discharge [27,28]. Rojstaczer et al. [29] highlighted that changes in the river regime can persist over time, indicating that the GW circulation feeding the river is changed; i.e., earthquakes can breach the seals between neighboring compartmented aquifers [30]. Aquifer breaching allows water mixing, which can be faster than other mechanisms (e.g., release of deep-seated fluids) and the anomaly can be long or even permanent [31].

A pre-and post-earthquake river recession curves analysis can help understand changes in the groundwater reservoir feeding rivers. Di Matteo et al. [24] reported that the 2016 seismic sequence changed the recession processes of the Nera River discharge recorded at the Visso gauge station. During the two years after the 2016 mainshock, the aquifers feeding the Nera River emptied at a faster rate, causing some concern among water-using companies, as the phenomenon's evolution was not known. The persistence of a more rapid depletion than in the past could have significant implications on river management, especially during recessions related to prolonged drought periods. These considerations need to be examined in more detail. Therefore, the present study aims to update the knowledge of the Nera River hydrogeological system by considering the analysis of river recession curves of the 2019 and 2020 periods. The Nera River is very suitable for such studies, as the river flows during recession periods are entirely supported by groundwater mainly fed by the basal aquifer. Consequently, the study of flow rates during recession phases provides information on the processes taking place within the hydrogeological system. The comparison of pre-and post-seismic recession curves contributes to understanding the dynamic of fractured carbonate aquifers useful for the water management of the multipurpose system.

2. Study Area

2.1. Geological Structural and Hydrogeological Setting

The area investigated in this study is located in the mountainous region of the Sibillini Mountains (northern Apennines) in Central Italy. The stratigraphic sequence of this part of the Apennine consists of Meso-Cenozoic carbonate multilayer formations (Fms) of the Umbria-Marche succession and siliciclastic foredeep deposits of the Laga Fm [32]. Figure 1a shows the hydrogeological map of the upper part of the Nera River basin. It is based on a new geological survey, which checked and updated the geological map published by [32]; rocks are described considering the stratigraphic relationships and focusing on the hydrogeological properties, e.g., grouping them in hydrogeological complexes [33–36].

The sequence starts with Upper Triassic evaporites, including anhydrides and dolomites (evaporites Fm). These rocks are not exposed in the study area and are characterized by a general low permeability (Evaporites Complex, EC). Upward, the sequence continues with shallow-water carbonates (Calcare Massiccio Fm) with a variable thickness of 600 to 700 m. In the Jurassic structural lows, a thick prevalent pelagic sequence was deposited. Micritic limestones represented the basal deposition of structural lows with nodules of chert (Corniola Fm). Being high fractured/karstified limestones, the Calcare Massiccio and Corniola host the basal aquifer with a maximum thickness of ca. 900 m (Basal Limestones Complex, BLC). The pelagic sequence continues with Calcareous Siliceous Marly Complex (CSMC, including Rosso Ammonitico, Marne del Serrone, Calcari a Posidonia, and Calcari Diasprigni Fms), characterized by low relative permeability. The subsequent deposition of Maiolica Fm occurred on both structural highs and lows, with a variable thickness from 150 to a maximum of 400 m, respectively (Maiolica Complex, MAC). Stratified and fractured limestones characterize the hydrogeological complex, with high relative permeability. Due to high permeability, it represents the middle aquifer. The sequence continues upward with the Marne a Fucoidi Fm, a marly rock with low relative permeability (Marne a Fucoidi Complex, MFC). Marly-limestones (about 400 m thick) were then deposited above the Marne a Fucoidi Fm; it includes the Scaglia Calcarea Complex (SCC, Scaglia Rossa and Scaglia Bianca), characterized by a moderate relative permeability. Above the SCC, Scaglia Variegata and Scaglia Cinerea Fms are deposited, composed of marls and marly limestones with low permeability (Calcareous Marly Complex, CMC). The subsequent Miocene marly units (Schlier and Bisciaro) and siliciclastic Laga Fm are mainly exposed in the eastern sector of the studied area. Most of these Fms are characterized by moderate or low permeability (Terrigenous Units Complex, TUC).

Formations belonging to the hydrogeological complexes were involved in distinct tectonic phases since the Jurassic. First, extensional tectonics was associated with the thinning of the Adriatic continental margin during Late Jurassic [32], involving the Basal Limestones Complex. Since the late Miocene, the whole limestone multilayer was affected by a compressional phase that produced important shortening of this part of the Apennines with a typical fold and thrust belt [37,38]. The Sibillini Mountains thrust (MST, Figure 1a) represented the main regional compressional structure [39], marking the tectonic boundary between the Mesozoic–Paleogene limestone sequence at its hanging-wall and the Late Miocene–Early Pliocene Laga sequence (Flysch della Laga Fm, [40]) at its footwall. A more internal compressional structure is represented by the Pizzo Tre Vescovi thrust (PTV, Figure 1a).

The compressional structures were subsequently crosscut by the NNW-SSE trending normal faults due to the prevalent extensional Quaternary tectonic phase [41,42]. These faults are still active and responsible for historical seismicity and represent preferential drainage paths of groundwater in the basal aquifer, which usually hinder the transversal groundwater exchanges [35,43].

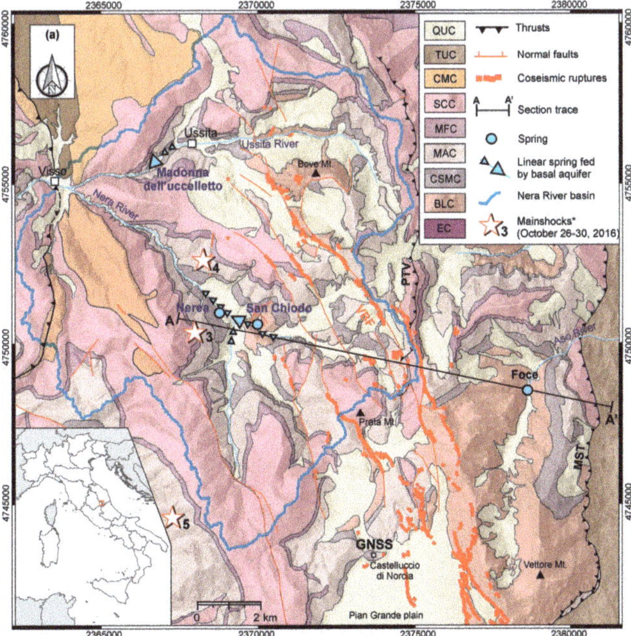

Figure 1. *Cont.*

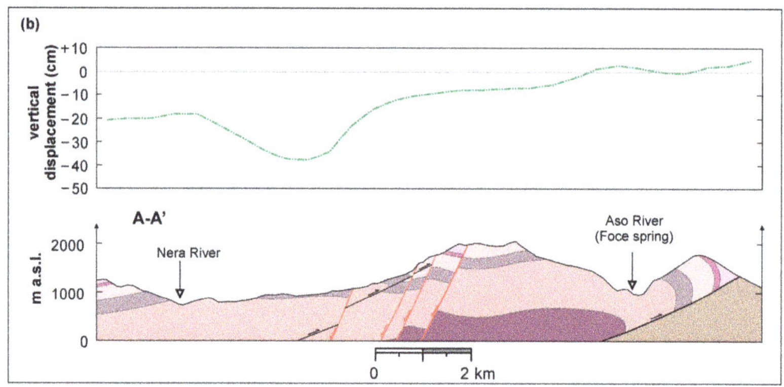

Figure 1. (**a**) Hydrogeological map of the upper part of the Nera River catchment with the mainshocks of 26–30 October 2016, seismic sequence (information on earthquakes are in Table 1). EC—Evaporitic-dolomitic Complex; BLC—Basal Limestones complex; CSMC—Calcareous Siliceous Marly Complex; MAC—Maiolica Complex; MFC—Marne a Fucoidi Complex; SCC—Scaglia Calcarea Complex; CMC—Calcareous Marly Complex; TUC—Terrigenous units complex; QUC—Quaternary Units Complex; VBF—Mt. Vettore-Mt. Bove fault systems; PTV—Pizzo Tre Vescovi thrust; SMT—Sibillini Mts. Thrust; GNSS—Global Navigation Satellite System station [44]. (**b**) Cross section roughly oriented WE (co-seismic displacements produced by the 30 October 2016 seismic sequence are taken from [45]).

2.2. Seismic Sequence and Co-Seismic Effects

The study area was struck by a long and intensive seismic sequence during 2016 and 2017 (Figure 1a), characterized by the occurrence of nine mainshocks with moment magnitude, Mw > 5.0 (Table 1). The seismic sequence was due to a system of mainly NNW-SSE striking and WSW dipping normal faults, as shown by the focal mechanism of the main shocks, the distribution of the aftershocks, and significant co-seismic ruptures [46–53].

Table 1. Mainshocks of the 2016–2017 seismic sequence (modified from [54]).

N°	Localization	Data	Mw	Hypocentral Depth (km)
1	Accumoli (42.70–13.70)	24 August 2016	6.0	4.65
2	Norcia (42.79–13.15)	24 August 2016	5.4	4.87
3	Castel Santangelo sul Nera (42.88–13.12)	26 October 2016	5.4	3.46
4	Visso (42.91–13.09)	26 October 2016	5.9	2.47
5	Norcia (42.83–13.11)	30 October 2016	6.5	5.78
6	Capitignano (42.56–13.29)	18 January 2017	5.1	7.87
7	Capitignano (42.55–13.28)	18 January 2017	5.5	7.72
8	Capitignano (42.52–13.29)	18 January 2017	5.4	8.38
9	Barete (42.48–13.28)	18 January 2017	5.0	9.43

The main seismogenic normal faults activated during the seismic sequence were the Vettore Mt.–Bove Mt. fault systems (VBF) to the north and the Laga (LAF) fault systems to the south [55]. This study is focused on the northern sector affected by the VBF segment.

In particular, the southern portion of the VBF and the northern portion of the LAF (Figure 1a) were activated during the 24 August Mw 6.0 Accumoli event, which enucleated at their overlapping area, at a depth of 8 km [51,56]. The 26 October Mw 5.9 Visso event enucleated at a depth of 4 km [57] and activated the northern part of the VBF, while during the 30 October Mw 6.5 Norcia event, the VBF was activated to the south and center. The average co-seismic throw of the entire sequence is ~0.3 m [52,53] and the activation of the VBF during the largest events produced important surface ruptures developed along major and minor fault segments as shown in Figure 1a [51,53]. Figure 1b shows a cross section roughly oriented WE with co-seismic displacements produced by the October 30, 2016, seismic sequence as taken from Valerio et al. [45]. Geodetic measurements, using the Global Navigation Satellite System (GNSS) technique together with the available surface deformation field obtained on the basis of the DInSAR technique, registered the co-seismic and immediately post-seismic deformation of the two major October shocks surrounding the VBF [44]. One of the GNSS stations (Figure 1a) recorded westward horizontal displacements of 419 mm and subsidence of 707 mm (with 95% confidence errors), with a total off-fault vertical displacement between footwall and hanging-wall blocks of 736 mm in correspondence of the northern sector of the VBF.

3. Materials and Methods

3.1. Meteorological Data and Techniques of Analysis

The study of river discharge fed by regional aquifers requires reliable meteorological data. According to Cambi et al. [13], as for other mountain regions in Central Italy, often few data are available to define the hydrogeological scheme. Figure 2 shows all the thermo-pluviometric stations located within the Nera River catchment at Visso section.

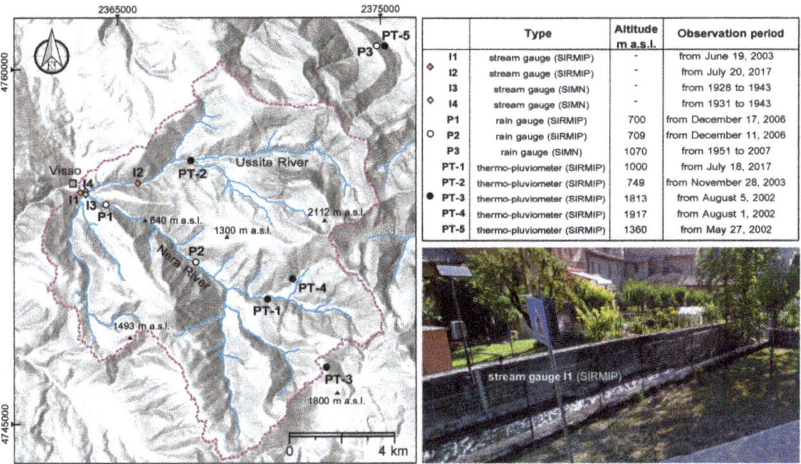

	Type	Altitude m a.s.l.	Observation period
I1	stream gauge (SIRMIP)	-	from June 19, 2003
I2	stream gauge (SIRMIP)	-	from July 20, 2017
I3	stream gauge (SIMN)	-	from 1928 to 1943
I4	stream gauge (SIMN)	-	from 1931 to 1943
P1	rain gauge (SIRMIP)	700	from December 17, 2006
P2	rain gauge (SIRMIP)	709	from December 11, 2006
P3	rain gauge (SIMN)	1070	from 1951 to 2007
PT-1	thermo-pluviometer (SIRMIP)	1000	from July 18, 2017
PT-2	thermo-pluviometer (SIRMIP)	749	from November 28, 2003
PT-3	thermo-pluviometer (SIRMIP)	1813	from August 5, 2002
PT-4	thermo-pluviometer (SIRMIP)	1917	from August 1, 2002
PT-5	thermo-pluviometer (SIRMIP)	1360	from May 27, 2002

Figure 2. Nera River catchment at Visso section (A = 130 km^2) with the location and description of the hydro-meteorological stations.

As for other locations in the Mediterranean region, the upper part of the Nera River catchment has had changes in the data acquisition network over time; the overall monitoring network has moved from the Servizio Idrografico e Mareografico Nazionale (SIMN) to the Civil Protection Agency of Marche Region (SIRMIP) during the beginning of the 2000s, resulting in lack of data often coupled with the relocation of some stations. To check the consistency of rainfall data, the double-mass analysis was used [57]. This method compares cumulated rainfall data of a single station with that of other stations in the area, allowing the individuation of abrupt deviations produced by instrument malfunctions and/or relocations. As recently reported by Caloiero et al. [19], lack of spatially distributed data can be overcome due to the increased availability of meteorological satellites. The Global Precipitation Measurement (GPM) provides monthly rainfall data on an approximately 10 × 10 km grid across the globe from April 2014 to the present. The latest release of the Integrated Multi-satellitE Retrievals for Global Precipitation Measurement product (IMERG, Version 6B-Final) fused GPM estimates with those collected by the TRMM satellite (Tropical Rainfall Measuring Mission), operating during 2000–2015 (https://gpm.nasa.gov/data/imerg, accessed on 25 May 2021). The Nera River catchment is included in a cell having the following lat-long coordinates: 42.45–13.05; and 42.55–13.15. Rainfall monthly data are downloaded from https://disc.gsfc.nasa.gov/datasets/GPM_3IMERGM_06/summary?keywords=%22IMERG%20final%22 (accessed on 25 May 2021). Due to the complex topography in mountain regions, checking the performance of satellite data is necessary to achieve analyses that are as realistic as possible [58–60]. To quantify the performance of GPM-IMERG data, Pearson's Correlation Coefficient (CC, Equation (1)), Nash-Sutcliffe efficiency (NSE, Equation (2)), Root Mean Square Error (RMSE, Equation (3)), and the relative bias (rBias, Equation (4)) are used; the latter reflects the systematic bias between the satellite and the ground observations [61].

$$CC = \frac{\sum_{i=1}^{n} \left(G_i - \overline{G}\right) \cdot \left(S_i - \overline{S}\right)}{\sqrt{\sum_{i=1}^{n} \left(G_i - \overline{G}\right)^2} \cdot \sqrt{\sum_{i=1}^{n} \left(S_i - \overline{S}\right)^2}} \tag{1}$$

$$NSE = 1 - \frac{\sum_{i=1}^{n} \left(S_i - G_i\right)^2}{\sum_{i=1}^{n} \left(G_i - \overline{G}\right)^2} \tag{2}$$

$$RMSE = \sqrt{\frac{\sum_{i=1}^{n} \left(S_i - G_i\right)^2}{n}} \tag{3}$$

$$rBias = \frac{\sum_{i=1}^{n} \left(S_i - G_i\right)^2}{\sum_{i=1}^{n} G_i} \cdot 100 \tag{4}$$

where:

S_i = satellite precipitation estimate (mm/month)
G_i = rain gauge observation (mm/month)
$\overline{S}$ = mean satellite precipitation estimate (mm/month)
$\overline{G}$ = mean gauge precipitation observation (mm/month)

Reliable monthly rainfall datasets are useful to check the frequency and length of droughts, which affect the groundwater recharge of hydrogeological systems. Worldwide, the Standardized Precipitation Index (SPI, [62]) is used as an effective drought index for detecting and characterizing meteorological droughts. According to WMO [63], for the computation at least 20–30 years of monthly rainfall data are needed. In the SPI computation the data are fitted to a gamma probability distribution and then transformed into a normal distribution. For a given data time series X_i as X_1, X_2 ... .X_n, the SPI_i is defined by Equation (5).

$$SPI_i = \frac{X_i - \overline{X}}{S_x} \tag{5}$$

where $\overline{X}$ is the arithmetic mean of rainfall and S_x is the standard deviation. According to McKee et al. [62], wet periods occur when SPI values are higher than 1 while drought periods occur when SPI is lower than −1. SPI equal to zero implies that there is no deviation from the mean. Table 2 shows the classification for the SPI values.

Table 2. Classification scale for the SPI values [63].

Class	Condition	SPI Values
1	Extremely wet	SPI > 2
2	Very wet	1.5 ≤ SPI < 2
3	Moderately wet	1.0 ≤ SPI < 1.5
4	Near normal	−1.0 ≤ SPI < 1.0
5	Moderately dry	−1.5 ≤ SPI < −1.0
6	Severely dry	−2.0 ≤ SPI < −1.5
7	Extremely dry	SPI ≤ −2.0

Monthly precipitation (P) and temperature (T) have been used to compute the monthly evapotranspiration (ETR) using the Thornthwaite and Mather [64] method. This method is still useful when no data concerning radiation, air humidity, etc.—necessary to apply more modern methods—are available, as occurred for the study area [59,65]. The computation of ETR was carried out by using the software developed by Čadro [66]. Surplus monthly data (S = P-ETR) were then computed to individuate the period with no recharge (P = ETR), i.e., the river discharge is fed only by the groundwater stored in the hydrological system, which empties as the depletion phase proceeds.

3.2. Discharge Data

Daily stream levels (H) of the Nera River were collected from the SIRMIP on-line monitoring network on the I1 river section (Figure 2), which also includes its main tributary (Ussita stream, I2 stream gauge). By using the rating curve published by SIRMIP (http://app.protezionecivile.marche.it/sol/indexjs.sol?lang=it, accessed on 25 May 2021), daily stream discharge values (Q) were computed from January of 2016. No substantial anthropic modification occurred in the catchment in the last decades that could have influenced SW–GW interactions through the river system. Apart from the construction of a fluent hydroelectric plant close to the P1 station (Figure 2), a drinking water intake was built near Castelsantangelo sul Nera village in the 1980s (San Chiodo spring, Figure 1). As the water is piped out of the Nera River catchment, the amount of drinking water withdrawn was added to the river discharge data; it corresponds to 0.150 m^3/s from January of 2016 to February of 2018 and 0.200 m^3/s from March of 2018 to today [67]. As reported by Di Matteo et al. [24], in addition to recent monitoring data, some historical discharges from the SIMN monitoring network (I3 and I4 in Figure 2) are also available (1928–1943 period).

3.3. Models for River Recession Curves

The study of the discharge during periods with no recharge (recession curve) provides insightful hydrologic information about the hydrogeological properties of the system feeding the river, at least in terms of average or equivalent values. Several conceptual models for recession curves have been developed to describe the discharge during the recession period. In the hydrogeological practice the exponential formula (EF) is extensively used (Equation (6), [68,69]). Although EF was developed for porous media (Darcian laminar flow), it has also been widely applied to fractured aquifers characterized by low hydraulic gradients and velocity [70–72]. When these conditions are not met (turbulent flow), the straight-line equation (SL) can describe the recession process in fractured and karst aquifers (Equation (7), [73,74]). The rate of change of the discharge (RCD) is computed making the first derivative of Equations (6) and (7), obtaining Equations (8) and (9) for EF and SL, respectively,

$$Q_t = Q_o \cdot e^{-\alpha_E \cdot t} \tag{6}$$

$$Q_t = Q_o - \alpha_{TL} \tag{7}$$

$$RCD_{EF} = -\alpha_E \cdot Q_t \tag{8}$$

$$RCD_{SL} = -\alpha_{ETL} \tag{9}$$

where:

Q_t and Q_o = discharge at time t and at the beginning of recession period (m^3/d)

α_E = Maillet recession coefficient (d^{-1})

α_{TL} = Torricelli-like recession coefficient (m^3/d^2)

RCD_{EF} = rate of change of the discharge of EF equation (m^3/d^2)

RCD_{TL} = rate of change of the discharge of SL equation (m^3/d^2)

According to Rorabaugh [75], and Kovacs and Perrochet [76], α_E is linked to the aquifer characteristics by Equation (10).

$$\alpha_E = \frac{\pi \cdot k \cdot H}{S \cdot L^2} \tag{10}$$

where:

k = aquifer' hydraulic conductivity (m/d)
H = aquifer thickness (m)
L = length of the one-dimensional domain (m)
S = storativity (dimensionless)

4. Results

4.1. Rainfall-Water Surplus Analysis

The recharge of aquifers is affected by prolonged drought periods, the evaluation of number and frequency of which require reliable rainfall data. Figure 3 shows the results of the double mass analysis; data of Ussita station (PT-2) are compared with those of ENDESA (P1) and M. Prata (PT-3) stations. The cumulative rainfall of PT-2 and P1—located at similar altitude—is aligned along a straight line with no abrupt deviations. On the contrary, an abrupt change between the years 2008 and 2010 is observed by comparing the cumulated rainfall data of PT-2 with PT-3. The latter shows lower cumulative rainfall despite being located 1050 m higher than PT-1. The double-mass analysis results for the 2002–2020 period found the Ussita station to be the most reliable in the study area.

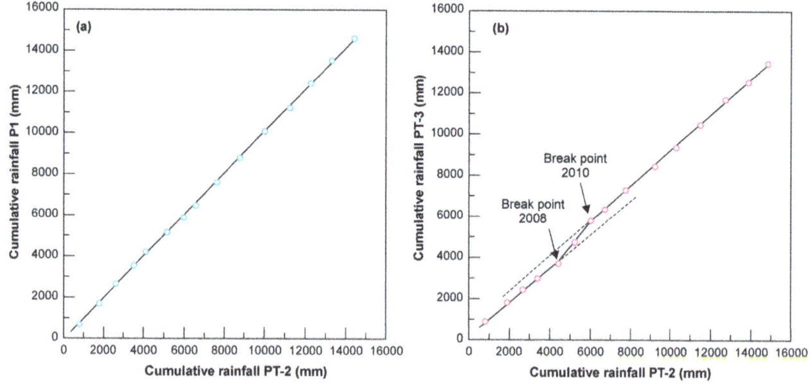

Figure 3. Double-mass curves for detecting the consistency of rainfall data. (**a**) Comparison between PT-2 and P1. (**b**) Comparison between PT-2 and PT-3. The location of rainfall stations is in Figure 2.

As illustrated in Section 2.1, the computation of SPI requires at least 20–30 years of monthly rainfall data [63]. The Ussita station does not meet this requirement, i.e., it is not suitable for SPI analysis. Among the available stations (Figure 2), only the Bolognola station could be useful for this analysis (P3 and PT-5). Unfortunately, in 2007, the historical rain gauge (P3, SIMN) was removed, and the new station was located at a higher elevation than the historical one (PT-5, SIRMIP), i.e., the rain gauge was moved from 1070 m a.s.l. to 1370 m a.s.l. Under these conditions, the aggregation of the two-time periods (1951–2007 and 2007–2020) was not successful; therefore, the data were not used for the SPI computation. In this framework, satellite data can be a valid alternative for this analysis, especially in data-scarce regions. The results of the application of some accuracy indices show that the correlation among monthly GPM-IMERG satellite data and rainfall at the Ussita gauge reaches a CC value equal to 0.76, a NSE of 0.99, a RMSE of 4.5 mm, and a rBias of about 24%, confirming that satellite data tend to slightly overestimate rainfall data as already recorded by Navarro et al. [60] along the Apennine chain of the Italian peninsula. The results of statistical evaluations indicate that—within the limits of use of satellite data in mountain areas (e.g., effects of topography, etc.)—the performance of GPM-IMERG allows the use of this dataset to compute the SPI during the 2000–2020 period. Similar results were recently obtained by Caloiero et al. [19], who used the GPM-IMERG data to calculate SPI values for the Italian peninsula. Figure 4a shows the SPI-12 computed over the last two decades;

prolonged drought events are detected in 2001–2003, 2007, 2012, and 2018. Three moderate to near normal wet periods are interposed between these events. Similar findings—even if with slightly different intensities and with some difference for 2018—were obtained by Valigi et al. [25] for the Pescara di Arquata Spring recharge area, located about 20 km southeast of the upper part of the Nera River catchment.

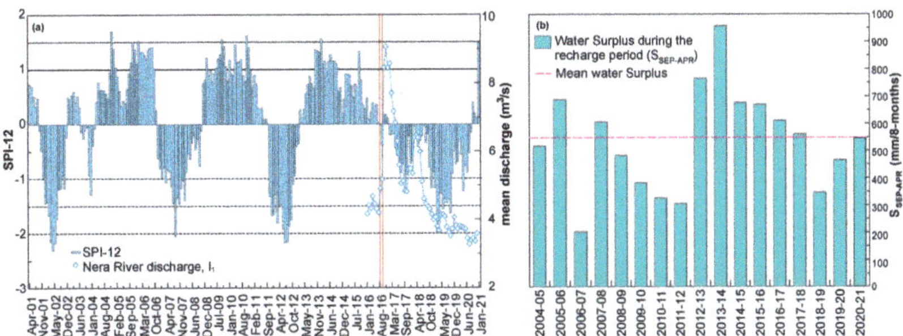

Figure 4. (**a**) A period of 12-months of SPI computed by using GMP-IMERG rainfall data with mean monthly discharge of the Nera River at section I1 (Figure 2). The two vertical red bars indicate the August 24 and the 30 October 2016, seismic shocks. (**b**) Cumulated water surplus (S)—based on the Thornthwaite and Mather [64] method—during the recharge period (8 months from September to April).

Figure 4b shows the cumulated water surplus (S) during the recharge periods (September–April); weather data of the Ussita station are used for the computation. During the observation period, the mean S was about 550 mm/8-months with the highest values from 2013–14 to 2017–18, confirming the results of SPI analysis. On the contrary, in the 2018–19 and 2019–20 periods, S values were 100–200 mm lower than the mean, corresponding to the moderate/severe dry periods highlighted with the SPI-12 (Figure 4a).

4.2. River Hydrograph Analysis

The flow regime of the Nera River was studied considering the available daily discharge data. According to the SIRMP monitoring system, reliable data were recorded during the 2016–2021 period (Figure 5). These data include both pre- and post-seismic phases.

The 30 October 2016, earthquake (n. 5 in Figure 5) produced the abrupt discharge rising observed during November–January of 2016 with an increase of about 4.8 m³/s. A small peak flow rate had already been recorded on 24 August 2016, connected with the Accumoli-Norcia earthquakes (ns. 1–2 in Figure 5). Figure 5 also shows the monthly water surplus, computed according to the Thornthwaite and Mather [64] method, and daily discharge data of the Ussita River. The discharge hydrograph of this sub-catchment overlaps the Nera River one with some differences during the late August–September periods of 2017 and 2018. In these two periods, the recession phase of the Ussita River continued while that of the Nera River stopped after receiving some recharge that did not occur in the Ussita catchment.

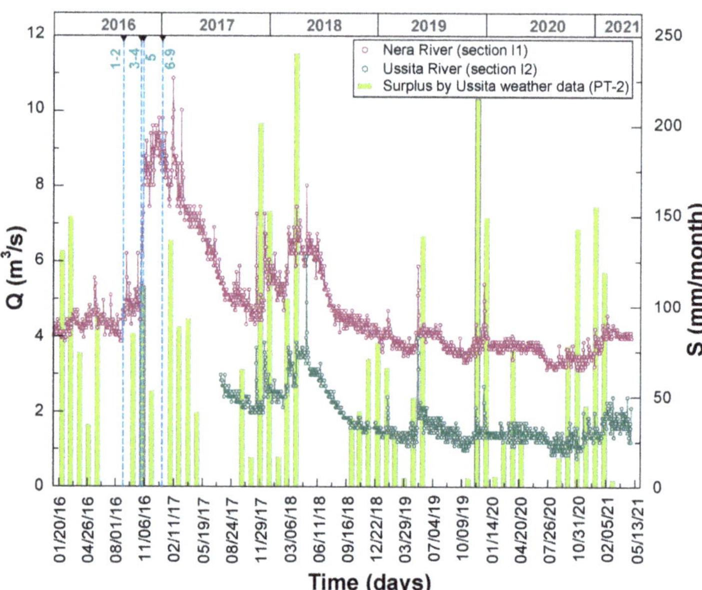

Figure 5. Hydrograph of the Nera River and Ussita River with monthly surplus calculated by using meteorological data of the Ussita station. The blue numbers adjacent to the dashed lines indicate the main earthquakes with moment magnitude (M_w) higher than 5.0 that occurred during the 2016–2017 seismic sequences in Central Italy (information about the earthquakes are reported in Table 1).

Considering periods with no aquifer recharge (water surplus S = 0 in Figure 5) the pre- and post-seismic recession curves were collected. After the individuation of months with S = 0, a deep analysis was carried out to separate consecutive days characterized by a decrease in discharge values, avoiding days with discharge increases in response to the previous effective rainfalls. These curves help to investigate the relationship between aquifer hydrogeological properties and groundwater discharge to the river. The results of the analysis of the 2019 and 2020 depletion phases are presented in Figure 6, moving to the discussion section for the comparison with previous studies published in the study area in the two years following the earthquake. The EF recession model describes both 2019 and 2020 recession curves as a result of the best-fit analysis, indicating that the hydrogeological system feeding the river empties with a Darcian laminar flow. The two curves show similar a recession coefficient α_E, -2.1×10^{-3} day^{-1} and -2.3×10^{-3} day^{-1} for 2019 and 2020, respectively.

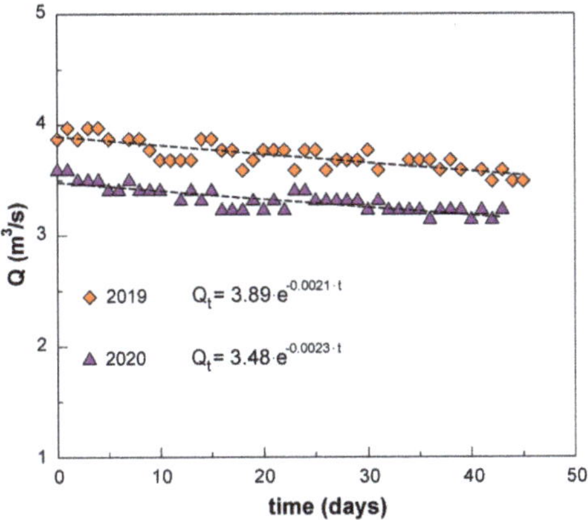

Figure 6. Recession curves for 2019 and 2020. Discharge data of the Nera River are recorded at section I1 (Figure 2).

5. Discussion

Updates on the response of the Nera River discharge to the 2016 seismic sequence are discussed, taking into consideration the meteorological conditions during pre- and post-seismic phases. During the two years after the 2016 seismic sequence (2017 and 2018), the peak of river discharge after the autumn–spring recharge period was very high compared with 2019 and 2020 (Figure 5). The high discharge value recorded in January of 2017 was mainly linked to co-seismic effects (e.g., increased aquifer permeability and pore water pressure). Moreover, as reported by Petitta et al. [22], the impressive increase in spring and river discharge observed in the upper Nera River may be correlated with the subsidence induced by the toe of VBF—as already recorded by GNSS station (Figure 1a)—which might have created an additional "squeezing effect" in the core of the Sibillini Mountains aquifer. Barberio et al. [77] reported some changes in hydrochemical characteristics of water of the basal aquifer feeding the San Chiodo spring (Figure 1a), such as an increase in the content of three metals (Cr, Fe, and V) and one metalloid (As) four months before or roughly in conjunction with the 2016 seismic sequence, respectively. Rosen et al. [78] reported no change in most of the major ions and in $\delta^{18}O_{H2O}$ and $\delta^{2}H_{H2O}$ of waters of the Nerea spring (Figure 1a), the recharge area of which is characterized by the basal aquifer extending towards SE including the Vettore Mountain faulting systems deeply affected by co-seismic ruptures (Figure 1). In detail, concentrations of SO_4 showed no change after the 30 October 2016, earthquake and trace element concentrations returned to pre-earthquake concentrations by the end of November, 2016. The authors concluded that the hydrochemical dynamics suggested within-aquifer changes instead of mixing with another aquifer, geothermal fluids, or aquifer breaching. Recently, Fronzi et al. [79] presented updated hydrogeochemical results for the Castelsantangelo sul Nera area— also fed by basal aquifer and located slightly upstream to the Nerea spring—showing some sampling points with an increase in SO_4 content and some others with no increase after the mainshock. The different results reported by [78,79] highlight the complexity of processes that occurs in a hydrogeological system highly influenced by normal faults and co-seismic fracturing systems. In other words, sampling points collected along the normal faults interacting with Triassic evaporites aquiclude below the Mt. Vettore–Pian Grande Plain show increases in SO_4, which is not evident in other zones—also fed by basal aquifer—where normal faults and co-seismic fracturing did not play a significant role.

Mastrorillo et al. [23], by a mechanism named "Aquifer Fault Rupture", hypothesized a shift eastward of the piezometric divide, severely penalizing the aquifers in the eastern side of the Sibillini Mountains. The authors left open whether or not the modified piezometric divide would recover to its original position. Our results on the Upper Nera River catchment (western part of the Apennine chain) show that co-seismic effects on GW lasted up to at least 2018 by analyzing river discharges at the Visso gauge. The peak discharge that occurred in 2018 is preceded by a recharge period characterized by a water surplus in line with or slightly above the historical mean. The maximum discharge was 1.3 times higher than that recorded in 2016 (pre-seismic conditions); the latter was preceded by about four years of very high water surplus values, which did not produce the high spring flows recorded in 2018 (Figure 4b). After about five years from the seismic sequence, the comparison of pre- and post-seismic recession curves can contribute to understanding what processes are taking place and whether or not permanent changes have taken place. Table 3 shows the recession processes and recession coefficients computed on the available recession periods. Figure 7 shows all the available pre- and post-seismic recession curves, sorted from the highest discharge to the lowest.

Table 3. Recession models that best describe the Nera River discharges (Section I1 in Figure 2) during the available recession periods.

Year	Recession Period	Recession Model	Recession Coefficient
1933	7 July–15 September	EF	-2.5×10^{-3} d^{-1}
1935	4 June–23 August	EF	-2.2×10^{-3} d^{-1}
1936	1 June–21 June	EF	-2.9×10^{-3} d^{-1}
1938	22 June–12 July	EF	-3.2×10^{-3} d^{-1}
2016	19 July–18 August	EF	-2.8×10^{-3} d^{-1}
2017	13 June–6 August	SL	-2274 m^3/d^2
2018	27 June–27 July	SL	-2267 m^3/d^2
2019	16 August–29 September	EF	-2.1×10^{-3} d^{-1}
2020	10 July–14 August	EF	-2.3×10^{-3} d^{-1}

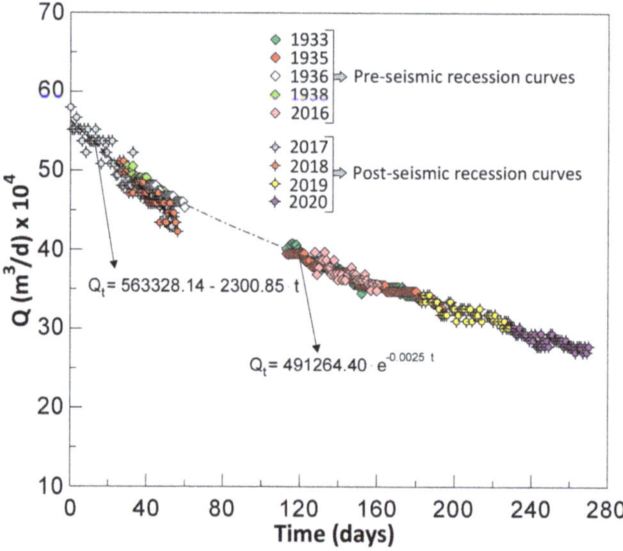

Figure 7. Pre- and post-seismic recession curves for the Nera River at section I1. Discharge data are in the supplementary materials (Table S1).

Curves with similar recession processes and recession coefficients have been merged using the strip method [80–82]. This approach allows obtaining a Master Recession Curve (MRC); the Root Mean Square Error (RMSE) criterion has been used to choose the best time-shifting to join the curves. A recent study carried out by Di Matteo et al. [24] highlighted that the SL equation described the recession curves of the two years following the seismic sequence (2017 and 2018). A permeability enhancement of the aquifer feeding the Nera River—due to cleaning of fractures and the co-seismic fracturing in the recharge area—coupled with an increase in groundwater flow velocity can explain this process, which lasted up to 2018. Fronzi et al. [26] also recorded consistent variations in the hydraulic conductivity distribution throughout carbonate aquifers related to the VBF system activation. Figure 7 also includes the most recent recession curves (2019 and 2020), which fit very well with the pre-seismic ones (Equation (11)), represented by the EF equation (Darcian flow).

$$Q_t = Q_o \cdot e^{-\alpha_E \cdot t} = 491{,}264.40 \cdot e^{-0.0025 \cdot t} \quad \left(R^2 = 0.98\right) \tag{11}$$

Considering that the recession process recovered to the same pre-earthquake conditions—at least for two consecutive years (2019 and 2020)—and α_E values are similar to the pre-seismic ones (Table 1), it can be deduced that hydrogeological changes induced by the seismic sequence seem to be almost recovered. As Equation (10) shows, co-seismic effects could have changed other parameters than aquifer hydraulic conductivity. It is interesting to point out that if the length of the one-dimensional domain (L) had increased (i.e., the recharge area had enlarged due to aquifer breaching), we should have expected much lower αE than pre-earthquake conditions. The permeability recovery after high magnitude earthquakes has been documented in a wide range of hydrogeological settings; most documented recovery times are about two years as obtained for the Nera River [27,28,83]. As Manga et al. [84] discussed, the mechanism of recovery is not clear because of the complexity and inaccessibility of the subsurface; a combination of mechanical and geochemical processes can cause the return of the enhanced permeability to approximately the pre-seismic value. Aben et al. [85] reported a list of possible recovery mechanisms of earthquake-related fracture damage including time-dependent mechanical recovery of micro and macro-cracks, sealing by mineral precipitation (e.g., calcite), resulting in fracture sealing, and fracture closure by pressure solution creep, the latter two act on longer time scales than mechanical ones. Based on the discharge at Visso gauge (I1), regardless of the mechanism that led to the recovery of the hydrogeological properties of the aquifer feeding the Nera River, it is important to emphasize again that, the emptying process seems to be restored to the same as it was before the 2016 earthquake. Since the reservoir in 2019 and 2020 empties at rates lower than 2017 and 2018, the hydrogeological system—as it was in the past—is less vulnerable to prolonged droughts such as those that are increasingly affecting the Apennine area of Central Italy.

6. Conclusions

The 2016 seismic sequence deeply affected the hydrogeological system of the upper part of the Nera River catchment—the waters of which are of strategic importance to the territory's economy. Through analysis of the geological-structural and hydrogeological data about five years after the seismic sequence, the following conclusions can be drawn:

- The peak river flow recorded in 2017 and 2018 did not occur in 2019 and 2020—the latter was preceded by a drought period well detected by the SPI, which affected the recharge of groundwater, and consequently the river discharge;
- The rapid depletion observed during the 2017 and 2018 recession periods (straight line equation and turbulent flow) were not detected in 2019 and 2020, where both recession process and recession coefficients seem to be restored to the same as prior to the 2016 earthquake (exponential equation and Darcian flow);
- Co-seismic effects on the hydrogeological system (e.g., increased aquifer permeability and pore water pressure) appear to have recovered after two years from the 2016 seismic sequence, as documented in other systems around the world affected by strong earthquakes.

Based on the discharge data representing the whole hydrogeological system feeding the Nera River, the findings reported here contribute to understanding processes and dynamics of fractured carbonate aquifers located in geologically and climatically complex regions, which are useful for water management. The results need to be further refined by continuing the monitoring and also with consideration of the response of hydrogeological systems located east of the Apennine chain.

Supplementary Materials: The following are available online at https://www.mdpi.com/article/10.3390/hydrology8030097/s1, Table S1: discharge data for the selected pre-and post-recession periods.

Author Contributions: Conceptualization, L.D.M. and C.P.; methodology, L.D.M., A.C., M.P. and C.P.; validation, L.D.M., A.C., M.P. and C.P.; formal analysis, L.D.M. and A.C.; investigation, L.D.M., A.C., M.P. and C.P.; data curation, L.D.M., A.C., M.P. and C.P.; writing—original draft preparation, L.D.M., A.C., M.P. and C.P.; writing—review and editing, L.D.M., A.C., M.P. and C.P.; visualization, L.D.M., A.C., M.P. and C.P.; supervision, L.D.M.; project administration, L.D.M. All authors have read and agreed to the published version of the manuscript.

Funding: This research received no external funding.

Institutional Review Board Statement: Not applicable.

Informed Consent Statement: Not applicable.

Data Availability Statement: Data are contained in the supplementary material.

Acknowledgments: The authors wish to thank the *Marche Region—Civil Protection Service* for discharge data.

Conflicts of Interest: The authors declare no conflict of interest.

References

1. Winter, T.C.; Harvey, J.W.; Franke, O.L.; Alley, W.M. *Ground Water and Surface Water: A Single Resource*; US Geological Survey: Denver, CO, USA, 1999; Volume 1139, pp. 1–88. Available online: https://pubs.usgs.gov/circ/circ1139 (accessed on 15 May 2021).
2. Sophocleous, M. Interactions between groundwater and surface water: The state of the science. *Hydrogeol. J.* **2002**, *10*, 52–67. [CrossRef]
3. Jolly, I.D.; McEwan, K.L.; Holland, K.L. A review of groundwater–surface water interactions in arid/semi-arid wetlands and the consequences of salinity for wetland ecology. *Ecohydrology* **2008**, *1*, 43–58. [CrossRef]
4. Fleckenstein, J.H.; Krause, S.; Hannah, D.M.; Boano, F. Groundwater-surface water interactions: New methods and models to improve understanding of processes and dynamics. *Adv. Water Res.* **2010**, *33*, 1291–1295. [CrossRef]
5. Li, M.; Liang, X.; Xiao, C.; Cao, Y. Quantitative Evaluation of Groundwater–Surface Water Interactions: Application of Cumulative Exchange Fluxes Method. *Water* **2020**, *12*, 259. [CrossRef]
6. Guzman, P.; Batelaan, O.; Wyseure, G. Comparative analysis of base flow recession curves for different Andean catchments. *Geophys. Res. Abs.* **2012**, *14*, 8318. Available online: https://meetingorganizer.copernicus.org/EGU2012/EGU2012-8318.pdf (accessed on 18 May 2021).
7. Berghuijs, W.; Woods, R.; Hrachowitz, M. A precipitation shift from snow towards rain leads to a decrease in streamflow. *Nat. Clim. Chang.* **2014**, *4*, 583–586. [CrossRef]
8. Boni, C.; Bono, P.; Capelli, G. Schema idrogeologico dell'Italia Centrale. *Mem. Soc. Geol. It.* **1986**, *35*, 991–1012. (In Italian). Available online: https://www.idrogeologiaquantitativa.it/wordpress/wp-content/uploads/2009/11/Pubb_1986_Schema_Italia_Centrale.pdf (accessed on 18 May 2021).
9. Cencetti, C.; Dragoni, W.; Nejad Massoum, M. Contributo alle conoscenze delle caratteristiche idrogeologiche del Fiume Nera (Appennino centro-settentrionale). *Geol. Appl. Idrogeol.* **1989**, *24*, 191–210.
10. Preziosi, E.; Romano, E. From a hydrostructural analysis to the mathematical modelling of regional aquifers (Central Italy). *It. J. Eng. Geol. Environ.* **2009**, *1*, 183–198. [CrossRef]
11. Boni, C.; Baldoni, T.; Banzato, F.; Cascone, D.; Petitta, M. Hydrogeological study for identification, characterization and management of groundwater resources in the Sibillini Mountains National Park (Central Italy). *It. J. Eng. Geol. Environ.* **2010**, *2*, 21–39. [CrossRef]
12. Altava-Ortiz, V.; Llasat, M.C.; Ferrari, E.; Atencia, A.; Sirangelo, B. Monthly rainfall changes in Central and Western Mediterranean basins, at the end of the 20th and beginning of the 21st centuries. *Int. J. Climatol.* **2011**, *31*, 1943–1958. [CrossRef]
13. Cambi, C.; Valigi, D.; Di Matteo, L. Hydrogeological study of data-scarce limestone massifs: The case of Gualdo Tadino and Monte Cucco structures (Central Apennines, Italy). *Bollet. Geofisica Teorica Appl.* **2010**, *51*, 345–360.
14. Longobardi, A.; Villani, P. Trend analysis of annual and seasonal rainfall time series in the Mediterranean area. *Int. J. Climatol.* **2010**, *30*, 1538–1546. [CrossRef]

15. Di Matteo, L.; Valigi, D.; Cambi, C. Climatic characterization and response of water resources to climate change in limestone areas: Some considerations on the importance of geological setting. *J. Hydrol. Eng.* **2013**, *18*, 773–779. [CrossRef]
16. Diodato, N.; Büntgen, U.; Bellocchi, G. Mediterranean winter snowfall variability over the past millennium. *Int. J. Climatol.* **2019**, *39*, 384–394. [CrossRef]
17. Gentilucci, M.; Barbieri, M.; Lee, H.S.; Zardi, D. Analysis of rainfall trends and extreme precipitation in the Middle Adriatic Side, Marche Region (Central Italy). *Water* **2019**, *11*, 1948. [CrossRef]
18. Caporali, E.; Lompi, M.; Pacetti, T.; Chiarello, V.; Fatichi, S. A review of studies on observed precipitation trends in Italy. *Int. J. Climatol.* **2021**, *41*, E1–E25. [CrossRef]
19. Caloiero, T.; Caroletti, G.N.; Coscarelli, R. IMERG-Based Meteorological Drought Analysis over Italy. *Climate* **2021**, *9*, 65. [CrossRef]
20. Valigi, D.; Luque Espinar, J.A.; Di Matteo, L.; Cambi, C.; Pardo Iguzquiza, E.; Rossi, M. Analysis of drought conditions and their effects on Lake Trasimeno (Central Italy) levels. *It. J. Groundwater* **2016**, *17–215*, 39–47. [CrossRef]
21. Diodato, N.; Bellocchi, G. Climate control on snowfall days in peninsular Italy. *Theor. Appl. Climatol.* **2020**, *140*, 951–961. [CrossRef]
22. Petitta, M.; Mastrorillo, L.; Preziosi, E.; Banzato, F.; Barberio, M.D.; Billi, A.; Cambi, C.; De Luca, G.; Di Carlo, G.; Di Curzio, D.; et al. Water table and discharge changes associated with the 2016–2017 seismic sequence in central Italy: Hydrogeological data and a conceptual model for fractured carbonate aquifers. *Hydrogeol. J.* **2018**, *26*, 1009–1026. [CrossRef]
23. Mastrorillo, L.; Saroli, M.; Viaroli, S.; Banzato, F.; Valigi, D.; Petitta, M. Sustained post-seismic effects on groundwater flow in fractured carbonate aquifers in Central Italy. *Hydrol. Proc.* **2020**, *34*, 1167–1181. [CrossRef]
24. Di Matteo, L.; Dragoni, W.; Azzaro, S.; Pauselli, C.; Porreca, M.; Bellina, G.; Cardaci, W. Effects of earthquakes on the discharge of groundwater systems: The case of the 2016 seismic sequence in the Central Apennines, Italy. *J. Hydrol.* **2020**, *583*, 124509. [CrossRef]
25. Valigi, D.; Fronzi, D.; Cambi, C.; Beddini, G.; Cardellini, C.; Checcucci, R.; Mastrorillo, L.; Mirabella, F.; Tazioli, A. Earthquake-induced spring discharge modifications: The Pescara di Arquata spring reaction to the august–october 2016 Central Italy earthquakes. *Water* **2020**, *12*, 767. [CrossRef]
26. Fronzi, D.; Di Curzio, D.; Rusi, S.; Valigi, D.; Tazioli, A. Comparison between Periodic Tracer Tests and Time-Series Analysis to Assess Mid-and Long-Term Recharge Model Changes Due to Multiple Strong Seismic Events in Carbonate Aquifers. *Water* **2020**, *12*, 3073. [CrossRef]
27. Manga, M.; Rowland, J.C. Response of Alum Rock springs to the October 30, 2007 earthquake and implications for the origin of increased discharge after earthquakes. *Geofluids* **2009**, *9*, 237–250. [CrossRef]
28. Geballe, Z.M.; Wang, C.-Y.; Manga, M. A permeability-change model for water level changes triggered by teleseismic waves. *Geofluids* **2011**, *11*, 302–308. [CrossRef]
29. Rojstaczer, S.; Wolf, S.; Michel, R. Permeability enhancement in the shallow crust as a cause of earthquake induced hydrological changes. *Nature* **1995**, *373*, 237–239. [CrossRef]
30. Wang, C.Y.; Manga, M. New streams and springs after the 2014 Mw 6.0 South Napa earthquake. *Nat. Commun.* **2015**, *6*, 7597. [CrossRef] [PubMed]
31. Binda, G.; Pozzi, A.; Michetti, A.M.; Noble, P.J.; Rosen, M.R. Towards the understanding of hydrogeochemical seismic responses in karst aquifers: A retrospective meta-analysis focused on the Apennines (Italy). *Minerals* **2020**, *10*, 1058. [CrossRef]
32. Pierantoni, P.; Deiana, G.; Galdenzi, S. Stratigraphic and structural features of the Sibillini mountains (Umbria-Marche Apennines, Italy). *It. J. Geosci.* **2013**, *132*, 497–520. [CrossRef]
33. Boscherini, A.; Checcucci, R.; Natale, G.; Natali, N. Carta Idrogeologica della Regione Umbria (scala 1:100.000). *Giornale Geol. Appl.* **2005**, *2*, 399–404. [CrossRef]
34. Civita, M. *Idrogeologia Applicata e Ambientale*; Casa editrice ambrosiana: Milano, Italy, 2005; p. 800. (In Italian)
35. Viaroli, S.; Mirabella, F.; Mastrorillo, L.; Angelini, S.; Valigi, D. Fractured carbonate aquifers of Sibillini Mts. (Central Italy). *J. Maps* **2021**, *17*, 140–149. [CrossRef]
36. Valigi, D.; Cambi, C.; Checcucci, R.; Di Matteo, L. Transmissivity Estimates by Specific Capacity Data of Some Fractured Italian Carbonate Aquifers. *Water* **2021**, *13*, 1374. [CrossRef]
37. Mazzoli, S.; Pierantoni, P.P.; Borraccini, F.; Paltrinieri, W.; Deiana, G. Geometry, segmentation pattern and displacement variations along a major Apennine thrust zone, central Italy. *J. Struct. Geol.* **2005**, *27*, 1940–1953. [CrossRef]
38. Porreca, M.; Minelli, G.; Ercoli, M.; Brobia, A.; Mancinelli, P.; Cruciani, F.; Giorgetti, C.; Carboni, F.; Mirabella, F.; Cavinato, G.; et al. Seismic reflection profiles and subsurface geology of the area interested by the 2016–2017 earthquake sequence (Central Italy). *Tectonics* **2018**, *37*, 1116–1137. [CrossRef]
39. Koopman, A. Detachment Tectonics in the Central Apennines, Italy. Ph.D. Thesis, Instituut voor Aardwetenschappen RUU, 1983. Available online: http://dspace.library.uu.nl/bitstream/handle/1874/216947/Koopman-Anton-30-1983.pdf?sequence=1&isAllowed=y (accessed on 31 May 2021).
40. Centamore, E.; Adamoli, L.; Berti, D.; Bigi, G.; Bigi, S.; Casnedi, R.; Cantalamessa, G.; Fumanti, F.; Morelli, C.; Micarelli, A.; et al. Carta geologica dei bacini della Laga e del Cellino e dei rilievi carbonatici circostanti (Marche meridionali, Lazio nordorientale, Abruzzo settentrionale). *Stud. Geol. Camert.* **1992**, *2*, Tavola 1. Available online: http://193.204.8.201:8080/jspui/bitstream/1336/782/1/Vol.%2091-2%20Cap.%2018%20Allegato%202.pdf (accessed on 31 May 2021).

41. Lavecchia, G.; Brozzetti, F.; Barchi, M.; Menichetti, M.; Keller, J.V. Seismotectonic zoning in east-central Italy deduced from an analysis of the Neogene to present deformations and related stress fields. *GSA Bullet.* **1994**, *106*, 1107–1120. [CrossRef]
42. Porreca, M.; Fabbrizzi, A.; Azzaro, S.; Pucci, S.; Del Rio, L.; Pierantoni, P.P.; Giorgetti, C.; Roberts, G.P.; Barchi, M.R. 3D geological reconstruction of the M. Vettore seismogenic fault system (Central Apennines, Italy): Cross-cutting relationship with the M. Sibillini thrust. *J. Struct. Geol.* **2020**, *131*, 103938. [CrossRef]
43. Tarragoni, C. Determinazione della "quota isotopica" del bacino di alimentazione delle principali sorgenti dell'alta Valnerina. *Geol. Romana* **2006**, *39*, 55–62.
44. De Guidi, G.; Vecchio, A.; Brighenti, F.; Caputo, R.; Carnemolla, F.; Di Pietro, A.; Lupo, M.; Maggini, M.; Marchese, S.; Messina, D.; et al. Brief communication: Co-seismic displacement on 26 and 30 October 2016 (M_W = 5.9 and 6.5)–earthquakes in central Italy from the analysis of a local GNSS network. *Nat. Hazards Earth Syst. Sci.* **2017**, *17*, 1885–1892. [CrossRef]
45. Valerio, E.; Tizzani, P.; Carminati, E.; Doglioni, C.; Pepe, S.; Petricca, P.; De Luca, C.; Bignami, C.; Solaro, G.; Castaldo, R.; et al. Ground Deformation and Source Geometry of the 30 October 2016 Mw 6.5 Norcia Earthquake (Central Italy) Investigated Through Seismological Data, DInSAR Measurements, and Numerical Modelling. *Remote Sens.* **2018**, *10*, 1901. [CrossRef]
46. EMERGEO Working Group. Coseismic effects of the 2016 Amatrice seismic sequence: First geological results. *Ann. Geophys.* **2016**, *59*, Fast Track 5. [CrossRef]
47. Aringoli, D.; Farabollini, P.; Giacopetti, M.; Materazzi, M.; Paggi, S.; Pambianchi, G.; Pierantoni, P.P.; Pistolesi, E.; Pitts, A.; Tondi, E. The August 24th 2016 Accumoli earthquake: Surface faulting and Deep-Seated Gravitational Slope Deformation (DSGSD) in the Monte Vettore area. *Ann. Geophys.* **2016**, *59*, Fast Track 5. [CrossRef]
48. Lavecchia, G.; Castaldo, R.; de Nardis, R.; De Novellis, V.; Ferrarini, F.; Pepe, S.; Brozzetti, F.; Solaro, G.; Cirillo, D.; Bonano, M.; et al. Ground deformation and source geometry of the 24 August 2016 Amatrice earthquake (Central Italy) investigated through analytical and numerical modeling of DInSAR measurements and structural-geological data. *Geophys. Res. Lett.* **2016**, *43*, 389–398. [CrossRef]
49. Livio, F.; Michetti, A.M.; Vittori, E.; Gregory, L.; Wedmore, L.; Piccardi, L.; Tondi, E.; Roberts, G.; and Central Italy Earthquake Working Grooup. Surface faulting during the August 24, 2016, Central Italy earthquake (Mw 6.0): Preliminary results. *Ann. Geophys.* **2016**, *59*, Fast Track 5. [CrossRef]
50. Galli, P.; Galadini, F.; Pantosti, D. Twenty years of paleoseismology in Italy. *Earth Sci. Rev.* **2008**, *88*, 89–117. [CrossRef]
51. Pucci, S.; De Martini, P.M.; Civico, R.; Villani, F.; Nappi, R.; Ricci, T.; Azzaro, R.; Brunori, C.A.; Caciagli, M.; Cinti, F.R.; et al. Coseismic ruptures of the24 August 2016, Mw 6.0 Amatrice earthquake (central Italy). *Geophys. Res. Lett.* **2017**, *44*. [CrossRef]
52. Civico, R.; Pucci, S.; Villani, F.; Pizzimenti, L.; De Martini, P.M.; Nappi, R.; Wedmore, L. Surface ruptures following the 30 October 2016 Mw 6.5 Norcia earthquake, central Italy. *J. Maps* **2018**, *14*, 151–160. [CrossRef]
53. Villani, F.; Pucci, S.; Civico, R.; De Martini, P.M.; Cinti, F.R.; Pantosti, D. Surface faulting of the 30 October 2016 Mw 6.5 central Italy earthquake: Detailed analysis of a complex coseismic rupture. *Tectonics* **2018**, *37*, 3378–3410. [CrossRef]
54. Michele, M.; Chiaraluce, L.; Di Stefano, R.; Waldhauser, F. Fine-Scale Structure of the 2016–2017 Central Italy Seismic Sequence From Data Recorded at the Italian National Network. *J. Geophys. Res. Solid Earth* **2020**, *125*, e2019JB018440. [CrossRef]
55. Barchi, M.R.; Carboni, F.; Michele, M.; Ercoli, M.; Giorgetti, C.; Porreca, M.; Azzaro, S.; Chiaraluce, L. The influence of subsurface geology on the distribution of earthquakes during the 2016—2017 Central Italy seismic sequence. *Tectonophysics* **2021**, *807*, 228797. [CrossRef]
56. Chiaraluce, L.; Di Stefano, R.; Tinti, E.; Scognamiglio, L.; Michele, M.; Casarotti, E.; Cattaneo, M.; De Gori, P.; Chiarabba, C.; Monachesi, G.; et al. The 2016 central Italy seismic sequence: A first look at the mainshocks, aftershocks, and source models. *Seismol. Res. Lett.* **2017**, *88*, 757–771. [CrossRef]
57. Searcy, J.K.; Hardison, C.H. *Double-Mass Curves*; US Government Printing Office: Washington, DC, USA, 1960; Volume 1541, p. 66.
58. Vernimmen, R.R.E.; Hooijer, A.; Aldrian, E.; Van Dijk, A.I.J.M. Evaluation and bias correction of satellite rainfall data for drought monitoring in Indonesia. *Hydrol. Earth Syst. Sci.* **2012**, *16*, 133–146. [CrossRef]
59. Di Matteo, L.; Dragoni, W.; Maccari, D.; Piacentini, S.M. Climate change, water supply and environmental problems of headwaters: The paradigmatic case of the Tiber, Savio and Marecchia rivers (Central Italy). *Sci. Total Environ.* **2017**, *598*, 733–748. [CrossRef] [PubMed]
60. Navarro, A.; García-Ortega, E.; Merino, A.; Sánchez, J.L.; Kummerow, C.; Tapiador, F.J. Assessment of IMERG precipitation estimates over Europe. *Remote Sens.* **2019**, *11*, 2470. [CrossRef]
61. Saouabe, T.; El Khalki, E.M.; Saidi, M.E.M.; Najmi, A.; Hadri, A.; Rachidi, S.; Jadoud, M.; Tramblay, Y. Evaluation of the GPM-IMERG precipitation product for flood modeling in a semi-arid mountainous basin in Morocco. *Water* **2020**, *12*, 2516. [CrossRef]
62. McKee, T.B.; Doesken, N.J.; Kleist, J. The relationship of drought frequency and duration to time scales. In Proceedings of the 8th Conference on Applied Climatology, Anaheim, CA, USA, 17–22 January 1993; Volume 17, pp. 179–183.
63. WMO-World Meteorological Organization. Standardized precipitation index user guide. M.; Svoboda, M. Hayes and D. Wood. (WMO-No. 1090), Geneva. Available online: www.wamis.org/agm/pubs/SPI/WMO_1090_EN.pdf (accessed on 25 May 2021).
64. Thornthwaite, C.W.; Mather, J.R. The water balance. Centerton: Drexel institute of technology, laboratory of climatology. *Publ. Climatol.* **1955**, *8*, 104.
65. Shuttleworth, W.J. Putting the "vap" into evaporation. *Hydrol. Earth Syst. Sci.* **2007**, *11*, 210–244. [CrossRef]

66. Čadro, S. Excel sheet for Potential Evapotranspiration (PET) and Soil Water Balance Calculation Based on Thornthwaite Method (1948). Available online: https://www.researchgate.net/profile/Sabrija_Cadro/publication/309740661_Thornthwaite_Potential_Evapotranspiration_PET_and_Water_Balance_1948/data/582187e808aeccc08af8d4eb/Thornthwaite-Evapotranspiration-PET-and-Water-Balance-1948.xlsx (accessed on 25 May 2021).
67. D.R. 35/2018. Fiume Nera–Derivazione Dalle Opere di Captazione Presso la Sorgente San Chiodo. AATO 3 Marche Centro–Società Acquedotto del Nera (MC). Marche Region Law with Subsequent Modifications and Additions. Available online: http://www.ato3marche.it/assemblea-di-ambito/atti-e-documenti-assemblea-di-ambito/decreti-del-presidente/2018-2/145 5-decreto-del-presidente-n-12-2018-del-08-06-2018/file (accessed on 25 May 2021).
68. Boussinesq, J. Essai sur la théorie des eaux courantes du movement non permanent des eaux souterraines. *Acad. Sci. Inst. Fr.* **1877**, *23*, 252–260.
69. Maillet, E. *Essais D'hydraulique Souterraine et Fluviale*; Librairie Sci.: Paris, France, 1905; p. 218.
70. Scanlon, B.R.; Mace, R.E.; Barrett, M.E.; Smith, B. Can we simulate regional groundwater flow in a karst system using equivalent media models? Case study Barton springs Edwards Aquifer, USA. *J. Hydrol.* **2003**, *276*, 137–158. [CrossRef]
71. Rehrl, C.; Birk, S. Hydrogeological characterisation and modelling of spring catchments in a changing environment. *Aust. J. Earth Sci.* **2010**, *103*, 106–117. [CrossRef]
72. Dragoni, W.; Mottola, A.; Cambi, C. Modeling the effects of pumping wells in spring management: The case of Scirca spring (Central Apennines, Italy). *J. Hydrol.* **2013**, *493*, 115–123. [CrossRef]
73. Coutagne, A. Les variations de dèbit en pèriode non influencèe par les prècipitations. In *Le dèbit d'Inflitration (Corrèlations Fluviales Internes)–2me Partie, Meteorologie et Hydrologie*; La Houille Blanche: Grenoble, France, 1948; pp. 416–436.
74. Bonacci, O. *Karst Hydrology With Special Reference to the Dinaric Karst*; Springer: Heidelberg, Germany, 1987; p. 184.
75. Rorabough, M.I. Estimating changes in bank storage and grounwater contribution to streamflow. *Int. Assoc. Sci. Hydro. Publ.* **1964**, *63*, 432–441.
76. Kovács, A.; Perrochet, P.A. quantitative approach to spring hydrograph decomposition. *J. Hydrol.* **2008**, *352*, 16–29. [CrossRef]
77. Barberio, M.D.; Barbieri, M.; Billi, A.; Doglioni, C.; Petitta, M. Hydrogeochemical changes before and during the 2016 Amatrice-Norcia seismic sequence (central Italy). *Sci. Rep.* **2017**, *7*, 11735. [CrossRef]
78. Rosen, M.R.; Binda, G.; Archer, C.; Pozzi, A.; Michetti, A.M.; Noble, P.J. Mechanisms of earthquake-induced chemical and fluid transport to carbonate groundwater springs after earthquakes. *Water Res. Res.* **2018**, *54*, 5225–5244. [CrossRef]
79. Fronzi, D.; Mirabella, F.; Cardellini, C.; Caliro, S.; Palpacelli, S.; Cambi, C.; Valigi, D.; Tazioli, A. The Role of Faults in Groundwater Circulation before and after Seismic Events: Insights from Tracers, Water Isotopes and Geochemistry. *Water* **2021**, *13*, 1499. [CrossRef]
80. Singh, V.P. *Hydrologic Systems: Watershed Modeling*; Prentice-Hall: Denver, CO, USA, 1989; p. 448.
81. Tallaksen, L.M. A review of baseflow recession analysis. *J. Hydrol.* **1995**, *165*, 349–370. [CrossRef]
82. Posavec, K.; Parlov, J.; Nakić, Z. Fully automated objective-based method for master recession curve separation. *Groundwater* **2010**, *48*, 598–603. [CrossRef]
83. Elkhoury, J.E.; Brodsky, E.E.; Agnew, D.C. Seismic waves increase permeability. *Nature* **2006**, *441*, 1135–1138. [CrossRef] [PubMed]
84. Manga, M.; Beresnev, I.; Brodsky, E.E.; Elkhoury, J.E.; Elsworth, D.; Ingebritsen, S.E.; Mays, D.C.; Wang, C.Y. Changes in permeability caused by transient stresses: Field observations, experiments, and mechanisms. *Rev. Geophys.* **2012**, *50*, 1–24. [CrossRef]
85. Aben, F.M.; Doan, M.L.; Gratier, J.P.; Renard, F. Experimental postseismic recovery of fractured rocks assisted by calcite sealing. *Geophys. Res. Lett.* **2017**, *44*, 7228–7238. [CrossRef]

 *hydrology*

Article

Water Budget Analysis Considering Surface Water–Groundwater Interactions in the Exploitation of Seasonally Varying Agricultural Groundwater

Sun Woo Chang and Il-Moon Chung *

Department of Land, Water and Environmental Research, Korea Institute of Civil Engineering and Building Technology, Goyang 10223, Korea; chang@kict.re.kr
* Correspondence: imchung@kict.re.kr; Tel.: +82-31-910-0334

Abstract: In South Korea, groundwater intended for use in greenhouse cultivation is collected from shallow riverside aquifers as part of agricultural activities during the winter season. This study quantified the effects of intensive groundwater intake on aquifers during the winter and examined the roles of nearby rivers in this process. Observation data were collected for approximately two years from six wells and two river-level observation points on the study site. Furthermore, the river water levels before and after the weir structures were examined in detail, because they are determined by artificial structures in the river. The structures have significant impacts on the inflow and outflow from the river to the groundwater reservoirs. As a result, a decline in groundwater levels owing to groundwater depletion was observed during the water curtain cultivation (WCC) period in the winter season. In addition, we found that the groundwater level increased owing to groundwater recharge due to rainfall and induced recharge by rivers during the spring–summer period after the end of the WCC period. MODFLOW, a three-dimensional difference model, was used to simulate the groundwater level decreases and increases around the WCC area in Cheongwon-gun. Time-variable recharge data provided by the soil and water assessment tool model, SWAT for watershed hydrology, was used to determine the amount of groundwater recharge that was input to the groundwater model. The groundwater level time series observations collected from observation wells during the two-year simulation period (2012 to 2014) were compared with the simulation values. In addition, to determine the groundwater depletion of the entire demonstration area and the sustainability of the WCC, the quantitative water budget was analyzed using integrated hydrologic analysis. The result indicated that a 2.5 cm groundwater decline occurred on average every year at the study site. Furthermore, an analysis method that reflects the stratification and boundary conditions of underground aquifers, hydrogeologic properties, hydrological factors, and artificial recharge scenarios was established and simulated with injection amounts of 20%, 40%, and 60%. This study suggested a proper artificial recharge method of injecting water by wells using riverside groundwater in the study area.

Keywords: water curtain cultivation; surface–groundwater interaction; MODFLOW; water budget analysis

Citation: Chang, S.W.; Chung, I.-M. Water Budget Analysis Considering Surface Water–Groundwater Interactions in the Exploitation of Seasonally Varying Agricultural Groundwater. *Hydrology* **2021**, *8*, 60. https://doi.org/10.3390/hydrology8020060

Academic Editor: Okke Batelaan

Received: 21 February 2021
Accepted: 24 March 2021
Published: 2 April 2021

Publisher's Note: MDPI stays neutral with regard to jurisdictional claims in published maps and institutional affiliations.

1. Introduction

Water is one of the most important resources on Earth and water used for irrigation accounts for a significant portion of the global water demand [1]. For stable food production, water supply methods are being diversified, and water usage is gradually increasing. Recently, the rate at which groundwater is consumed for irrigation has been increasing rapidly. Moreover, the continuous consumption of water resources has worsened the drought condition worldwide [2]. In many countries, excessive use of groundwater from shallow aquifers is perceived as the main cause of reduced available surface water supply [3]. Consequently, groundwater changes attributed to aquifer depletion have been investigated, and various predictions have been made in sites with large extraction wells [4–6]. Groundwater withdrawal from deep aquifers leads to permanent depletion except in cases where the groundwater is recharged by high-intensity rainfall, which occurs occasionally [7]. In contrast, groundwater in shallow aquifers has different characteristics, because it can be renewed by nearby rivers or rainfall. Moreover, continuous large-scale depletion varies near the river environment owing to groundwater–surface-water interaction. Thus, shallow aquifers undergo spatiotemporal changes owing to various factors; seasonality is an indispensable element in understanding the underground water resources of shallow aquifers. In monsoon or semi-arid climates, rainfall is concentrated during one period in a year. During these periods, aquifers are recharged. Furthermore, anthropogenic activities, which are the main causes of water depletion, also exhibit seasonality. In many areas, the abstraction of groundwater may increase during the dry season [8–10]. Two opposite activities (groundwater recharge and abstraction from shallow aquifers) occur in accordance with the seasonal cycle. This is observed in large agricultural areas in monsoon climates. For example, the groundwater level of the Ganges–Brahmaputra–Meghna (GBM) delta in Bangladesh is decreasing at an annual rate of 0.1–0.5 m/year owing to intensive irrigation. In the Dhaka area, where the degree of groundwater usage is high, the groundwater is decreasing at a maximum speed of 1 m/year [9]. The recharge of shallow aquifers is indispensable for the sustainable groundwater use. However, Perrin [10] found that the locally heterogeneous pattern of groundwater depletion intensifies owing to seasonal characteristics, and that the imbalance worsens during low-rainfall years. Even in monsoon areas where precipitation is relatively abundant, the annual pattern of groundwater use varies considerably depending on the degree of yearly precipitation. For example, Pavelic et al. [8] investigated the upper Bhima River basin in India and reported that less than 10% of pumping wells became dry during the off-monsoon season dry if there was large precipitation the previous year, whereas 40% of pumping wells became dry if the prior year had low rainfall. The seasonal precipitation pattern causes short-term reliability issues rather than long-term groundwater depletion, as they influence food production. It has been found that extreme water shortage caused by extreme droughts in conjunction with climate changes also causes problems in shallow aquifers [11]. Thus, even rechargeable groundwater areas are influenced by factors that cause or interfere with recharging (i.e., adjacent geographic characteristics, climate, relationship with rivers, and use of public water). Therefore, a highly precise analysis that considers multiple hydrological and geologic parameters, as well as site-specific conditions, is required to determine whether the groundwater in the study area is a sustainably available resource.

Groundwater resources located along rivers are influenced considerably by hydrometeorological effects, such as changes in precipitation and natural recharge amounts, use of surrounding lands, various river environments, maintenance status of riverside facilities, and agricultural water intake methods. To identify the water circulation state of riverside groundwater under the complex effects of natural and artificial environments and to prepare for water resource depletion and pollution of the water environment, the behavior of each hydrological component should be identified. In addition, an unbiased quantitative evaluation of each hydrological component of the corresponding groundwater system must be obtained using raw data collected from the site of interest.

Ruud et al. [12] named the semi-arid basin where intensive groundwater development is performed as an "active basin" and pointed out analytical errors caused by unmeasurable or inaccurate data when analyzing the water budget of an active basin. In particular, it may be difficult to acquire data, because usage data for agricultural groundwater or private pumping wells do not exist or cannot be obtained due to privacy issues [2]. Given that the measurement of groundwater usage in a large basin-scale area increases the economic burden of research, groundwater usage is an unknown factor in many studies [12].

The seasonality of groundwater use and the problem of groundwater usage measurements are similar in South Korea to other monsoon and semi-arid climates. In 2015, the agricultural water usage of South Korea was 2.08 billion m^3. This accounted for 51% of the total groundwater usage [13]. Since 2009, domestic and industrial water usage has not changed significantly; thus, the use of agricultural water has become a major reason for the increase in total groundwater usage [14]. Greenhouse cultivation methods for farming using groundwater are becoming increasingly popular in South Korea. The demand for fresh vegetables and fruits that are difficult to store for long periods is increasing, as the demand is high throughout the year and the income level of people is rising. Although the demand and supply of off-season agricultural products throughout the year is increasing, these products are only being produced by greenhouse cultivations in most areas, because growth conditions, such as the climate, soil, and moisture, are not suitable during the winter months (November to February) in South Korea. Moreover, the greenhouse cultivation method has attracted considerable attention as a profitable method for farmers, because stable cultivation can be achieved by avoiding weather disasters and pests. Farmers can grow fresh vegetables and fruit in a short period and sell them at much higher prices. However, greenhouse cultivation requires much higher artificial energy consumption than field cultivation, because special heating or insulation are required to artificially create a suitable environment for the growth of agricultural products.

Recently, tremendous growth has been documented in the area of plastic film-covered greenhouse cultivations in most agricultural lands near rivers in South Korea. In particular, the winter facility cultivation method, which uses groundwater to maintain the warmth of the greenhouse in winter, has been adopted by many agricultural lands in South Korea. This greenhouse warming method is called water curtain cultivation (WCC). Since the first introduction of WCCs in South Korea in 1984, WCCs have been used extensively. Indicatively, in 2006, WCC facilities spanned 10,746 ha [15]. WCC is a cultivation method whereby groundwater, which maintains a constant temperature of 15 °C even in winter, is extracted and sprayed on the roof of the greenhouse to form a water curtain. The heat emitted from the water maintains the warmth of the greenhouse. WCC technology is advantageous in that it dramatically reduces the fuel cost, because no fuel is required to heat the facility. The nationwide area under WCC cultivation is growing constantly, owing to its advantages. It is not affected by oil prices, and a stable indoor temperature can be maintained. For WCC, considerable volumes of groundwater are pumped during the winter and discharged into rivers through agricultural waterways after use [16]. Most of the WCC facilities in South Korea are installed in alluvial areas near a river in anticipation of a sufficient supply of groundwater from the river. However, because WCC facilities are concentrated in a large-scale complex near rivers, excessive use of water curtain water within a limited area causes groundwater depletion, thus making it difficult to continually supply water for water curtains. Using a large amount of groundwater for 4–5 months during the winter and discharging it to nearby rivers has resulted in riverside groundwater depletion. In large WCC complexes that use large amounts of groundwater, disruptions in the water curtain water supply often occur in January every year.

This study aimed to examine the surface–groundwater interaction near a small stream called Musimcheon within the Mihocheon basin, and to quantitatively diagnose the groundwater shortage phenomenon in aquifers. Several studies have been conducted on this watershed. Chung et al. [17] analyzed the spatiotemporal distribution of groundwater recharge in a watershed based on the application of the combined surface-water–groundwater model SWAT–MODFLOW [18] for an area that spanned approximately 1868 km². Chung et al. [19] proposed a sustainable intake amount along the stream within the Mihocheon basin based on an analysis of the effect of groundwater pumping. The SWAT–MODFLOW model used in these studies was an integrated hydrological analysis model and was capable of calculating the heterogeneous recharge characteristics, simulating the hydrogeological characteristics and groundwater pumping, and evaluating the interaction between the river and groundwater. Chung et al. [17] and Chung et al. [19] used a relatively large grid size of 300 × 300 m to study the entire watershed area. The present study conducted precise groundwater level and water budget analyses on a small demonstration area of ~4 km² in the southeastern part of the Mihocheon basin in fine grids using MODFLOW. The hydrologically analyzed spatiotemporal groundwater recharge amount was used as the input component of the MODFLOW. In South Korea, the status of groundwater development and use of nearby rivers are still not adequately understood; groundwater development is not officially measured and reported in many WCC facilities that are operated by small farmers. Hence, the groundwater pumping pattern in the WCC area was identified using the correlation between the daily minimum temperature and groundwater usage of the WCC area. Precise groundwater modeling was performed in the demonstration area using the three-dimensional (3D) groundwater flow model MODFLOW [20] and the watershed hydrology model soil and water assessment tool (SWAT) [21]. To our knowledge, this is the first attempt to study the response of the surface water, as well as the shallow aquifer, to the impact of seasonal WCC activity. For this purpose, integrated hydrological analyses using the MODFLOW and SWAT models were performed to determine the long-term water budget depletion at the demonstration site based on an expansion of the range of the water budget analysis.

2. Site Description

Figure 1 shows the boundaries of the study area, the cross-section, locations of rivers, artificial structures in the rivers, and pumping wells and observation wells installed for groundwater development. Sangdae-ri, Gadeok-myeon, Cheongwon-gun, and Chungcheongbuk-do, which are demonstration study areas, belong to the Mihocheon basin in the Geum River area. Most of this watershed comprises farmlands used for agricultural activities; greenhouse areas mixed with rice fields and other crop cultivations. The target areas for this study are located in the middle to the upstream part of the Musimcheon, which is a farming area in which paddy fields and greenhouse cultivation areas (represented by recent vinyl houses) are mixed. Pumping wells for WCC and other agricultural activities are scattered throughout the cultivation area; strawberries are mainly cultivated in the vinyl houses. During the summer, the water irrigated from the river flows through the agricultural waterway in the demonstration area; almost all the water that flows through the agricultural waterway during the winter is discharged to the waterway after being used for WCC. The greenhouse cultivation sites in this area suffer temporary reductions in groundwater intake volume owing to continuous groundwater level decreases during the winter.

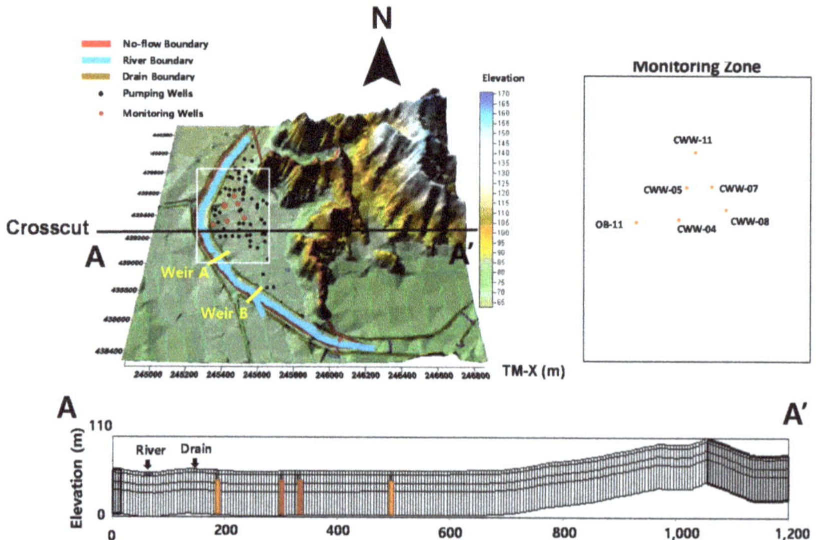

Figure 1. Model boundary, weirs, and wells within the studied area (modified from Chang and Chung [22]).

The geology of Cheongwon-gun is composed of igneous rocks, metasedimentary rocks, and alluvial layers. When the geological age, rock formation, pore shape, and hydraulic characteristics of groundwater are considered, the hydrogeological unit of this study site corresponds to intrusive igneous rock [14]. Although geological heterogeneity exists, on average, the strata are generally composed of (from the top down) a sedimentary layer, weathered soil layer, weathered rock, and soft rock layer. The soft rock layer is usually 35 m below the surface [15]. From a geological perspective, there is an alluvial layer deposited by the flooding of the river above the Cheongju granite, which is connected to the east.

The WCC demonstration area located in the most upstream part of the Musimcheon watershed is located on riverside land to the east of the stream flow created by the meandering stream of the Musimcheon from the northeast. The land near Musimcheon is relatively flat; the northern area of the study area is a low-mountainous terrain with an altitude of 100 m. There is a small reservoir at the most upstream part of the river. Sudden water discharge was not observed from the reservoir. Most rivers in South Korea are not pristine rivers, and artificial structures are installed in most of them. Small weirs are also installed at intervals spanning several kilometers along the Musimcheon. Thus, the river water level is relatively constant during most of the year, with the exception of rainy seasons. Reductions in the water flow of the Musimcheon were not observed during the study period, even during intensive groundwater pumping in the winter.

3. Study Method

3.1. Acquisition of Field Measurement Data

3.1.1. Groundwater Observation Data

According to field survey results, there were 56 motor-activated groundwater pumping wells that were active during the WCC period. The observation wells for groundwater level measurements were mainly located in the west and north, wherein many WCC houses existed among the simulated areas. The black dots in Figure 1 indicate the 56 groundwater pumping wells, and the red dots indicate the observation wells. Observation wells are located along the Musimcheon embankment in the south–north direction and seven observation wells in the WCC area that correspond to the riverside land. In this study, the calibration and validation of the numerical analysis were performed based on daily data from these observation wells equipped with sensors and data loggers. To measure the groundwater amount used by WCC, Moon et al. [23] calculated the total amount of groundwater taken up for WCC before it was discharged to the Musimcheon from the agricultural waterway in the studied site.

3.1.2. Collection of Surface Water Observation Data

The Musimcheon River in the simulation area has a relatively slow flow rate because of the effects of weirs, and its water-level changes are insignificant except during the summer rainy season. To examine the water level changes at the center of the river, the structures of the upstream and downstream weirs were identified.

The preliminary modeling results indicated that the water level and conductance of the river at the river boundary condition had a significant effect on the inflow and outflow from the river to groundwater. When setting the River Boundary as the boundary condition, the Conductance variable was used to estimate stream leakage near the aquifer. The stream leakage was calculated from the conductance multiplied by the difference in the water levels between the river and the groundwater in the aquifer. Conductance is expressed as a combination of the vertical hydraulic conductivity of the riverbed material, the length of the river, the thickness of the floor, and the width of the river. The conductance also can be interpreted as the flow resistance between the river and the aquifer. Hence, it was necessary to examine the weir structures and prepare river water levels and other data related to the river. Furthermore, because it was found that the model had a large sensitivity to most of the parts of the upstream river, Musimcheon was investigated to input the river water level data at the boundary area in addition to the river water level data at two other locations.

Figure 2 shows the locations of the major weirs. According to the time series data of the river water level measurements, upstream and downstream of weir A had an upstream elevation of 67 m and a downstream elevation of 65 m, which were verified by observations. Moreover, weir A exhibits elevation differences downstream, in addition to the one shown in this figure. The structure of weir B was also measured. Figure 2 shows a full view and conceptual crosscut of the weir.

Figure 2. Photographs of weirs (**a**) and (**b**) and their conceptual crosscut schematics.

According to the measurements, the upstream weir maintains an elevation of 67 m and the weir downstream maintains an elevation of 65 m throughout the WCC period. The identification of the structure of weir B upstream of weir A was useful in numerical modeling. The structure of weir B is vertically higher and larger than that of weir A. As a result, the river becomes wider upstream of weir B.

3.1.3. Estimation of the Usage Pattern of Groundwater for WCC

Figure 3 show the precipitation, temperature, and the estimated total pumping rate during the study period from July 2012 to July 2014.

The estimation of the groundwater usage in the WCC region in the demonstration area was conducted for approximately three months (mid-November to mid-February). During this period, which corresponds to the winter period in South Korea, the groundwater recharge rate due to precipitation was low.

The average excavation depth of the groundwater wells in the survey area was 50 m. The motors for the pumping wells in the survey area were mostly ground pumps for agricultural water used to collect groundwater in shallow aquifers. According to the Groundwater Act of South Korea, if the daily pumping capacity for agricultural and fishing purposes is less than 100 m^3/day, the usage amount may not be reported, and groundwater impact investigation is not conducted. Most motors used in the WCC site of this study corresponded to unreported pumping wells, because their pumping capacities were less than 100 m^3/day. Therefore, as conducted by Ruud et al. [12], the locations of pumping wells in the test area were investigated individually because the location and development amount of groundwater pumping for the modeling study were unknown. Furthermore, the method of Moon et al. [23] was used to estimate the groundwater usage. Moon et al. [23]

estimated that the number of days of pumping for WCC was approximately 100, and also estimated the groundwater usage for WCC during the winters of 2011 and 2012 as 53,138 m³ per 1 ha. This estimation was based on an analysis of the motor specifications of representative pumping wells and the record of the daily lowest temperatures below 0 °C for 94 days in the WCC area in Cheongwon. By using this method, we assumed a daily average discharge volume of 68 m³/day, with the assumption that the operation time of the motor and on/off periods were influenced by the daily minimum temperature. In addition, the operation time was determined to be equal to 0.68 d immediately after sunset until the sunrise of the subsequent day. This time was used as an input for the model. Furthermore, 68 pumping wells were classified into two groups. In one of the groups, the motor was operated when the temperature fell below 0 °C, and the other group was operated when the temperature fell below −5 °C. Only the motors of the first group were operated between 0 and −5 °C. When the WCC operations began in late November, only some motors were operated, because the minimum temperature was not low. At other times, all the motors installed in the demonstration site were operated, because the daily minimum temperatures were consistently lower than −5 °C. In addition, in early March, when WCC was terminated, the minimum temperature slowly increased. All of the pumping wells were no longer used when the WCC stopped.

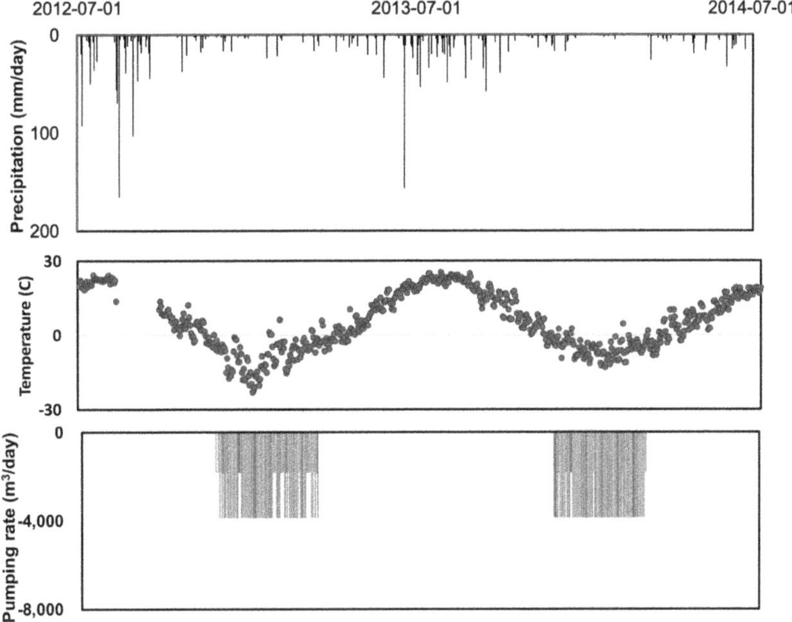

Figure 3. Precipitation, temperature, and the estimated total pumping rate from July 2012 to July 2014.

3.2. Model Description

3.2.1. Calculation of Time-Varying Groundwater Recharge Amount using Watershed Hydrology Model

This study calculated the groundwater recharge amount of the Cheongwon area during the model application period using the SWAT model [21] developed by the Agricultural Research Service (ARS) of the United States Department of Agriculture. Recently, SWAT-related models for this area have been successfully calibrated. The SWAT–MODFLOW [18], an integrated surface-water–groundwater model was applied to the Mihocheon basin (area of 1868 km²), which included the area evaluated in this study. The results were calibrated using six stream gauging stations and ten national groundwater observation wells. In a follow-up study, Chung et al. [19] calibrated the Mihocheon basin (area of 1602 km^2) with SWAT–MODFLOW to examine and ensure the development of a sustainable groundwater level based on simulations of the groundwater development scenario.

The target basin was the Musimcheon watershed in the southeast of the Mihocheon basin, which had a basin area of 198 km^2. Preprocessing was performed using the ESRI's ArcView 3.3 software. The digital elevation map was processed to 100 m grids and divided into 34 sub-basins in total, reflecting the terrain elevations. Furthermore, a land use map was created with the land cover map of the Ministry of Environment, and a soil distribution map was created based on the soil data. One hydrologic response unit was determined based on the combination of the features of these two maps, and hydrological components considering land use and land composition were determined accordingly.

The input precipitation, temperature, wind speed, solar radiation, and relative humidity were calculated for three years (from 2012) based on meteorological data from the Cheongju weather station (Korea Meteorological Administration database). The Cheongju watermark data were used for the river runoff as hydrological data. All hydrological data were collected for 23 years from 1990 to 2015.

In general, the watershed hydrological model was calibrated based on the observed runoff amount. In this study, the water flow data (five years from 2010 to 2014) of the Cheongju gauging station were downloaded from the Water Resources Management Information System (WAMIS, http://www.wamis.go.kr (accessed on 9 March 2021)).

Based on the aforementioned hydrologic component analysis, the evapotranspiration amount of the demonstration site sub-basin, groundwater recharge amount, and groundwater runoff were estimated. Given that the demonstration site of this study was relatively small, the spatial distribution of the amount of groundwater recharge in the groundwater flow model was not considered.

The calculated groundwater recharge amount was expressed in the form of daily data that could be used to simulate the unsteady groundwater flow. Additionally, the model values were verified based on comparisons with the time series data of the groundwater for the same period.

3.2.2. Analysis of Spatiotemporal Distribution of Groundwater Level with a Groundwater Flow Model

To examine the presence of groundwater in aquifers in the study area and to analyze the local water circulation mechanism, a distributed 3D groundwater flow model was constructed. The modeling area of the analysis basis was set, the values and ranges of hydrogeological parameters were determined from various data and measurements, and the boundary conditions were established by inferring the inflow/outflow methods of various water bodies and aquifers. In particular, the physical shape of the aquifer, including the area of the aquifer, thicknesses of the aquifer and pressurized layer, locations of surface water and rivers, boundary conditions of the aquifer, and hydrologic information, such as the hydraulic conductivity of the aquifer, the retention coefficient of the aquifer, locations of permeability and specific storage coefficient of pressurized bed, and the hydrologic connectivity between aquifer and surface water are required.

Figure 4 shows the boundary conditions of MODFLOW model. The model area for numerical analysis was divided into 360 rows and 240 columns made of 5 × 5 m grids through a convergence test for an area of 1200 × 1800 m (see Figure 4a). The model stratum (with an approximate depth of 80 m) was divided into three layers based on the geological characteristics. The input values of the hydrogeological parameters of each layer are listed in Table 1.

The 2–3 m thick topsoil was included in the unconfined aquifer, which is the first layer of the model. The first layer of the model represents weathered soil, and its spatial distribution was estimated based on a statistical interpolation method in conjunction with the hydraulic conductivity obtained by the pumping test.

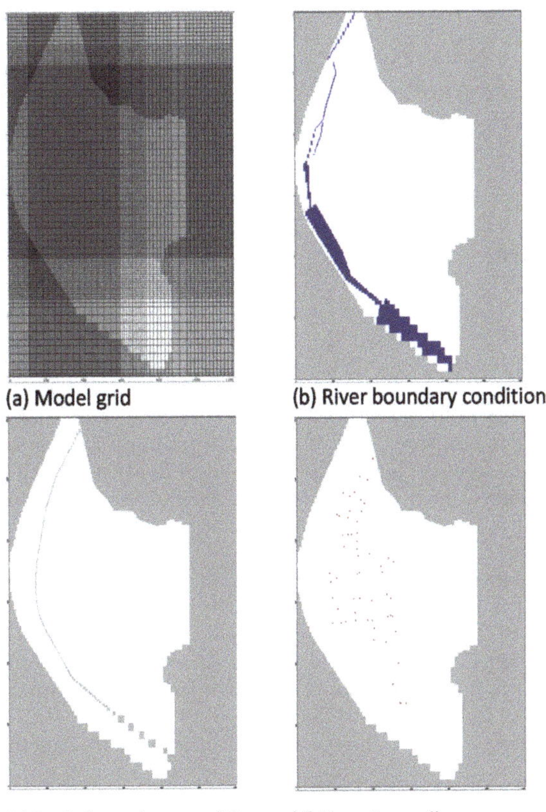

(a) Model grid   (b) River boundary condition

(c) Drain boundary condition   (d) Pumping wells

Figure 4. Model grids and the boundary conditions of the study site.

Table 1. MODFLOW input data for layer properties (modified from Chang and Chung [22]).

	Thickness	Aquifer Type	Hydraulic Conductivity (m/sec)		Specific Yield	Storage Coefficient
	(m)		K_H	K_Z	(Sy)	(Ss, m^{-1})
Layer 1 (Weathered rock)	12	Unconfined	2.7×10^{-5}~7.6×10^{-5}	2.7×10^{-5}~7.6×10^{-5}	0.09	
Layer 2 (Weathered rock)	20	Confined	2×10^{-6}	1×10^{-6}		1×10^{-5}
Layer 3 (Soft rock)	50	Confined	1×10^{-5}	1×10^{-5}		1×10^{-4}

In the Figure 4b, the boundary condition for the small stream developed near the aquifer was set using the MODFLOW river package. The characteristics of the river were distinguished based on the weir at the center of the river. The river boundary was set by referring the time series data for the river water level downstream of the weir (CWR-01) and the measurement data in the upstream region (CWR-02). There were no large variations in the river water level upstream of the weir, and the upstream water level variations in the summer are smaller than those downstream. The downstream time series data from 18 April to 16 November 2013, could not be acquired, because the equipment was lost due to the flooding.

The restoration of the lost data was performed by adding 1.5 m to the river level measured at the Geum River Flood Control Center. The river water level measurement data at the center (located ~4 km downstream) was compared to the CWR-01 time series data from the study site. We found that the two water levels varied in similar patterns when comparing the three peak water levels. In addition, there was no significant lag in the time of peak occurrence. The agricultural waterway was set as the drain boundary condition in the model (see Figure 4c). The input values for the drain package were calibrated based on the evaluated depth of the bottom of the agricultural waterway located in the south–north direction, parallel to the agricultural waterway and river shapes at the demonstration site. Pumping wells were distributed as shown in Figure 4d.

To identify the terrain elevation of the area where a research facility exists and the boundary of the watershed, the elevation contour lines were extracted from the digital map (scale: 1:25,000) received from the National Geographic Information Institute (30707071_008, 30707071_009 from http://map.ngii.go.kr/ (accessed on 9 March 2021)), and were converted to grid data. A digital elevation map was then created and used as input data for MODFLOW. The minimum elevation was 63 m near the river downstream in the north, and the maximum elevation was 170 m in the mountainous terrain in the east. In general, the slope of the terrain was along the southeast–northwest direction from the mountainous terrain to the river. The outer parts of the simulation area, such as the impermeable bedrock and watershed in the demonstration site, were set as no-flow boundaries, and the time series values calculated using SWAT were used for the boundary of water recharge in the soil layer which flowed to the aquifer according to the percolation process.

We conducted simulations by extending the simulation period to 30 July 2014, based on the simulation results of Chang and Chung obtained from 10 July 2012 to 2 March 2014, for the Cheongwon WCC site [22]. The hydraulic conductivity distribution values of the first layer used by Chang and Chung [22] were adjusted with an additional calibration process. As a result, hydraulic conductivity values of 3.2×10^{-5}–9.1×10^{-5}, 2.0×10^{-6}, and 1.0×10^{-5} m/s were used for the first, second, and third layers, respectively.

4. Results and Discussion

4.1. Analysis of Groundwater Level

The groundwater flow decreased from the south toward the north. Groundwater flows from the south to the north, and rapid changes in the groundwater level were observed around the weir in the Musimcheon. The flow analysis results exhibited characteristics of a loss river upstream of the Musimcheon until early November during the full-scale WCC period, whereas the downstream of the Musimcheon in the north exhibited characteristics of a gain river. However, unlike the discharge of groundwater downstream of the Musimcheon before WCC, the entire section of the Musimcheon acted as a loss river owing to the large-scale water intake during the winter period. Because of the difference in elevation (of up to 2 m) in the upstream and downstream parts of the weir, the distribution of groundwater levels was dense. This suggests that the groundwater level variations near a weir were caused by the surrounding groundwater and river water levels. The pattern of the groundwater level decline during the WCC period set as the operation time of the water intake well does not affect the Musimcheon watershed during the first ten days, and variations in the groundwater level only occur near the water intake site, but the radius of the influence region gradually widened. However, the demonstration basin in this study was a narrow basin blocked by rivers and mountains, and a relatively fast overall groundwater level decline could be observed.

For model calibration, the validity of the characteristic hydraulic values was examined based on the field measurement results for water budget simulation, such as the water level contour map, groundwater level time series, and recharge and discharge amounts. In addition, input variables, such as the hydraulic conductivity of the model and retention coefficient were adjusted so that the groundwater level time series values measured in the field would be matched within a reasonable range. Chang and Chung [24] previously showed a graph that compared the data collected from the observation wells on August 3 before the onset of the WCC and the simulation results of MODFLOW. A good correlation with a coefficient of determination (R^2) of 0.97 was obtained.

Figure 5 shows six different groundwater level decline–recovery curves, which compare the simulation results obtained from the six observation wells. The gray line in this figure indicates the groundwater level based on actual well measurements, and the black line indicates the precise modeling results using MODFLOW. The blue box represents the WCC pumping periods. The six graphs were based on the values of the six observation wells of CWW-04, CWW-05, CWW-07, CWW-08, CWW-11, and OB-11 in the bank, used in the model calibration. When the WCC began, the pattern of the groundwater level decline exhibited an abrupt change (increased slope) at the early stage. As the slope decreased, the rate of change of the downward trend gradually decreased. When WCC ended, the water level recovered for 3–4 months. Examining the graphs in detail, the periods could be divided into the following parts: (a) changes in aquifer groundwater levels owing to rainfall in the summer, (b) groundwater decline owing to WCC for four months between 2012 and 2013, (c) aquifer water level recovery, groundwater level fluctuations owing to precipitation during the summer in 2013, (d) the re-decline of groundwater level owing to WCC, which started in November 2013, and (e) the re-recovery of groundwater level owing to precipitation during the summer in 2014. The examination of each observation well showed that the groundwater level, which ranged from 63 to 68 m in steady state, declined after the operation of the water pumping well. As the well was farther in the northeast region, which is inland of the river's downstream, the clustering of the pumping wells was denser. Additionally, as the distance from the river increased, the degree of groundwater declines increased. By contrast, the simulation results showed that in the upstream of the river, a high-water level was maintained because of the river, and an induced recharge in the river resulted in a small water level decline. For example, CWW 8, which is located at a distance farthest from the river, exhibited a large groundwater level decline; however, CWW 11, which was located at a short distance from the river, exhibited a relatively small groundwater decline.

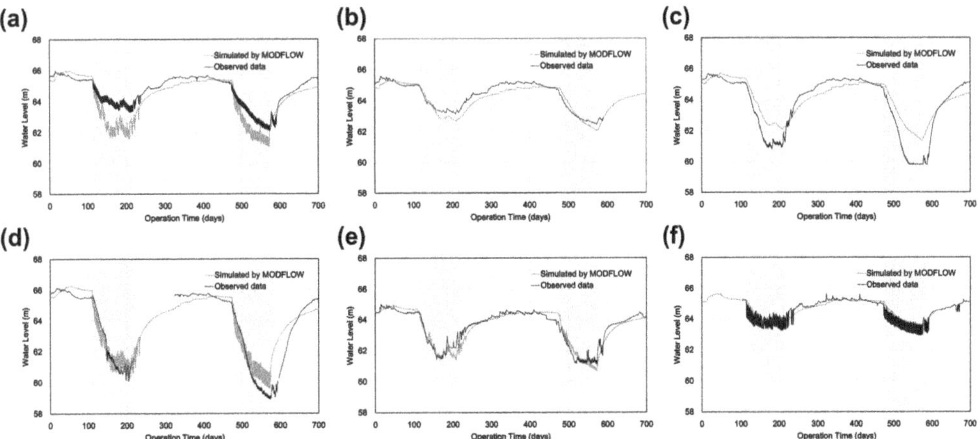

Figure 5. Comparisons between observation and simulated results by MODFLOW for a two-year seasonal groundwater exploitation owing to water curtain cultivation (WCC) for monitoring wells of (**a**) CWW-04, (**b**) CWW-05, (**c**) CWW-07, (**d**) CWW-08, (**e**) CWW-11, and (**f**) OB-11 (blue boxes represent WCC period).

Herein, the simulation results are also discussed, including the variations of the aquifer groundwater level owing to precipitation during the summer period, in addition to the groundwater decline during the WCC period in the simulation period (total of 700 days). In the graph, the groundwater levels during the two cycles of groundwater depletion–recovery were observed. In this period, one year in which the following events occurred was recognized as the WCC cycle: (1) aquifer recharge owing to rainfall in the summer, (2) groundwater depletion owing to WCC, and (3) induced recharge by the river after the termination of the WCC. The analysis onset time and the graph of this study also followed this cycle. As shown in the Figure 5, the modeling results indicated that most of the demonstrations were influenced by sporadically scattered groundwater pumping. When the two cycles of groundwater level decline and the respective recoveries were compared with the observations, the simulated results were reproduced and were found to be close the observed values. Notably, even though the winter temperatures were not significantly different between the first and second water curtain periods, the groundwater declined more in the second water curtain period.

4.2. Water Budget Analysis

Figure 6 shows a graph of the groundwater disturbance caused by the WCC. The inflow and outflow rates (m^3/day) of aquifer, drain, and river boundary were analyzed throughout the simulation period. Examining the graphs in detail, in the summer, outflows from the aquifer to the river owing to increased precipitation were observed intermittently. During the WCC period, the groundwater depletion in the aquifer increased, whereas the aquifer recharge owing to the induced recharge of the river also increased significantly. The phenomenon of groundwater loss of the aquifer during the WCC period can be expressed as the storage out rate in the graph. The yellow line in this graph shows the recharge induced by the river that increased continually during WCC and that decreased during the replenishment of groundwater in the aquifer during the spring period. Hence, the abrupt change pattern of the aquifer and river can be referred to as an aquifer disturbance by the WCC or surface–groundwater (SW–GW) interaction, showing the direct river-response to the pumping pattern during the WCC in winter.

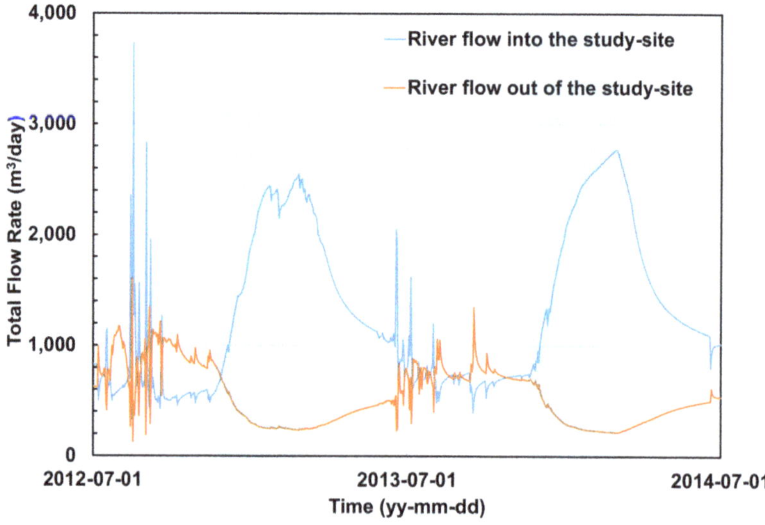

Figure 6. Temporal changes of water budget of the Mushimcheon river owing to the seasonal water use in the study-site (blue shaded boxes represent WCC period).

In addition, the water budget at the center of the aquifer was analyzed. The quantitative values for water budget analysis are expressed as the cumulative volume of groundwater inflow/outflow in the aquifer during a specific period. The changes in the water budget inflow/outflow that an aquifer undergoes are influenced by the pumping pattern during the WCC period. The water budget analysis for this study shows that the aquifer groundwater depleted by pumping was recharged naturally by the recharge induced by rivers and rainfall. When analyzing the groundwater storage of the aquifer, which corresponds to the difference between the cumulative outflow and inflow of the aquifer, the recharge of the aquifer attributed to the rainfall or induced recharge by the river was expressed as a positive value, whereas the depletion of aquifer was expressed as a negative value. These values were used in an effort to show the depletion and recovery patterns of the aquifer.

Figure 7 shows a graph that expresses the aquifer net storage, which is determined by dividing the volume [L^3] with a water budget analysis value based on the aquifer owing to the area of the demonstration site in the vertical direction [L^3/L^2]. The volume-based water budget analysis value for a large cultivation area is larger than that for a small area, even if the actual degrees of depletion may be different. Correspondingly, direct comparisons often become difficult. In this case, one-dimensional water budget analysis can be used to compare target sites with different areas. The water budget analysis method expressed in one dimension in this way has the advantage in that it reduces the number of factors to be considered by removing the basin area from the result analysis. It is thus easy to intuitively understand groundwater depletion compared with volume analyses.

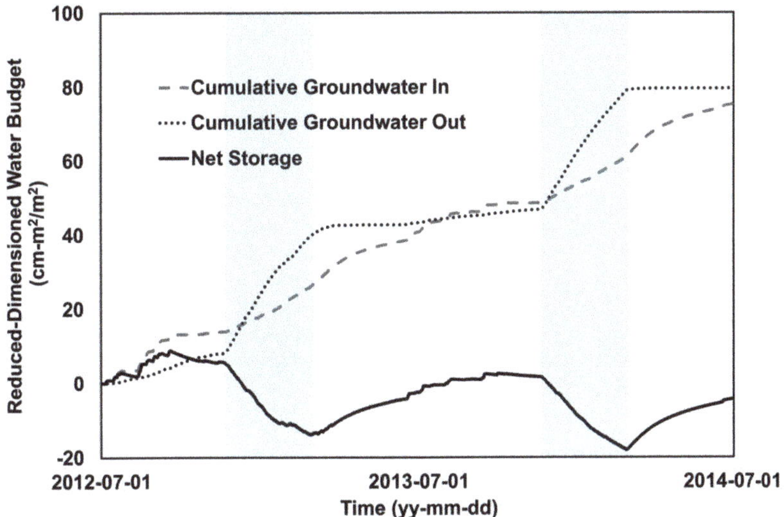

Figure 7. Cumulative water budget plot for aquifer from September 2012 to June 2014 (blue boxes represent WCC period).

Note that the shape of net storage in Figure 7 is similar to the groundwater level decline–recovery curve. The lowest point of storage owing to the first cycle of WCC in 2012–2013 was −13.7 cm-m²/m². The amount of groundwater recharged by rainfall and by the induced recharge from the river until 1 July 2013 (when the one-year analysis cycle was completed), did not recover to the level of the previous year (−2.5 cm-m²/m²). Consequently, the groundwater depletion owing to the WCC in the previous year was not recovered by the aquifer recharge over a year, and the storage value at the end of the WCC period for 2013 and 2014 decreased to −18.1 cm-m²/m², which was lower than the lowest point during the previous year. A comparison of the two lowest points verified quantitatively that it is difficult to completely recover the aquifer depletion owing to recharge from a river or precipitation, and the degree of depletion can therefore worsen every year. We estimate that if this pattern continues, the depletion of the demonstration site could continue for a long time in an irreversible direction.

When the results of July 2012, July 2013, and July 2014 were compared, the groundwater shortage across the aquifer worsened by 2.5 cm per year. This value can be used as an indicator that conceptually represents groundwater depletion, rather than a factor that indicates the actual depth of groundwater decline. This is because the actual groundwater depletion occurs within several tens of centimeters around the pumping well, and almost no groundwater decline occurs in a place where the pumping well is far and where the river is nearby.

4.3. Application of Artificial Recharge Scenario

To predict the groundwater level recovery and water budget change caused by artificial recharge, an artificial recharge scenario was established. This scenario was applied to the model. In this scenario, artificial recharge was performed for the period of 2013–2014, which corresponds to the second WCC period in the study period. The artificial recharge started in December 2013 when groundwater pumping was most active and the temperature was low. Each pumping well performed pumping every day for 16 h and 30 min (which corresponds to 0.68 days on average). Groundwater re-injection was simulated for 7 h and 30 min (which corresponds to 0.32 days), which is the remaining time in each day if the pumping time is excluded. In addition, artificial recharge was performed for 24 h when pumping was not performed owing to high temperatures.

Figure 8 shows the results in the form of a groundwater level distribution map for February 2014, 100 days after the onset of the WCC in November 2013. Figure 5a shows the groundwater level distribution map representing the spatial distribution of groundwater when WCC was performed using the existing model, while Figure 5b shows the result of artificial recharge. These results are represented on the plane of the model based on the same data. A distinct recovery of groundwater level can be observed in the north of the cultivation area that corresponds to the downstream section of the river. Specifically, in the cultivation area to the northeast, the water level decreased farther from the river and closer to the mountainous area. Given that a large water level recovery occurred in this area compared with the existing (precise) model outcome, the application of a scenario for WCC water reinjection of 10% can be expected to achieve a maximum water level recovery of 0.8 m. The water level not only recovered at the groundwater injection point, but the largest degree of water level recovery was achieved in the background when the groundwater level decreased considerably owing to the WCC.

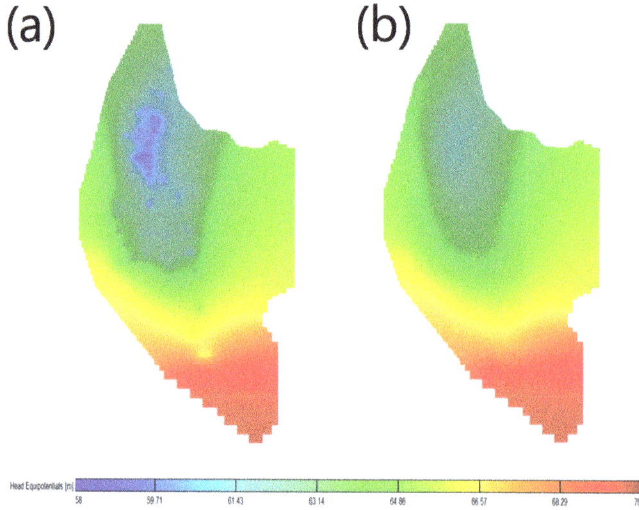

Figure 8. Spatial head distribution owing to (**a**) WCC and (**b**) at 100 days after artificial recharge.

The effect of groundwater recharge by the recovery of water level yields different values depending on the observation location. Hence, in order to determine the overall effect of the aquifer changing over time, we performed a water budget analysis. Figure 9 shows a graph plotted using the results of the model that introduced the 20% of artificial recharge scenario for the water budget analysis of groundwater aquifers. A 20% artificial recharge scenario means that freshwater was injected into the aquifer by all wells with an amount equivalent to 20% of the total groundwater withdrawn during WCC. The graph that shows the slope change between the scenarios when the artificial recharge started in December 2013 previously described that groundwater pumping was most active and the temperature was low. The black solid line indicates the cumulative groundwater inflow and the dots indicate the cumulative groundwater outflow from the aquifer as a result of the WCC in the existing model (Figure 7). The solid red line and dots indicate the cumulative water budget inflow and outflow amounts that changed when 20% of the freshwater was injected for artificial recharge. When the artificial recharge technique was applied, the amount of cumulative inflow became larger than the existing WCC result, as shown in the red solid line compared to the black solid line in the figure. The change in the slope of the solid line indicates the rapid recovery of groundwater storage in the aquifer, while the dotted line indicates relatively little change in outflow from the aquifer. This means that the groundwater lost from the aquifer during this period was reduced owing to artificial recharge.

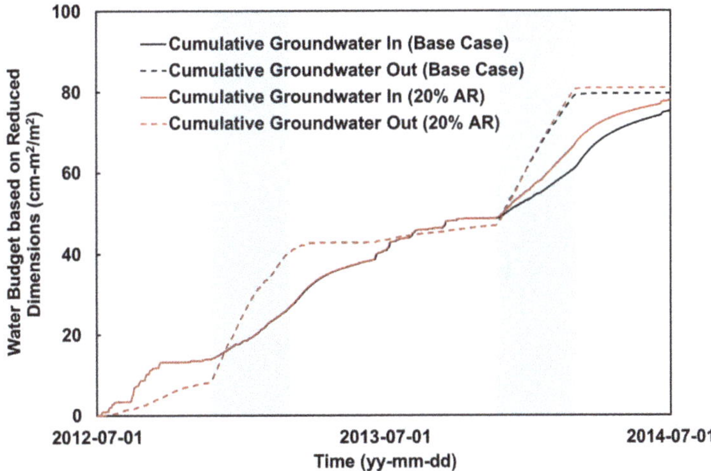

Figure 9. Cumulated water budget of the aquifer storage model from September 2012 to June 2014 (blue boxes represent WCC period).

In previous study, Chang and Chung [24] explained the aquifer changed according to various AR scenarios using the volumetric water balance [L^3], and this study modified the values into the reduced dimensions using cm-cm^2/m^2 shown in the Figure 10.

Scenarios that accounted for increasing artificial recharges owing to groundwater reinjection (20%, 40%, and 60%) were applied to the model. The goal of the AR scenarios was to maintain the sustainable development of WCC. The 2012–2013 were simulated without AR scenario to be compared the AR application for 2014. As the result, it was found that a 20% artificial recharge of the aquifer is a condition that allows the same level of the lowest decline of previous year, 2013. Meantime, the 40% injection AR scenario showed the similar level of water budget at the end of simulation time that July 2014 shows. Therefore, sustainability solutions can be different depending on whether the groundwater level indicator or the water balance indicator is selected.

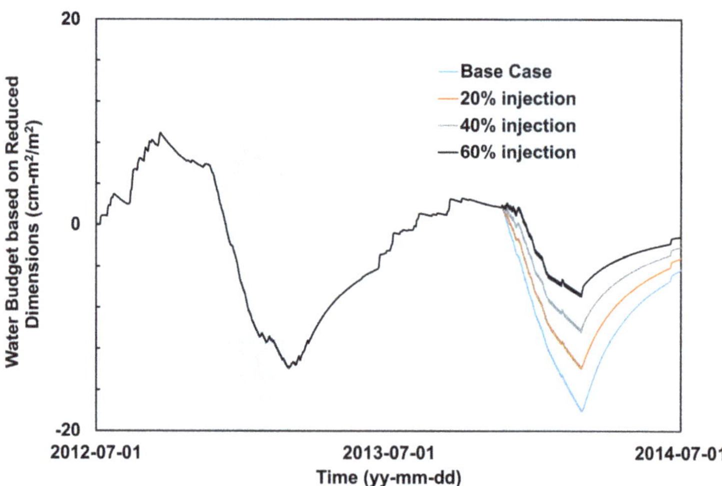

Figure 10. Comparison between the artificial recharge scenarios based on the aquifer storage model (Modified from Chang and Chung [24]) from September 2012 to June 2014 (blue boxes represent WCC period).

5. Summary and Conclusions

According to the United Nations Food and Agriculture Organization, water, energy, and food security are closely correlated. WCC can be a slightly different concept from the water–energy–food nexus, given that water is used directly as an energy source. However, WCC is closely related to water, energy, and food security in South Korea. It uses groundwater as a seemingly infinite energy resource depending on the induced river recharge. However, we found that it causes groundwater depletion of aquifers in a very subtle way. Thus, WCC has a negative aspect (groundwater depletion), and guidelines for appropriate groundwater usage are needed, instead of allowing unrestricted groundwater development. This is because there are many cultivation areas that face difficulties owing to the lack of groundwater.

This study introduced a quantitative analytical method and attempted to present water management directions by establishing a precise water budget analysis method for the WCC demonstration area in Cheongwon-gun, wherein long-term groundwater depletion was expected owing to seasonal groundwater use.

The aquifer disturbance and recovery attributed to the WCC facilities were analyzed precisely for different periods based on the introduction of a method for daily inflow/outflow and cumulative water budget analysis. In addition, a precise investigation was conducted to examine the role of rivers in the aquifer and the model boundary conditions. Furthermore, this study suggested a quantitative method to determine the possibility of sustainable development of an aquifer at the demonstration site from a long-term perspective, and verified the precise water budget analysis method using actual measurements and simulation research for the WCC period of two years. In addition, the appropriate injection range that can prevent long-term aquifer depletion by artificial recharge was explored based on a scenario that re-injected 10% of WCC water with the use of the precise water budget analysis method. The results proved that the degree of aquifer disturbance decreased as the injection amount owing to the artificial recharge increased. However, a 20% injection could maintain the aquifer environment of the previous year's level in the current situation of the demonstration site. Bredehoeft [25] considered that when the system enters a new steady state, sustainable development is maintained. At this study site, a groundwater level declines of 2.5 cm in the total area occurred every year. In fact, intense local groundwater depletion occurred at the center of the study site.

If the regulation of some areas was enforced with an appropriate optimization method, instead of 20% artificial recharge of the total usage, sustainable development of the total area could be possible if the development amount was reduced to values less than 20% of the usage. Based on these arguments, a follow-up study is required. Furthermore, given that groundwater recharge by river water is a major recharge method for groundwater, a follow-up study that simulates the effect of the river water temperature on aquifers during the winter season should be conducted.

Author Contributions: Conceptualization: S.W.C. and I.-M.C.; methodology: S.W.C.; validation: S.W.C. and I.-M.C.; formal analysis: S.W.C.; investigation: S.W.C.; resources: I.-M.C.; data curation: S.W.C.; writing—original draft preparation: S.W.C.; writing—review and editing: I.-M.C.; project administration: I.-M.C.; funding acquisition: I.-M.C. All authors have read and agreed to the published version of the manuscript.

Funding: This research was supported by the Development program of Minimizing of Climate Change Impact Technology through the National Research Foundation of Korea (NRF), funded by the Korean government. (Ministry of Science and ICT (MSIT)). (No. NRF-2020M3H5A1080735).

Institutional Review Board Statement: Not applicable.

Informed Consent Statement: Not applicable.

Conflicts of Interest: The authors declare no conflict of interest.

References

1. Siebert, S.; Burke, J.; Faures, J.M.; Frenken, K.; Hoogeveen, J.; Döll, P.; Portmann, F.T. Groundwater use for irrigation—A global inventory. *Hydrol. Earth Syst. Sci.* **2010**, *14*, 1863–1880. [CrossRef]
2. Wada, Y.; Van Beek, L.P.H.; Van Kempen, C.M.; Reckman, J.W.T.M.; Vasak, S.; Bierkens, M.F.P. Global depletion of groundwater resources. *Geophys. Res. Lett.* **2010**, *37*, 37. [CrossRef]
3. Winter, T.C.; Harvey, J.W.; Franke, O.L.; Alley, W.M. *Ground Water and Surface Water: A single Resource*; United States Geological Survey: Reston, VA, USA, 1998.
4. Barlow, P.M.; Leake, S.A. *Streamflow Depletion by Wells—Understanding and Managing the Effects of Groundwater Pumping on Streamflow*; U.S. Geological Survey: Reston, VA, USA, 2012; p. 84.
5. Kollet, S.J.; Zlotnik, V.A. Stream depletion predictions using pumping test data from a heterogeneous stream–aquifer system (a case study from the Great Plains, USA). *J. Hydrol.* **2003**, *281*, 96–114. [CrossRef]
6. Nyholm, T.; Christensen, S.; Rasmussen, K.R. Flow Depletion in a Small Stream Caused by Ground Water Abstraction from Wells. *Ground Water* **2002**, *40*, 425–437. [CrossRef] [PubMed]
7. Thomas, B.F.; Behrangi, A.; Famiglietti, J.S. Precipitation Intensity Effects on Groundwater Recharge in the Southwestern United States. *Water* **2016**, *8*, 90. [CrossRef]
8. Pavelic, P.; Patankar, U.; Acharya, S.; Jella, K.; Gumma, M.K. Role of groundwater in buffering irrigation production against climate variability at the basin scale in South-West India. *Agric. Water Manag.* **2012**, *103*, 78–87. [CrossRef]
9. Shamsudduha, M.; Chandler, R.E.; Taylor, R.G.; Ahmed, K.M. Recent trends in groundwater levels in a highly seasonal hydrological system: The Ganges-Brahmaputra-Meghna Delta. *Hydrol. Earth Syst. Sci.* **2009**, *13*, 2373–2385. [CrossRef]
10. Perrin, J.; Ferrant, S.; Massuel, S.; Dewandel, B.; Maréchal, J.; Aulong, S.; Ahmed, S. Assessing water availability in a semi-arid watershed of southern India using a semi-distributed model. *J. Hydrol.* **2012**, *460–461*, 143–155. [CrossRef]
11. Ferrant, S.; Caballero, Y.; Perrin, J.; Gascoin, S.; Dewandel, B.; Aulong, S.; Dazin, F.; Ahmed, S.; Maréchal, J.-C. Projected impacts of climate change on farmers' extraction of groundwater from crystalline aquifers in South India. *Sci. Rep.* **2014**, *4*, 3697. [CrossRef] [PubMed]
12. Ruud, N.; Harter, T.; Naugle, A. Estimation of groundwater pumping as closure to the water balance of a semi-arid, irrigated agricultural basin. *J. Hydrol.* **2004**, *297*, 51–73. [CrossRef]
13. MOLIT. *2015 Groundwater Annual Report*; 11-1611000-000155-10; MOLIT: Daejeon, Korea, 2015.
14. Lee, J.-Y.; Kwon, K.D. Current Status of Groundwater Monitoring Networks in Korea. *Water* **2016**, *8*, 168. [CrossRef]
15. KIGAM. *Groundwater Restoration Technology for Riverside Area*; Report on the Advanced Technology for Groundwater Development and Application in Riversides (Geowater+) in Water Resources Management Program; 11-1613000-001827-01; KIGAM: Daejeon, Korea, 2017; p. 549.
16. Chang, S.; Chung, I.-M. Analysis of Groundwater Variations using the Relationship between Groundwater use and Daily Minimum Temperature in a Water Curtain Cultivation Site. *J. Eng. Geol.* **2014**, *24*, 217–225. [CrossRef]
17. Chung, I.-M.; Kim, N.-W.; Lee, J.; Sophocleous, M. Assessing distributed groundwater recharge rate using integrated surface water-groundwater modelling: Application to Mihocheon watershed, South Korea. *Hydrogeol. J.* **2010**, *18*, 1253–1264. [CrossRef]

18. Kim, N.W.; Chung, I.M.; Won, Y.S.; Arnold, J.G. Development and application of the integrated SWAT–MODFLOW model. *J. Hydrol.* **2008**, *356*, 1–16. [CrossRef]
19. Chung, I.-M.; Lee, J.; Kim, N.W.; Na, H.; Chang, S.W.; Kim, Y.; Kim, G.-B. Estimating exploitable amount of groundwater abstraction using an integrated surface water-groundwater model: Mihocheon watershed, South Korea. *Hydrol. Sci. J.* **2014**, *60*, 863–872. [CrossRef]
20. Harbaugh, A.; Banta, E.; Hill, M.; McDonald, M. *MODFLOW-2000, the U.S. Geological Survey Modular Ground-Water Model User Guide to Modularization Concepts and the Ground-Water Flow Process*; United States Geological Survey: Reston, VA, USA, 2000; Volume Open-File Rep 00-92.
21. Arnold, J.G.; Williams, J.R.; Srinivasan, R.; King, K.W. *The Soil and Water Assessment Tool (SWAT) User's Manual*; Texas Water Resources Institue: College Station, TX, USA; Temple, TX, USA, 1996.
22. Chang, S.W.; Chung, I.-M. Analysis of Groudwater Budget in a Water Curtain Cultivation Site. *J. Korean Soc. Civ. Eng.* **2015**, *35*, 1259–1267. [CrossRef]
23. Moon, S.-H.; Ha, K.; Kim, Y.; Yoon, P. Analysis of Groundwater Use and Discharge in Water Curtain Cultivation Areas: Case Study of the Cheongweon and Chungju Areas. *J. Eng. Geol.* **2012**, *22*, 387–398. [CrossRef]
24. Chang, S.; Chung, I.-M. Long-term groundwater budget analysis based on integrated hydrological model for water curtain cultivation site: Case study of Cheongweon, Korea. *J. Geol. Soc. Korea* **2016**, *52*, 201–210. [CrossRef]
25. Bredehoeft, J.D. The Water Budget Myth Revisited: Why Hydrogeologists Model. *Ground Water* **2002**, *40*, 340–345. [CrossRef] [PubMed]

Article

Importance of the Induced Recharge Term in Riverbank Filtration: Hydrodynamics, Hydrochemical, and Numerical Modelling Investigations

Rudy Rossetto [1,*], Alessio Barbagli [1], Giovanna De Filippis [1], Chiara Marchina [1], Thomas Vienken [2] and Giorgio Mazzanti [3]

1   Institute of Life Sciences, Scuola Superiore Sant'Anna, 56127 Pisa, Italy; barbagli@cgt-spinoff.it (A.B.); giovanna.df1989@libero.it (G.D.F.); chiara.marchina@unipd.it (C.M.)
2   Helmholtz Centre for Environmental Research GmbH—UFZ, Permoserstraße 15, 04318 Leipzig, Germany; thomas.vienken@ufz.de
3   Genio Civile Toscana Nord, Regione Toscana, 55100 Lucca, Italy; giorgio.mazzanti@regione.toscana.it
*   Correspondence: rudy.rossetto@santannapisa.it

Received: 31 October 2020; Accepted: 5 December 2020; Published: 8 December 2020

Abstract: While ensuring adequate drinking water supply is increasingly being a worldwide challenging need, managed aquifer recharge (MAR) schemes may provide reliable solutions in order to guarantee safe and continuous supply of water. This is particularly true in riverbank filtration (RBF) schemes. Several studies aimed at addressing the treatment capabilities of such schemes, but induced aquifer recharge hydrodynamics from surface water bodies caused by pumping wells is seldom analysed and quantified. In this study, after presenting a detailed description of the Serchio River RBF site, we used a multidisciplinary approach entailing hydrodynamics, hydrochemical, and numerical modelling methods in order to evaluate the change in recharge from the Serchio river to the aquifer due to the building of the RBF infrastructures along the Serchio river (Lucca, Italy). In this way, we estimated the increase in aquifer recharge and the ratio of bank filtrate to ambient groundwater abstracted at such RBF scheme. Results highlight that in present conditions the main source of the RBF pumping wells is the Serchio River water and that the groundwater at the Sant'Alessio plain is mainly characterized by mixing between precipitation occurring in the higher part of the plain and the River water. Based on chemical mixing, a precautionary amount of abstracted Serchio River water is estimated to be on average 13.6 Mm^3/year, which is 85% of the total amount of water abstracted in a year (~16 Mm^3). RBF is a worldwide recognized MAR technique for supplying good quality and reliable amount of water. As in several cases and countries the induced recharge component is not duly acknowledged, the authors suggest including the term "induced" in the definition of this type of MAR technique (to become then IRBF). Thus, clear reference may be made to the fact that the bank filtration is not completely due to natural recharge, as in many cases of surface water/groundwater interactions, but it may be partly/almost all human-made.

Keywords: drinking water supply; water supply scheme; surface water/groundwater interactions; managed aquifer recharge; induced riverbank filtration; groundwater resource management

---

1. Introduction

Ensuring adequate drinking water supply is one of the most pressing needs for our societies [1]. However, finding reliable sources is more and more a difficult task because of the widespread deterioration of surface water resources (and the related costs for treatment) and groundwater deterioration and overexploitation [2–4]. In this sense managed aquifer recharge (MAR) schemes may provide reliable solutions in order to guarantee safe and continuous supply of water [5–7]. Managed aquifer recharge consists in the intentional recharge of aquifers using excess water in wet periods, or non-conventional water sources while at the same time assuring adequate protection of the environment and human-health [8]. Several techniques (spreading methods, Aquifer Storage and Recovery, Soil Aquifer Treatment systems, etc.) have been devised and widely adopted at global scale in the last 60 years [5,9]. Among these, riverbank filtration (RBF) has been in use since about two centuries, being a popular way to tap surface water ensuring at the same time adequate water treatment [10].

RBF is a technique in which the bed and bank of a river serve as treatment zone for the induced river water [10–14], when pumping wells are placed adjacently to a surface water body with continuous and adequate discharge or storage (such as rivers, lakes or basins). As such pumped groundwater in RBF schemes is constituted by a mix of ambient groundwater and induced surface water recharge. Several Authors (i.e., [11,15,16]) well-stated that while groundwater quality depends on the land use in the catchment area, bank filtrate quality is a function of river water quality and the efficiency of the purification processes during RBF. Moreover, other studies discuss the treatment capabilities of the RBF MAR systems in dealing with common surface water contaminants, but also with organics and emerging contaminants [17–20]. At the same time, while many examples exist in literature in evaluating surface- and groundwater interactions, at different spatial scales [21] with a number of methods [22], still induced recharge hydrodynamics is seldom analysed and quantified. Sottani and Vielmo (2014) [23] evaluated the extent and the rate of the recharge effects in groundwater due to weir realization in the middle Brenta River plain (Italy). Wei-shi et al. (2020) [24] evaluated the impact of a river reach restoration on the groundwater flow on bank filtration by using a transient flow and heat transport numerical model. Shankar et al. (2009) [15] quantified the contributing ratio of bank filtrate to ambient groundwater by means of a transient numerical model simulation at Grind well field (Germany).

The assessment of the induced recharge and of the ratio of bank filtrate to ambient groundwater is then important as it may provide relevant insights on the risk of groundwater contamination at the RBF wells. As such, in this study, we applied a multidisciplinary approach entailing hydrodynamics, hydrochemical and numerical modelling methods in order to evaluate the change in recharge from the Serchio river (Lucca, Italy) to the aquifer due to the building of the RBF infrastructures along the river. In this way, we estimated the increase in aquifer recharge and the ratio of bank filtrate to ambient groundwater abstracted at such RBF scheme.

2. Materials and Methods

The Serchio Riverbank Filtration scheme in Sant'Alessio (Lucca, Italy) supplies an average volume of 16 Mm^3/year of drinking water (data from years 2016–2019; source GEAL SpA, local water utility) to about 300,000 citizens of Lucca, Pisa, and Livorno. This plant can be considered an exemplary site for MAR RBF systems. As a blue infrastructure, it provides ecosystem services such as "water storage" and "water quality improvement" [25].

The RBF scheme is set along the Serchio River (Figure 1). Groundwater is tapped by Holocene coarse sand and gravel aquifer overlain by a silty surficial cover. Water is pumped by twelve vertical wells (four at the Sant'Alessio pumping station and eight of the Pisa-Livorno water supply scheme), located between 30 m and 100 m from the river reach, inducing river water into the high yield (10^{-2} m^2/s transmissivity on average [26]) sand and gravel aquifer. The aquifer storage is then increased by the presence of one weir downstream the well field (Figure 2) to raise the river head of about 1.5 m, therefore raising the saturated part of the aquifer along the river reach.

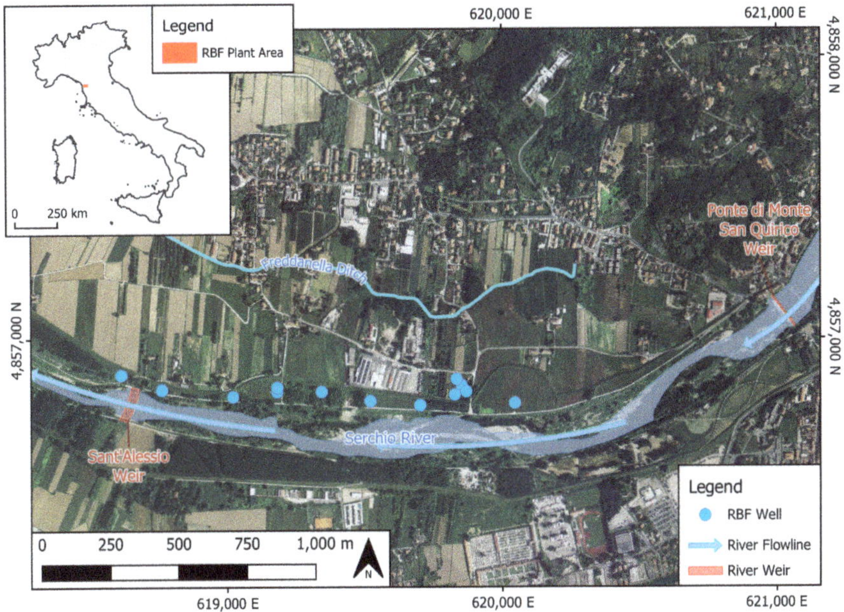

Figure 1. Layout of the Serchio River riverbank filtration scheme in Lucca, Italy. Coordinates are in WGS84-UTM zone 32N (EPSGr: 32632).

Figure 2. The Sant'Alessio weir on the Serchio River.

Since the 1960s, the Sant'Alessio area along the Serchio River was deemed suitable for groundwater abstraction for drinking water purposes and, initially, four vertical wells were set in operation in 1967. They were drilled about 100 m away from the Serchio River to supply the north western households of the Lucca town with about 0.1 m³/s [27]. Following hydrogeological investigations, in the 1980s a highly yielding sand and gravel aquifer capable of supplying the towns of Pisa and Livorno, about 20 km and 40 km away, respectively, was identified. At the end of the 1980s, an Expert Commission evaluated the potential impacts of increasing abstractions and of the building of a weir to increase aquifer storage.

During the following ten years, the full MAR scheme was completed with the building of the river weir in Sant'Alessio, and it reached a total of about 0.500 m³/s supply capacity. More information on the historical development and the socio-economic importance of the Serchio river MAR scheme may be found in Rossetto et al. 2020a [28].

2.1. Site Description

The Sant'Alessio plain surface geology is characterized by unconsolidated silty to sandy sediments covering a sandy-gravel aquifer (Figure 3). The aquifer is limited at the bottom by a level of silty clays of lacustrine origin, which outcrops as stiff clays in the northern part of the area. From the Serchio river (south) towards the San Quirico-Carignano hills (north) the silty-sandy cover of the aquifer becomes thinner, going from 6.5–7 m to less than 1 m (Figure 3 Bottom). According to Rossetto and Bockelmann-Evans (2007) [29], also the aquifer thickness follows the same trend, starting from about 30 m in the Serchio riverbed and pinching-out towards the San Quirico-Carignano hills, where the silty-sandy cover directly meets the clay sediments at the aquifer bottom (Figure 3 Top). The Serchio river shows variable discharge ranging from more than 1000 m³/s during extreme rainfall events down to few m³/s during the summer season in extreme dry years. In the Serchio riverbed the aquifer is covered by coarse and clean gravels outcrops, as well as sand bars.

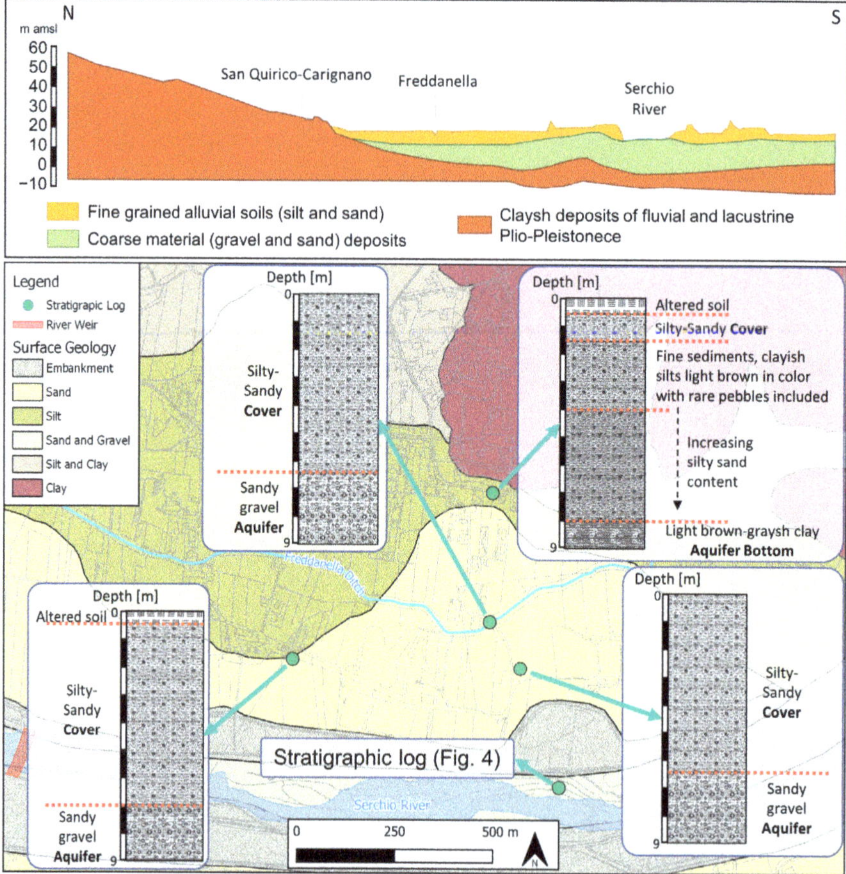

Figure 3. Simplified stratigraphic cross section (top) and detailed stratigraphic logs around the Serchio River RBF area (bottom).

Outcrops along the Serchio river show a strong contrast in the sedimentological composition of the river deposits which may lead to changes in hydraulic properties over several orders of magnitude on short vertical distances. The vertical heterogeneity of the aquifer was further investigated by a 17 m deep core drilled directly in the riverbed during a low-flow period (December 2015; Figure 4). According to the drilled core, the river bed stratigraphy was then reconstructed as follows (from the top of the river bed): 0–2 m gravels and sand, 2–5 m sand with gravel in a weakly silty matrix, 5–9.5 m gravel and medium sand in fine matrix, 9.5–12 m gravel in a silty-sandy matrix, 12–15.50 m gravel and sand, 15.50–17.3 m silty compacted clay.

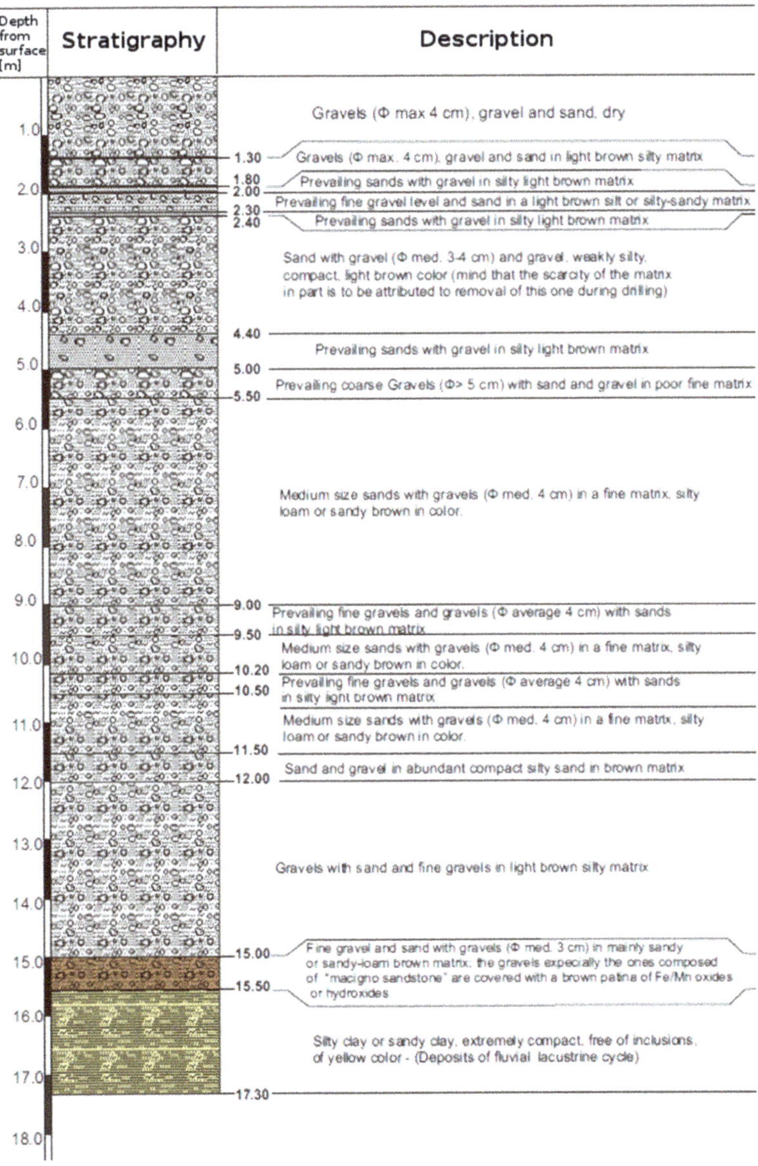

Figure 4. Stratigraphic log from the top of the Serchio river riverbed down to about 17 m.

2.2. Hydrodynamic Investigations

Investigations of the aquifer properties were performed during different campaigns. The first campaign was performed to characterize the aquifer structure in vertical and horizontal extent around the most western RBF well (Figure 1), with a special focus placed on the characterization of subsurface heterogeneity. The successive field campaigns aimed at characterizing the flow field in the vicinity of this pumping well. Direct Push technology, which describes an approach of pushing, hammering, and/or vibrating small-diameter hollow steel rods into the ground [30], was employed during the site investigation to install groundwater monitoring wells and to collect vertical high resolution subsurface information. The latter are obtained using specialized probes that are attached to end of the Direct Push rod string [31–36].

In this study, the Direct Push Injection Logger (DPIL) was used together with Direct Push electrical conductivity logging. The DPIL is used for vertical in-situ hydrostratigraphic profiling as it allows vertical differentiation of units with different hydraulic properties, measured as relative hydraulic conductivity (KDPIL), a parameter that can be linked to absolute hydraulic conductivity [35]. DPIL measurements were performed in vertical steps of 0.4 m; see [37] for detailed information on the tool and data analysis routines. In contrast, Direct Push electrical conductivity was employed for vertical high resolution identification of clay containing, confining layers or lenses as an increase in subsurface electrical conductivity can be related to an increase of clay mineral content under non-saline conditions; see [33,38] for further information. Thirteen DPIL profiles as well as 10 Direct Push electrical conductivity logs were collected during the initial site investigation campaign.

Several tracer tests were additionally performed to characterize the flow field in the vicinity of the pumping well. Tracer test included heat tracer as well as salt tracer testing; the latter was combined with Direct Push electrical conductivity logging for enhanced in-situ monitoring, following the approach of Vienken et al. (2017) [39].

From the hydrodynamic point of view, three main flow directions are converging to the RBF plant (Figure 5): (i) a flow from the alluvial plain northern boundary (San Quirico-Carignano hills) given by the local meteoric recharge infiltrating through low-permeability deposits (from north to south flow) and then infiltrating in the aquifer; (ii) from east to west and from south to north flows. During the year, no seasonal changes are affecting these main flows directions given by the Serchio River filtration into the aquifer. The presence of vertical upward groundwater flow was tested by means of multi-level piezometers, but no relevant vertical gradient was highlighted. The piezometric level at the well field shows a direct relationship with the Serchio river stage, with the range between the maximum and minimum groundwater head no larger than 2 m between 2011 and 2016 (Figure 6).

River discharge measurements were performed in May 2015 in the Serchio river at two cross-sections corresponding to the 2 weirs (Figure 1), measuring the flow speed with an acoustic flow meter (acoustic digital current meter, OTT Messtechnik GmbH) every 50 cm along three verticals.

2.3. Hydrochemical Characterisation

Surface water and groundwater sampling campaigns for hydrochemical characterization were repeated both in dry (2 campaigns from May 2015 to October 2015 and 3 from May 2016 to October 2016) and in wet (2 campaigns from November 2015 to April 2016) periods. Data were collected from 25 different sampling points (including multilevel piezometer clusters at point PR_01-02-03-04-05-06) to assess hydrochemical seasonal variability. Groundwater samples were mostly collected using a low-flow pump after purging at least three well volumes and reaching stable values of the chemical-physical parameters monitored during sampling (Temperature, Electric Conductivity, Redox Potential, and Dissolved Oxygen). Surface water samples were collected 30 cm below the water surface, where the flow was sufficiently high to guarantee a representative sample. Sampling campaigns for isotopic characterization were performed on a selected sub sample of the monitoring points (14 in total) and on 2 different periods: dry season (May 2015) and wet season (January 2016).

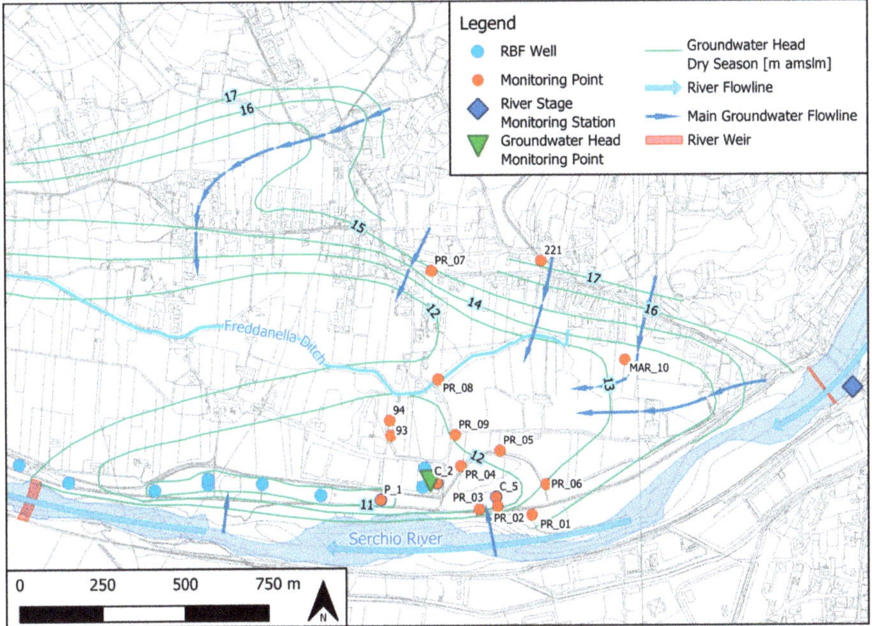

Figure 5. Monitoring network (groundwater head and hydrochemistry) in relation to the main groundwater flow directions. Piezometric head data from September 2003 (dry season; data from Studio Nolledi 2003).

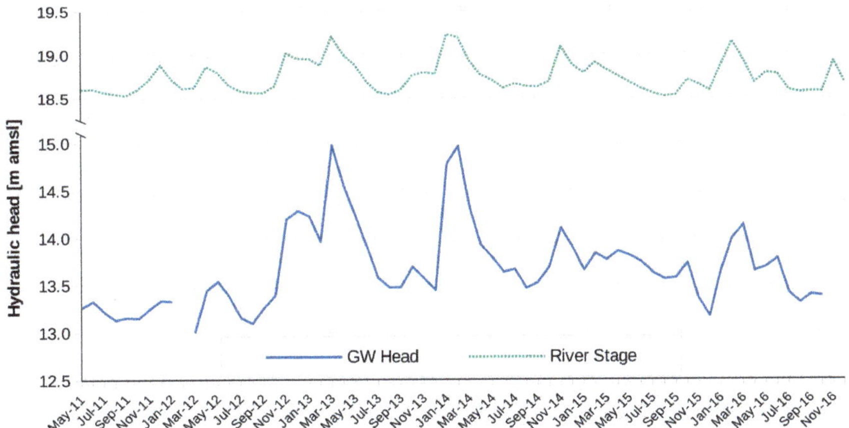

Figure 6. Monthly groundwater head data at the Serchio River RBF plant and Serchio river stage at the Ponte di Monte San Quirico monitoring station from May 2011 to December 2016 (data courtesy of Servizio Idrologico Regionale—Regione Toscana). The locations of the piezometer and the Serchio river monitoring station are marked in Figure 4.

Water samples were filtered by 0.45 μm filters (Minisart® NML syringe cellulose acetate filters) and separated in different aliquots. 50 mL PE bottles were taken for the analysis of anions and oxygen/hydrogen isotopes in each site. An additional 50 mL PE bottle (acidified with 0.5 mL of concentrated Suprapur HNO_3) was taken for the determination of cations and trace elements. After sampling all the aliquots were stored at 4 °C until analyzed.

Cations and trace elements were measured by inductively coupled plasma mass spectrometry (ICP-MS) using a Thermo-Scientific X Series instrument (Thermo, Bremen, Germany). The samples were previously diluted 1:10 by deionized Milli-Q water (resistivity of ca. 18.2 M$\Omega\times$cm) and a known amount of Re and Rh have been introduced as internal standard; in each analytical session. The analysis of samples was verified with that of the reference materials EU-L-1 and ES-L1 provided by SCP-Science (www.scpscience.com).

Anions were determined by ion chromatography using a Dionex ICS-1000 (Thermo, Bremen, Germany), calibrated by different dilutions of the Dionex "7-ion standard". The coherence of chemical data was verified by checking the ionic balance, as the sum of cation (expressed in meq/L) approaches that of anions with a relative error less than 5%. Hydrogen and oxygen isotope ratios were determined using the CRDS Los Gatos LWIA 24-d isotopic analyzer (Los Gatos Research, California, USA). The isotopic ratios of ^{2}H/^{1}H and ^{18}O/^{16}O are expressed as δ notation [δ = ((Rsample/Rstandard) $-$ 1) $\times$ 1000] with respect to the Vienna Standard Mean Ocean Water (VSMOW) international standard. Four bracketing standards were run throughout the analytical sessions, as indicated in Marchina et al. (2020) [40].

2.4. Groundwater Flow Numerical Modelling

A groundwater flow numerical model was built by means of the FREEWAT software [41,42] using the MODFLOW-2005 code [43]. FREEWAT, as its predecessor SID&GRID [44], is a free and open source platform, QGIS-integrated, for planning and management of ground- and interactions with surface–water resources. It is the main result of the HORIZON 2020 FREEWAT project (FREE and open source software tools for WATer resource management [45]). FREEWAT provides tools for archiving, analyzing and processing the groundwater data and information (e.g., sensor and field data on groundwater levels and hydrochemical data) along with post-processing tools for visualization and analysis of the model results [46,47]. MODFLOW is one of the codes most widely used in the world for numerical simulation of groundwater flow in aquifers. It solves the groundwater flow equation in three dimensions using a finite difference scheme. The FREEWAT software has been applied at several case studies; details may be found in, i.e., [48,49].

We simulated an active domain extending over 4.2 km^2 discretized using square cells 100 m^2 wide and two model layers, a silty-sandy superficial cover and the sandy-gravelly aquifer. The conceptual model for the simulated domain is shown in Figure 7. The northern boundary of the active domain, corresponding to the outcrop of impervious clayey sediments of the Monte S. Quirico-Carignano hills, was defined according to the geological setting of the area (Figure 3). Hydraulic connection with the Serchio river and exploitation of the aquifer through the Sant'Alessio well field were simulated. The model has been implemented to simulate two years, from 20 November 2014 until 30 November 2016. The simulation started with a steady-state stress period lasting 11 days, and then with 24 transient monthly stress periods. Calibration was run using a mixed trial and error and automatic parameter estimation approach [50]. Details and more information on the model implementation may be found in Rossetto et al. (2017) [51].

The calibrated model was then used to simulate the evolution of the RBF scheme in order to evaluate the change in recharge from the Serchio river to the aquifer. The following scenarios were simulated:

(a) Simulation of the aquifer river exchange prior to any riverbed modification or aquifer exploitation in order to evaluate pristine conditions: to this aim, we linearly reconstructed the riverbed profile prior to the weir construction;

(b) Simulation of the aquifer river exchange prior to any riverbed modification considering aquifer exploitation for 0.350 m^3/s;

(c) Simulation of the aquifer river exchange considering the Sant'Alessio weir in operation and aquifer exploitation for 0.350 m^3/s;

(d) Simulation of the aquifer river exchange considering the Sant'Alessio weir in operation and aquifer exploitation for 0.430 m^3/s.

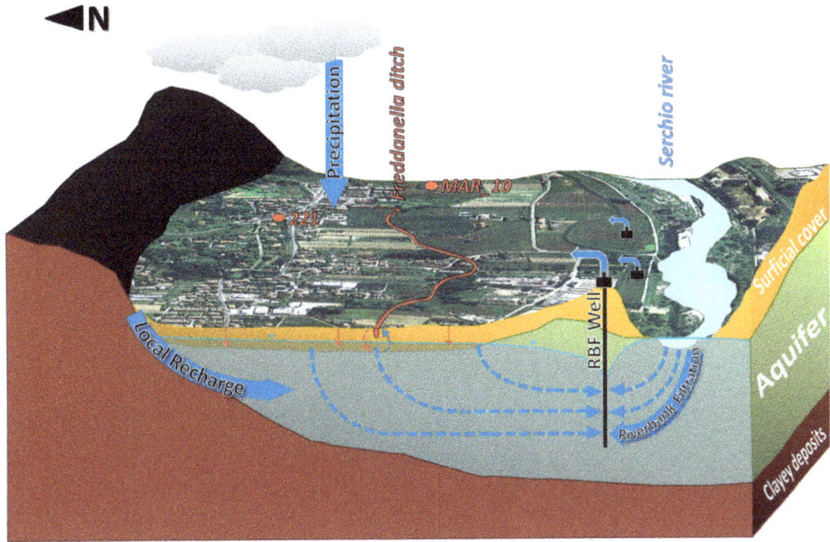

Figure 7. Conceptual model for the simulated domain; main hydrological elements are shown.

3. Result and Discussion

In order to answer the research questions, we jointly used the results of the performed analysis to provide data for the definition of the conceptual model and for the implementation of the numerical modelling (i.e., data coming from the hydrodynamic investigations) or to cross-validate the results of each approach.

Results from the hydrodynamics investigations informed the implementation of the numerical model and provided additional insights on the groundwater flow field. DPIL measurements revealed a strong variation of DPIL values on short vertical distances within the aquifer, while measured electrical conductivity values showed only minor variation, ranging between 10 mS/m and 20 mS/m. These are typical values for clean sands and gravels. Figure 8 depicts Direct Push logging results at an investigation point located approximately 90 m north-east of the pumping well C5 (Figure 5). Based on the results, the identification of hydrostratigraphic units or prevailing layers within the aquifer itself was not possible. Hence, Direct Push based pneumatic slug testing (see [52]) was performed at four locations in proximity of the pumping well C5 at different depth intervals between 3.6 and 10 m below ground surface to obtain further information on the distribution of hydraulic conductivity. Measured values ranged between 5.6×10^{-3} m/s and 3.3×10^{-4} m/s.

Thanks to the tracer tests a maximum bulk groundwater flow velocity of 26 m/d was estimated based on the first arrival times of the sodium-chloride tracer test breakthrough curve measurement. Despite the use of vertical high-resolution exploration, it was not possible to resolve the aquifer heterogeneity and to obtain in-depth understanding of the hydrodynamics using field investigations alone. This clearly highlights the necessity to approach experimental complex questions in groundwater hydrology by combining sophisticated site investigation and advanced modelling.

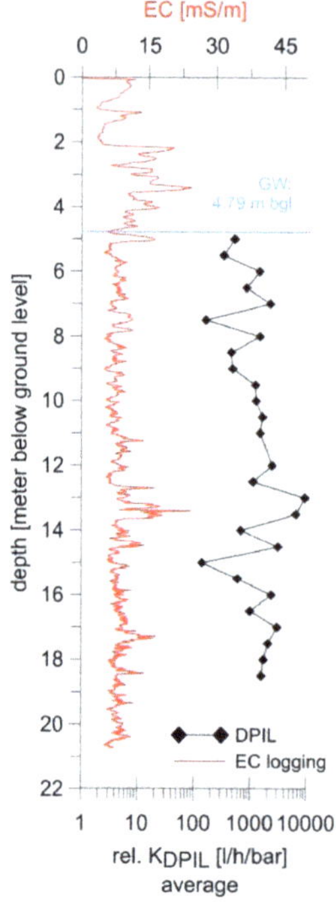

Figure 8. Direct Push profiling results (DPIL and Electrical Conductivity measurements).

The simulated water budget for one hydrological year (October 2015 to September 2016) is presented in Table 1. As far as the inflow terms, the Serchio river recharge constitutes the bulk of aquifer recharge in the Sant'Alessio plain with an overall yearly volume of about 58 Mm^3, in line with the loss of river water estimated by means of flow measurements (1.4 m^3/s). This recharge occurs within the reach Ponte di Monte San Quirico—Sant'Alessio weir (Figures 1 and 9). A smaller inflow term is constituted by meteoric recharge and inflow from the adjoining hills. Outflows from the systems are dominated by RBF pumping wells abstraction (about 16 Mm^3), outflow from the southern boundary (about 8 Mm^3), and drainage of the aquifer operated by the river south of the Sant'Alessio weir. The effect of the Freddanella ditch in draining the aquifer is negligible. The simulated head contour map (Figure 9) clearly shows this situation, presenting also a large drainage axis at about the middle of the plain.

Table 1. Simulated water budget for the Sant'Alessio aquifer for the hydrologic year October 2015 to September 2016.

	Inflow Terms		**Outflow Terms**	
	Cumulative Volume (m^3)	**% Over the Total**	**Cumulative Volume (m^3)**	**% Over the Total**
Storage	846	Negligible	938	Negligible
Inflow from the Monte S.Quirico—Carignano hills	207,089	0.3	15,640,341	26.4
Rainfall infiltration	673,287	1.1	658	Negligible
River leakage	58,154,744	98.2	38,439,232	64.9
Southern boundary of the domain	160,214	0.3	5,109,272	8.6
TOTAL	59,196,180	100.0	59,190,432	100.0

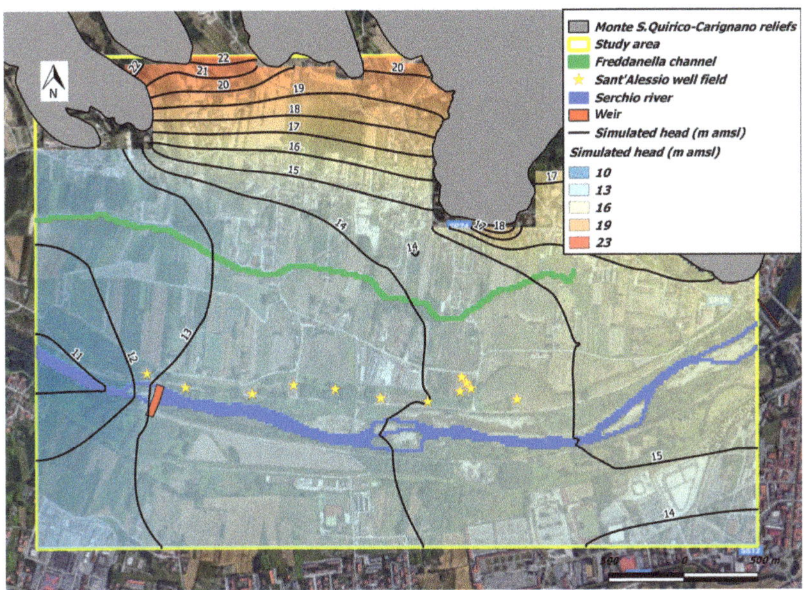

Figure 9. Simulated hydraulic head at the end of May 2016.

Table 2 shows the results of the analyses run in order to evaluate the change in recharge from the Serchio river to the aquifer induced during the evolution of the RBF scheme. In pristine conditions (riverbed not modified and wells not built), simulation results show that the natural recharge of the aquifer was of about 0.150 m^3/s. Construction of the first set of wells and their operation at an overall abstraction rate of 0.350 m^3/s would have caused an estimated recharge of about 0.490 m^3/s, hence inducing in the aquifer around 0.340 m^3/s. We then estimated in about 0.030 m^3/s the additional recharge provided as result of the construction in the 1990s of the Sant'Alessio weir. This recharge then creates an additional aquifer storage of about 1 Mm^3. Finally, increasing groundwater withdrawal up to 0.430 m^3/s (from initial 0.350 m^3/s) induces an additional recharge of about 0.09 m^3/s.

Table 2. Simulated changes in river recharge from pristine conditions to completion of the Serchio River RBF managed aquifer recharge scheme.

	No-Weir/No-Wells	No-Weir/Average Pumping (0.350 m^3/s)	Change in Recharge Rate (m^3/s)
Net Aquifer Recharge (m^3/s)	0.151	0.488	0.337
	No-Weir/Average Pumping (0.350 m^3/s)	Weir/Average Pumping (0.350 m^3/s)	Change in Recharge Rate (m^3/s)
Net Aquifer Recharge (m^3/s)	0.488	0.521	0.033
	Weir/Average Pumping (0.350 m^3/s)	Weir/High Pumping (0.430 m^3/s)	Change in Recharge Rate (m^3/s)
Net Aquifer Recharge (m^3/s)	0.521	0.609	0.088

These results show that the increase in recharge induced by pumping is larger than that caused by the weir construction. The latter brought a limited, yet important, change in storage. Figure 10 shows the groundwater head increase in the aquifer after the construction of the Sant'Alessio weir compared to the previous condition, with only the well field in operation. At the RBF scheme, the river/aquifer connection ensures, thanks to Serchio river discharge, stable recharge of the aquifer induced by the pumping wells and by the presence of the river weir. Hence, this also guarantees a limited drawdown of the water table.

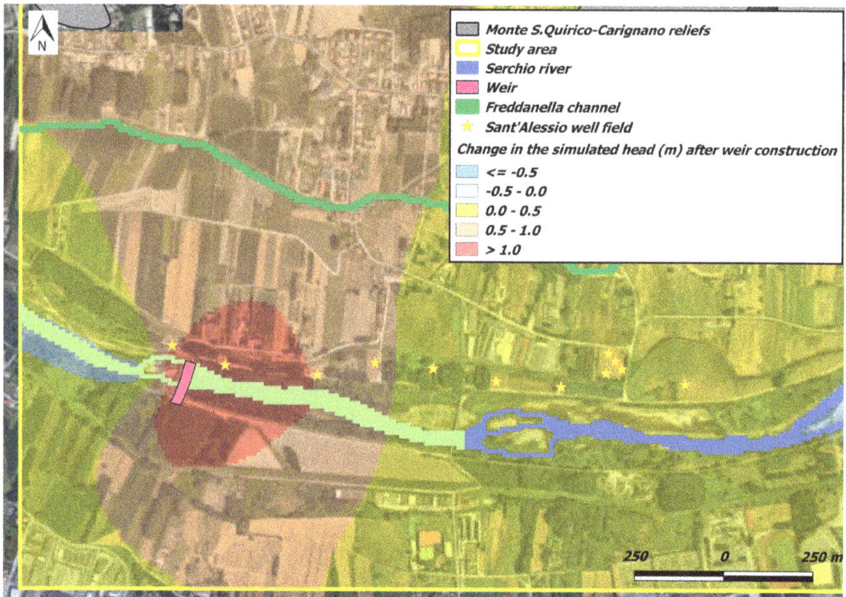

Figure 10. Simulated difference in groundwater head between pre- and post-operam of the Sant'Alessio weir.

As far as water hydrochemistry, according to the Piper classification diagrams (Figure 11 bottom), the samples mainly exhibit a Na-K-HCO$_3$ hydrochemical facies with a clear trend towards the Ca-Mg-SO$_4$ facies, perfectly aligned on a mixing line from the point 221 to the Serchio river (see Figures 5 and 7). The only exception to this trend is the sample at the "Freddanella" ditch, a canal draining the Sant'Alessio plain and also collecting untreated wastewater, hence to be considered as an end-member for potential groundwater pollution in the conceptual model [53]. This canal did not show any hydrochemical facies predominance during the dry season 2015.

The mixing process between surface water and ambient groundwater is also confirmed by the binary plots of conservative elements such as Cl$^-$ and Br$^-$ (Figure 11, top right). However, as for the geochemical facies, the mixing process of conservative elements is more pronounced during the dry season, when the chloride content of the Serchio river is not affected by dilution due to rainfall, and, at the same time, the Cl$^-$ concentrations are more uniform in the groundwater.

The δ^{18}O and δ^2H values of water are very uniform in most of the samples and they have isotopic values ranging between −8.2‰ and −4.9‰ for δ^{18}O, and −31.7‰ and −49‰ for δD. These values are plotted in Figure 11 (top left), together with the Global Meteoric Water Line (GMWL) calculated by Rozansky (1993) [54] and the Local Meteoric Water Lines calculated by Longinelli and Selmo (2003) [55] and La Ruffa and Panichi (2000) [56]. The results evidence a trend of precipitation typical for the west coast of Italy. This data is also confirmed by an average deuterium excess of 14.6‰, respect the GMWL, which is in the typical range for the northern Tuscan area (10.8–14.6‰; Longinelli and Selmo, 2003) [55].

Isotopic data describes a clear mixing process between the Serchio river water and the groundwater. The isotopic composition of the local recharge (point 221, in Figures 7 and 11) is the least depleted (in light isotopes, i.e., ^{16}O and ^{1}H), while the Serchio river reveals an isotopic signature similar to those of the groundwater pumped at the RBF wells. The groundwater in the Sant'Alessio plain and at the RBF plant shows a clear signature characterized by the Serchio river. This River originates in the Apuan Alps catchment (with maximum elevation of about 2000 m amsl, and average elevation of 717 m amsl) and therefore shows values more depleted (therefore richer in light isotopes) than the ones of the local recharge (originating in the San Quirico-Carignano hills, average elevation ~40 m amsl). In particular, the point sampled in the Sant'Alessio plain (MAR_10) is placed almost at the middle of the mixing line (Figure 11 left). The mixing processes highlighted by the $\delta^{18}O$ and δ^2H can be detected during the dry and, also, the wet season. However, during the wet period, the Serchio river shows an enrichment in heavy isotopes due to the impact of local precipitation and the influence of tributaries from lower sub-catchments in the upstream part of the basin and, at the same time, its influence on the groundwater of the Sant'Alessio plain (also at MAR_10) is higher due to its higher stage and discharge.

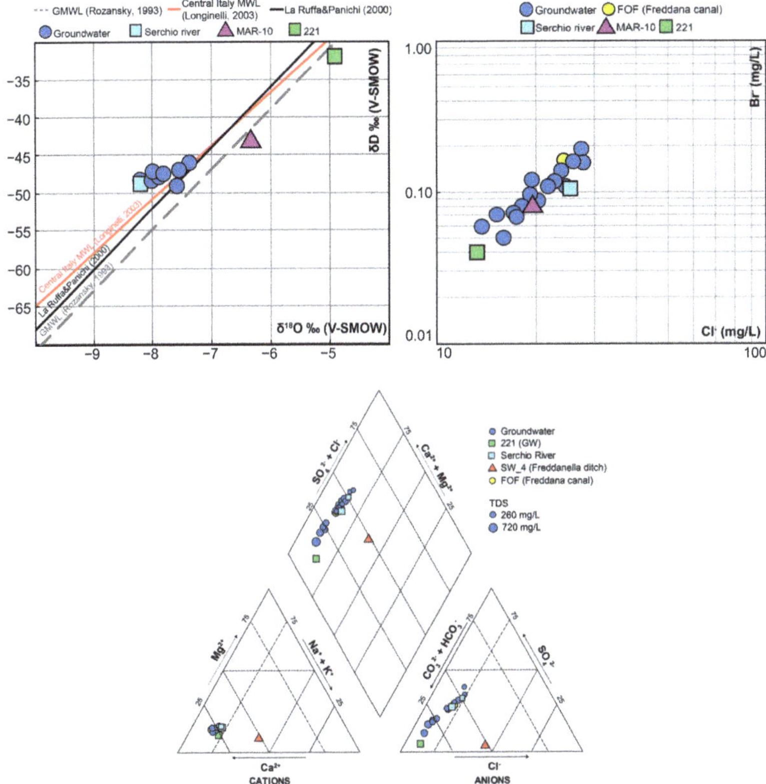

Figure 11. Piper diagram (**bottom**), Cl⁻ vs Br⁻ binary plot (**top right**) and $\delta^{18}O$ vs δ^2H binary plot (**top left**) of the samples collected during the dry season of 2015.

Therefore, as suggested both by the hydrodynamic and the hydrochemical data, the Serchio river and the local recharge from the northern part of the aquifer (groundwater entering the aquifer along the northern hilly border) can be considered end-member terms of the mixing system. Based on this assumption, different seasonal mixing fractions were computed using the distribution of lithium as conservative tracer. Lithium is the trace element showing the widest range of values among the end-members in the Sant'Alessio area (with high concentration in the Serchio river and values below the detection limit in the point 221; Table 3). The mixing fractions showed the extension of the Serchio river recharge within the Sant'Alessio aquifer and its importance at the RBF plant (Figure 12).

Table 3. Average values during dry and wet season of the presented analytes.

Sampling Point	Dry Season											Wet Season										
	Na^+ mg/L	K^+ mg/L	Ca^{2+} mg/L	Mg^{2+} mg/L	HCO_3^- mg/L	Cl^- mg/L	SO_4^{2-} mg/L	Br^- mg/L	Li µg/L	$\delta 18O$ ‰	δD ‰	Na^+ mg/L	K^+ mg/L	Ca^{2+} mg/L	Mg^{2+} mg/L	HCO_3^- mg/L	Cl^- mg/L	SO_4^{2-} mg/L	Br^- mg/L	Li µg/L	$\delta 18O$ ‰	δD ‰
93	13.97	2.23	71.42	9.47	234	16.83	55.79	0.06	7.33	-	-	16.19	2.09	104.90	11.34	191	29.04	103.31	0.16	7.63	-	-
94	14.66	2.86	79.81	10.41	228	15.10	47.70	0.07	10.26	-	-	-	-	-	-	-	-	-	-	-	-	-
221	13.35	3.80	52.67	4.59	249	20.85	14.67	0.11	0.77	−4.9	−31.7	14.45	3.85	69.84	5.42	270	22.67	10.48	0.14	0.74	−5.0	−30.6
C_2	15.67	1.60	68.32	9.54	164	26.15	97.05	0.12	11.66	-	-	19.05	1.62	83.55	10.53	150	27.00	98.88	0.13	13.30	−6.7	−46.6
C_5	19.30	1.55	69.45	9.44	155	31.54	101.77	0.11	12.29	−7.9	−47.6	17.82	1.55	74.93	9.07	144	27.38	103.87	0.13	13.13	−6.4	−43.7
MAR_10	15.16	2.18	63.30	10.09	320	24.42	91.97	0.08	7.93	−6.3	−43.3	21.39	2.81	117.27	13.84	224	28.93	80.76	0.10	10.32	−6.8	−42.3
P_1	17.73	1.65	71.88	9.73	162	29.58	116.05	0.14	13.21	-	-	22.63	1.42	86.18	12.01	140	42.87	129.96	0.28	15.22	-	-
PR_01a	17.77	1.98	74.06	9.24	162	33.49	101.05	0.16	13.25	−8.2	−48.1	16.62	1.37	79.39	10.26	120	27.80	102.04	0.18	12.31	−7.2	−44.0
PR_01b	15.84	2.36	61.09	8.40	87	17.30	60.80	0.07	9.29	−7.4	−45.7	-	-	-	-	-	-	-	-	-	-	-
PR_01c	16.06	1.65	73.14	8.60	159	19.95	62.98	0.07	12.22	−7.8	−47.3	20.40	1.37	83.29	10.53	174	30.66	110.71	0.21	12.02	−7.0	−44.0
PR_02a	13.76	1.48	59.21	8.52	209	20.25	62.65	0.13	10.91	−8.0	−47.1	19.77	1.48	80.87	10.79	155	32.91	126.16	0.20	14.78	−7.2	−43.7
PR_02b	15.77	1.70	67.90	9.08	133	25.69	99.80	0.12	13.95	−8.0	−48.2	21.02	1.78	78.75	11.63	127	33.23	125.07	0.34	13.89	-	-
PR_03b	13.80	2.34	68.65	7.61	144	25.50	67.20	0.16	11.51	−7.4	−46.5	-	-	-	-	-	-	-	-	-	-	-
PR_04a	14.70	1.74	57.40	7.66	181	21.67	69.99	0.11	9.98	−7.6	−46.9	18.97	1.64	79.04	10.37	143	32.99	119.66	0.17	9.99	-	-
PR_04b	14.41	1.54	61.04	8.51	159	24.05	72.62	0.12	11.11	−7.6	−49.0	20.14	1.65	83.55	10.78	153	31.09	122.80	0.14	13.02	−7.2	−46.1
PR_04c	14.75	1.36	59.65	8.05	123	27.24	81.39	0.17	11.18	−7.5	−46.7	-	-	-	-	-	-	-	-	-	-	-
PR_05a	13.89	1.34	52.22	8.03	135	20.72	58.03	0.07	10.10	-	-	19.43	1.51	77.18	10.38	231	31.89	121.68	0.18	12.20	−7.5	−46.6
PR_05b	14.03	1.24	59.17	8.91	178	18.88	68.47	0.05	8.08	-	-	19.98	1.99	87.29	10.97	163	31.12	118.68	0.33	13.42	-	-
PR_06a	11.11	1.18	48.91	7.01	160	25.31	75.40	0.10	8.86	-	-	19.54	1.91	76.68	10.85	133	31.73	125.62	0.17	15.10	−7.1	−44.6
PR_06b	11.40	1.86	52.51	7.68	129	23.50	83.93	0.08	11.40	-	-	-	-	-	-	-	-	-	-	-	-	-
PR_07	20.95	2.67	31.06	16.44	180	45.45	70.11	<LOD	1.72	-	-	24.92	3.56	137.75	20.02	264	39.40	43.00	0.09	2.99	−7.3	−43.4
PR_08	17.10	1.57	58.04	10.04	190	24.26	59.84	<LOD	7.60	-	-	17.38	2.50	110.06	11.99	196	32.95	100.60	0.18	9.20	−6.8	−42.4
PR_09	13.49	3.46	55.24	9.88	190	16.56	59.66	<LOD	4.71	-	-	16.56	3.84	102.96	11.09	214	27.03	95.95	0.16	7.50	−7.1	−43.0
Freddanella Ditch	40.02	6.39	62.35	6.77	294	112.21	57.11	0.11	5.43	-	-	47.04	4.45	92.45	8.43	205	94.42	41.95	0.11	6.14	-	-
Serchio River	20.92	1.59	64.95	9.35	144	29.06	97.60	0.19	11.27	−8.2	−48.1	15.15	1.35	58.69	8.10	153	23.03	67.27	0.11	12.30	−6.5	−44.2

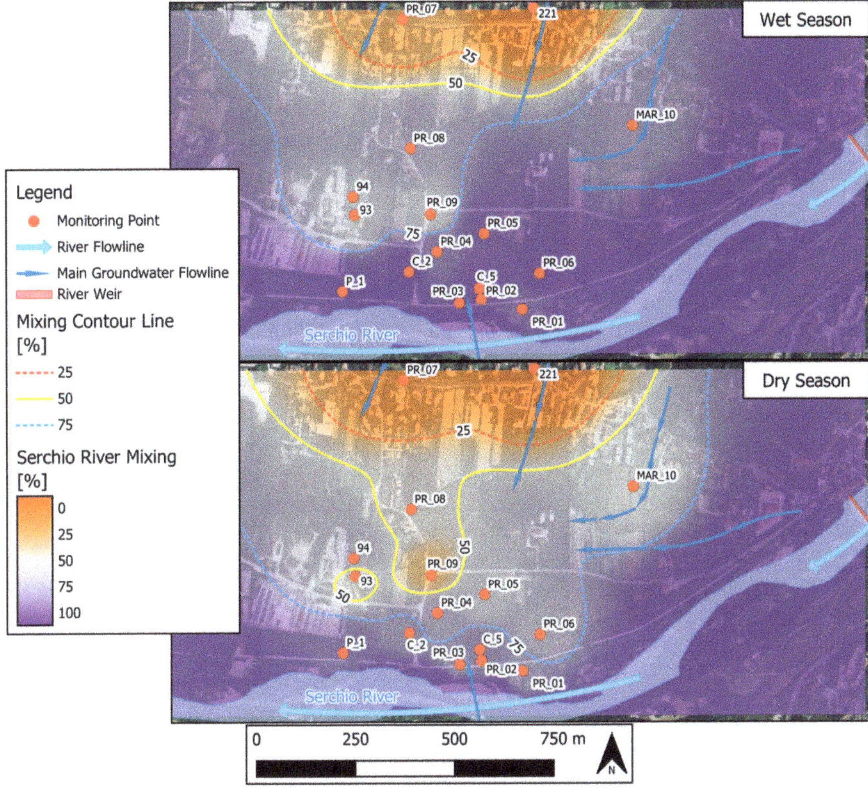

Figure 12. Interpolation showing the Serchio river influence on the Sant'Alessio aquifer and the Serchio River RBF, using lithium as tracer.

These results confirm that the main sources of the RBF pumping wells is the Serchio river water and that the groundwater at the Sant'Alessio plain is mainly characterized by mixing between precipitation occurring in the higher part of the plain and the river water. As an example, the percentage of Serchio river water at the RBF wells ranges from nearly 100% (during the wet season) to around 80% (in the dry period), while for the point MAR_10 sample fraction of about 50% and 70% of Serchio river water may be estimated for the dry and wet season, respectively (Figure 11). Based on chemical mixing, a precautionary amount of abstracted Serchio river water is estimated to be on average 13.6 Mm^3/year, this is 85% of the total amount of water abstracted in a year (~16 Mm^3). As such, only the 15% (2.4 Mm^3) of abstracted groundwater comes from ambient groundwater.

4. Conclusions

In this study, we presented a detailed description of the Serchio River RBF site, and we used a multidisciplinary approach entailing hydrodynamics, hydrochemical, and numerical modelling methods in order to evaluate the change in recharge from the Serchio river to the aquifer due to the building of the RBF infrastructures along the river, and the ratio of bank filtrate to ambient groundwater abstracted at such RBF scheme.

The use of Direct Push technology in combination with traditional site investigation approaches was feasible to provide critical information for reliable characterization of subsurface properties and to inform first the conceptual model and then the implementation of the numerical model. Thereby, the use of DPIL proved successful to produce quantitative information about vertical distribution of hydraulic conductivities in a high K environment. Direct Push slug testing as well as tracer testing was successfully used for the in-situ characterization of hydraulic conductivity variations and to capture high flow velocities at the river/aquifer interface in the vicinity of the most western pumping well that were in the order of m/day.

Our results show that when discussing surface water/groundwater interactions, the Serchio river water constitutes the main source of water for the pumping wells at the RBF scheme. By means of numerical modelling of groundwater flow we simulated the increase in induced recharge, starting from natural conditions, up to present conditions with the Sant'Alessio weir and the vertical wells in full operation. The amount of induced recharge caused by the pumping wells is by far larger than that related to the construction of the Sant'Alessio weir raising the riverbed of about 1.5 m. The weir is however important as it contributes to guarantee a small drawdown in the well-field area. Hydrochemical investigations confirmed and specified the results of the numerical modelling activities. Groundwater at the Sant'Alessio plain is mainly characterized by mixing between precipitation occurring in the higher part of the plain and the river water. This mix seasonally varies depending on the Serchio river discharge, with a marked decrease of the induced component during the Serchio low-flow periods. As an example, the percentage of abstracted Serchio river water at the RBF wells ranges from nearly 100% (in the wet season) to around 80% (in the dry season). Based on chemical mixing, a precautionary amount of abstracted Serchio river water is estimated to be on average 13.6 Mm^3/year, which is 85% of the total amount of water abstracted in a year (~16 Mm^3). Finally, we wish to stress that the assessment of the induced recharge and of the ratio of bank filtrate to ambient groundwater is very important at RBF sites as it may provide relevant insights on the risk of groundwater contamination. This is especially true for the Serchio River RBF as it was already affected by surface water related pollution at the beginning of the 2000 [29]. Disregarding the interactions between surface and groundwater brought to search the cause of contamination in the adjoining agricultural areas and to issue inappropriate local regulations for farming activities [29]. All these issues were discussed in Rodríguez-Escales et al. (2018) [57] in comparison with other MAR sites across the Mediterranean basin: the risk of failure of the Serchio River RBF scheme was defined low. Nowadays, the Serchio River RBF scheme is a reliable site for continuous water supply.

RBF is a worldwide recognized MAR technique for supplying good quality and reliable amounts of water. However, in many cases and countries (including Italy) government authorities do not appear to acknowledge that groundwater extraction is sustained by induced recharge through the river bank and it is not entirely from natural recharge/ambient groundwater. Additionally, several studies take this connection for granted and often only focus on bank filtration treatment capabilities disregarding the amount of recharge induced by the pumping wells. This "induced" component is even not taken into account in the definition of this MAR type in the literature (i.e., [58]). In fact, this remains one of the few MAR schemes where reference to the recharge mechanism is not adequately made in the definition, conversely to, i.e., spreading methods.

Because of this, we suggest to include the term "induced" to complete the definition of this type of MAR technique, hence defining it *"Induced RiverBank Filtration"*. This way we make reference to the fact that the bank filtration is not completely due to natural recharge, as in several cases of natural surface water/groundwater interactions, but it may be partly/almost all human-made.

Author Contributions: Conceptualization, R.R. and T.V.; methodology, R.R., T.V., A.B., C.M.; software, R.R. and G.D.F.; validation, R.R. and T.V.; formal analysis, R.R., A.B., T.V., C.M., G.D.F.; investigation, R.R., A.B., T.V., C.M., G.D.F.; data curation, R.R., A.B., T.V., G.D.F.; writing—original draft preparation, R.R. and A.B.; writing—review and editing, R.R., A.B., T.V., C.M., G.D.F.; visualization, A.B., C.M. and G.D.F.; supervision, R.R.; project administration, R.R. and G.M.; funding acquisition, R.R. All authors have read and agreed to the published version of the manuscript.

Funding: This research was funded by the European Union, Grant Number 619120.

Acknowledgments: The work presented here was performed during the EU co-funded project FP7 MARSOL (Grant Agreement No. 619120). This research also exploited results from the H2020 FREEWAT project, funded by the European Union within the Horizon 2020 research and innovation programme (Grant number 642224).

Alessio Barbagli (now Department of Physics and Earth Sciences, University of Ferrara, Italy) took part in the research activities during his three-year PhD in Agrobiosciences at the Institute of Life Sciences, Scuola Superiore Sant'Anna. Giovanna De Filippis (now AECOM URS, Milano, Italy) and Chiara Marchina (now Department of Land, Environment, Agriculture and Forestry, University of Padova, Italy) took part to the research activities during a post-doc scholarship at the Institute of Life Sciences, Scuola Superiore Sant'Anna. Thomas Vienken (now also Weihenstephan-Triesdorf University of Applied Sciences, TUM Campus Straubing for Biotechnology and Sustainability, Straubing, Germany) performed these activities while at UFZ. The authors thanks GEAL S.p.a. for the technical support provided and granting access to the well-field during the research activities.

Conflicts of Interest: The authors declare no conflict of interest.

References

1. WHO; UNICEF. Progress on Drinking Water, Sanitation and Hygiene: Joint Monitoring Programme 2017 Update and SDG Baselines. Available online: http://www.who.int/water_sanitation_health/publications/jmp-2017/en/ (accessed on 22 October 2020).

2. Meffe, R.; de Bustamante, I. Emerging organic contaminants in surface water and groundwater: A first overview of the situation in Italy. *Sci. Total Environ.* **2014**, *481*, 280–295. [CrossRef] [PubMed]

3. Paldor, A.; Shalev, E.; Katz, O.; Aharonov, E. Dynamics of saltwater intrusion and submarine groundwater discharge in confined coastal aquifers: A case study in northern Israel. *Hydrogeol. J.* **2019**, *27*, 1611–1625. [CrossRef]

4. Wanner, P. Plastic in agricultural soils—A global risk for groundwater systems and drinking water supplies?—A review. *Chemosphere* **2021**, *264*, 128453. [CrossRef] [PubMed]

5. Dillon, P.; Stuyfzand, P.; Grischek, T. Sixty years of global progress in managed aquifer recharge. *Hydrogeol. J.* **2019**, *27*, 1–30. [CrossRef]

6. Dillon, P.; Fernández Escalante, E.; Megdal, S.B.; Massmann, G. Managed Aquifer Recharge for Water Resilience. *Water* **2020**, *12*, 1846. [CrossRef]

7. Fernandez Escalante, E.; Henao Casas, J.D.; Vidal Medeiros, A.M.; San Sebastián Sauto, J. Regulations and guidelines on water quality requirements for Managed Aquifer Recharge. International comparison. *Acque Sotter. Ital. J. Groundw.* **2020**, *9*. [CrossRef]

8. Dillon, P. Future management of aquifer recharge. *Hydrogeol. J.* **2005**, *13*, 313–316. [CrossRef]

9. Stefan, C.; Ansems, N. Web-based global inventory of managed aquifer recharge applications. *Sustain. Water Resour. Manag.* **2018**, *4*, 153–162. [CrossRef]

10. Umar, D.A.; Ramli, M.F.; Aris, A.Z.; Sulaiman, W.N.A.; Kura, N.U.; Tukur, A.I. An overview assessment of the effectiveness and global popularity of some methods used in measuring riverbank filtration. *J. Hydrol.* **2017**, *550*, 497–515. [CrossRef]

11. Kühn, W.; Müller, U. Riverbank filtration. *JAWWA* **2000**, *92*, 60–69. [CrossRef]

12. Hiscock, K.M.; Grischek, T. Attenuation of groundwater pollution by bank filtration. *J. Hydrol.* **2002**, *266*, 139–144. [CrossRef]

13. Hunt, H.; Schubert, J.; Ray, C. Conceptual Design of Riverbank Filtration Systems. In *Riverbank Filtration: Improving Source Water Quality*; Springer: Berlin/Heidelberg, Germany, 2002; pp. 19–27.

14. Ray, C. Riverbank Filtration Concepts and Applicability to Desert Environments. In *Riverbank Filtration for Water Security in Desert Countries*; Springer: Berlin/Heidelberg, Germany, 2011; pp. 1–4.

15. Shankar, V.; Eckert, P.; Ojha, C. Transient three-dimensional modeling of riverbank filtration at Grind well field, Germany. *Hydrogeol. J.* **2009**, *17*, 321–326. [CrossRef]

16. Ray, C.; Grischek, T.; Schubert, J.; Wang, Z.; Speth, T.F. A perspective of riverbank filtration. *JAWWA* **2002**, *94*, 149–160. [CrossRef]

17. Hoppe-Jones, C.; Oldham, G.; Drewes, J.E. Attenuation of total organic carbon and unregulated trace organic chemicals in U.S. riverbank filtration systems. *Water Res.* **2010**, *44*, 4643–4659. [CrossRef] [PubMed]

18. Massmann, G.; Dünnbier, U.; Heberer, T.; Taute, T. Chemosphere Behaviour and redox sensitivity of pharmaceutical residues during bank filtration–Investigation of residues of phenazone-type analgesics. *Chemosphere* **2008**, *71*, 1476–1485. [CrossRef] [PubMed]

19. Sprenger, C.; Lorenzen, G.; Hülshoff, I. Vulnerability of bank filtration systems to climate change. *Sci. Total Environ.* **2011**, *409*, 655–663. [CrossRef]

20. Ahmed, A.K.A.; Marhaba, T.F. Review on river bank filtration as an in situ water treatment process. *Clean Technol. Environ. Policy* **2017**, *19*, 349–359. [CrossRef]

21. Barthel, R.; Banzhaf, S. Groundwater and Surface Water Interaction at the Regional-scale—A Review with Focus on Regional Integrated Models. *Water Resour. Manag.* **2016**, *30*, 1–32. [CrossRef]

22. Cook, P.G. Quantifying river gain and loss at regional scales (Review). *J. Hydrol.* **2015**, *531*, 749–758. [CrossRef]

23. Sottani, A.; Vielmo, A. Groundwater conservation and monitoring activities in the middle Brenta River plain (Veneto Region, Northern Italy): Preliminary results about aquifer recharge. *Acque Sotter. Ital. J. Groundw.* **2014**, *3*. [CrossRef]

24. Wang, W.-S.; Oswald, S.E.; Gräff, T.; Lensing, H.-J.; Liu, T.; Strasser, D.; Munz, M. Impact of river reconstruction on groundwater flow during bank filtration assessed by transient three-dimensional modelling of flow and heat transport. *Hydrogeol. J.* **2020**, *28*, 723–743. [CrossRef]

25. Barbagli, A.; Jensen, B.N.; Raza, M.; Schueth, C.; Rossetto, R. Assessment of soil buffer capacity on nutrients and pharmaceuticals in nature-based solution applications. *Environ. Sci. Pollut. Res.* **2019**, *26*, 759–774. [CrossRef] [PubMed]

26. Rossetto, R.; Barbagli, A.; Borsi, I.; Mazzanti, G.; Vienken, T.; Bonari, E. Site investigation and design of the monitoring system at the Sant'Alessio Induced RiverBank Filtration plant (Lucca, Italy). *Rend. Online Soc. Geol. Ital.* **2015**, *35*, 248–251. [CrossRef]

27. Studio Nolledi. Indagini Idrogeologiche Finalizzate All'individuazione Dell'area di Salvaguardia del Campo Pozzi di sant'Alessio—Lucca Art. 21 del D.Lgs. 152/99. Prima Fase Individuazione e Caratterizzazione Dell'area di Rispetto. 2003; (Comune di Lucca Unpublished Report). Available online: file:///C:/Users/mdpi/AppData/Local/Temp/D39.pdf (accessed on 7 December 2020).

28. Rossetto, R.; Barbagli, A.; De Filippis, G.; Marchina, C.; Mazzanti, G.; De Caterini, A. Case Study 19: The Serchio River Bank Filtration for Drinking Water Supply in Sant'Alessio area of Lucca, Italy. In *Managing Aquifer Recharge: A Showcase for Resilience and Sustainability*; Zheng, Y., Ross, A., Villholth, K.G., Dillon, P., Eds.; A Unesco-Iah-Gripp Publication, 2002; (to be Advised). Available online: https://recharge.iah.org/unesco-exemplary-mar-projects-booklet (accessed on 8 December 2020).

29. Rossetto, R.; Bockelmann-Evans, B. Modellazione numerica del flusso e del trasporto di soluti ai fini dell'investigazione dei processi di trasporto dell'erbicida terbutilazina nel sistema acquifero della pianura di S. Alessio (Lucca). *G. Geol. Appl.* **2007**, *5*, 29–44.

30. EPA. *Expedited Site Assessment Tools for underground Storage Tank Sites. A Guide for Regulators*; U.S. Government Printing Office: Washington, DC, USA, 1997.

31. Butler, J.J. Hydrogeological Methods for estimation of Spatial Variations in Hydraulic Conductivity. In *Hydrogeophysics*; Rubin, Y., Hubbard, S.S., Eds.; Springer: Berlin/Heidelberg, Germany, 2005; pp. 23–58.

32. Dietrich, P.; Leven, C. Direct Push-Technologies. In *Groundwater Geophysics*; Kirsch, R., Ed.; Springer: Berlin/Heidelberg, Germany, 2006; pp. 321–340.

33. McCall, W.; Nielsen, D.M.; Farrington, S.P.; Christy, T.M. Use of Direct-Push Technologies in Environmental Site Characterization and Ground-Water Monitoring. In *Practical Handbook of Environmental Site Characterization and Ground-Water Monitoring*; Nielsen, D.M., Ed.; CRC Press Taylor and Francis Group: Boca Raton, FL, USA, 2006; pp. 345–471.

34. Liu, G.S.; Butler, J.J.; Reboulet, E.; Knobbe, S. Hydraulic conductivity profiling with direct push methods. *Grundwasser* **2012**, *17*, 19–29. [CrossRef]

35. Vienken, T.; Leven, C.; Dietrich, P. Use of CPT and other direct push methods for (hydro-) stratigraphic aquifer characterization—A field study. *Can. Geotech. J.* **2012**, *49*, 197–206. [CrossRef]

36. Vienken, T.; Kreck, M.; Hausmann, J.; Werban, U.; Dietrich, P. Innovative strategies for high resolution site characterization: Application to a flood plain. *Ital. J. Groundw.* **2014**, *138*, 7–14. [CrossRef]

37. Dietrich, P.; Butler, J.J.; Faiss, K. A rapid method for hydraulic profiling in unconsolidated formations. *Ground Water* **2008**, *46*, 323–328. [CrossRef]

38. Sellwood, S.M.; Healey, J.M.; Birk, S.; Butler, J.J. Direct-push hydrostratigraphic profiling: Coupling electrical logging and slug tests. *Ground Water* **2005**, *43*, 19–29. [CrossRef]

39. Vienken, T.; Huber, E.; Kreck, M.; Huggenberger, P.; Dietrich, P. How to chase a tracer—Combining conventional salt tracer testing and direct push electrical conductivity profiling for enhanced aquifer characterization. *Adv. Water Resour.* **2017**, *99*, 60–66. [CrossRef]

40. Marchina, C.; Zuecco, G.; Chiogna, G.; Bianchini, G.; Carturan, L.; Comiti, F.; Engel, M.; Natali, C.; Borga, M.; Penna, D. Alternative methods to determine the δ^2H-δ^{18}O relationship: An application to different water types. *J. Hydrol.* **2020**, *587*, 124951. [CrossRef]

41. De Filippis, G.; Borsi, I.; Foglia, L.; Cannata, M.; Velasco Mansilla, V.; Vasquez-Suñe, E.; Ghetta, M.; Rossetto, R. Software tools for sustainable water resources management: The GIS-integrated FREEWAT platform. *Rend. Online Soc. Geol.* **2017**, *42*, 59–61. [CrossRef]

42. Foglia, L.; Borsi, I.; Mehl, S.; De Filippis, G.; Cannata, M.; Vasquez-Sune, E.; Criollo, R.; Rossetto, R. Freewat, A Free and Open Source, GIS-Integrated, Hydrological Modeling Platform. *Groundwater* **2018**, *56*, 521–523. [CrossRef] [PubMed]

43. Harbaugh, A.W. *MODFLOW-2005, The U.S. Geological Survey Modular Ground-Water Model—The Ground-Water Flow Process*; Techniques and Methods 6-A16; U.S. Geological Survey: Reston, VA, USA, 2005. [CrossRef]

44. Rossetto, R.; Borsi, I.; Schifani, C.; Bonari, E.; Mogorovich, P.; Primicerio, M. SID&GRID: Integrating hydrological modeling in GIS environment. *Rend. Online Soc. Geol. Ital.* **2013**, *24*, 282–283.

45. Rossetto, R.; Borsi, I.; Foglia, L. Freewat: Free and open source software tools for WATer resource management. *Rend. Online Soc. Geol. Ital.* **2015**, *35*, 252–255. [CrossRef]

46. Criollo, R.; Velasco, V.; Nardi, A.; Vries, L.M.; Riera, C.; Scheiber, L.; Jurado, A.; Brouyère, S.; Pujades, E.; Rossetto, R.; et al. AkvaGIS: An open source tool for water quantity and quality management. *Comput. Geosci.* **2020**, *127*, 123–132. [CrossRef]

47. Cannata, M.; Neumann, J.; Rossetto, R. Open source GIS platform for water resource modelling: FREEWAT approach in the Lugano Lake. *Spat. Inf. Res.* **2018**, *26*, 241–251. [CrossRef]

48. De Filippis, G.; Pouliaris, C.; Kahuda, D.; Vasile, T.A.; Manea, V.A.; Zaun, F.; Panteleit, B.; Dadaser-Celik, F.; Positano, P.; Nannucci, M.S.; et al. Spatial Data Management and Numerical Modelling: Demonstrating the Application of the QGIS-Integrated FREEWAT Platform at 13 Case Studies for Tackling Groundwater Resource Management. *Water* **2020**, *12*, 41. [CrossRef]

49. Joodavi, A.; Izady, A.; Karbasi Maroof, M.T.; Majidi, M.; Rossetto, R. Deriving optimal operational policies for off-stream man-made reservoir considering conjunctive use of surface- and groundwater at the Bar dam reservoir (Iran). *J. Hydrol. Reg. Stud.* **2020**, *31*, 100725. [CrossRef]

50. La Vigna, F.; Hill, M.C.; Rossetto, R.; Mazza, R. Parameterization, sensitivity analysis, and inversion: An investigation using groundwater modeling of the surface-mined Tivoli-Guidonia basin (Metropolitan City of Rome, Italy). *Hydrogeol. J.* **2016**, *24*, 1423–1441. [CrossRef]

51. Rossetto, R.; Barbagli, A.; De Filippis, G.; Marchina, C.; Di Bartolo, S.; Bonari, E.; Sabbatini, T.; Triana, F.; Tozzini, C.; Ercoli, L.; et al. Deliverable 8.4. Report on the Induced RiverBank Filtration MAR Plant at Sant'Alessio (Lucca, Italy). FP7 MARSOL (Demonstrating Managed Aquifer Recharge as a Solution to Water Scarcity and Drought). Available online: http://marsol.eu/ (accessed on 28 November 2020).

52. Butler, J.J., Jr.; Garnett, E.J.; Healey, J.M. Analysis of Slug Tests in Formations of High Hydraulic Conductivity. *Groundwater* **2003**, *41*, 620–631. [CrossRef] [PubMed]

53. Barbagli, A. Analysis of Water-Soil Interaction in Drainage Water Phyto-Treatment and in Aquifer Recharge Schemes. Ph.D. Thesis, Scuola Superiore Sant'Anna, Pisa, Italy. Available online: https://dta.santannapisa.it (accessed on 28 November 2020).

54. Rozanski, K.; Araguás-Araguás, L.; Gonfiantini, R. *Isotopic Patterns in Modern Global Precipitation, in Climate Change in Continental Isotopic Records*; Swart, P.K., Lohmann, K.C., Mckenzie, J., Savin, S., Eds.; American Geophysical Union: Washington, DC, USA, 1993. [CrossRef]

55. Longinelli, A.; Selmo, E. Isotopic composition of precipitation in Italy: A first overall map. *J. Hydrol.* **2003**, *270*, 75–88.

56. La Ruffa, G.; Panichi, C. *Caratterizzazione Chimico-Isotopica Delle Acque Fluviali: Il Caso Del Fiume Arno*; Istituti Editoriali Poligrafici Internazionali: Pisa, Italy, 2000; p. 101.

57. Rodríguez-Escales, P.; Canelles, A.; Sanchez-Vila, X.; Folch, A.; Kurtzman, D.; Rossetto, R.; Fernández-Escalante, E.; Lobo-Ferreira, J.P.; Sapiano, M.; San-Sebastián, J.; et al. A risk assessment methodology to evaluate the risk failure of managed aquifer recharge in the Mediterranean Basin. *Hydrol. Earth Syst. Sci.* **2018**, *22*, 3213–3227. [CrossRef]

58. Natural Resource Management Ministerial Council, Environment Protection and Heritage Council National Health and Medical Research Council. Australian Guidelines for Water Recycling, Managing Health and Environmental Risks, Vol 2C: Managed Aquifer Recharge. Available online: http://webarchive.nla.gov.au/gov/20130904195601/http://www.environment.gov.au/water/publications/quality/water-recycling-guidelines-mar-24.html (accessed on 29 September 2020).

Publisher's Note: MDPI stays neutral with regard to jurisdictional claims in published maps and institutional affiliations.

 *hydrology*

Article

Meeting SDG6 in the Kingdom of Tonga: The Mismatch between National and Local Sustainable Development Planning for Water Supply

Ian White [1,*], Tony Falkland [2] and Taaniela Kula [3]

[1] Fenner School of Environment and Society, Australian National University, Canberra, ACT 0200, Australia

[2] Island Hydrology Services, 9 Tivey Place, Hughes, Canberra, ACT 2605, Australia; tony.falkland@netspeed.com.au

[3] Natural Resources Division, Ministry of Lands and Natural Resources, Nuku'alofa, Tonga; tkula@naturalresources.gov.to

* Correspondence: ian.white@anu.edu.au; Tel.: +61-418-262-881

Received: 14 September 2020; Accepted: 12 October 2020; Published: 22 October 2020

Abstract: UN Sustainable Development Goal 6 challenges small island developing states such as the Kingdom of Tonga, which relies on variable rainwater and fragile groundwater lenses for freshwater supply. Meeting water needs in dispersed small islands under changeable climate and frequent extreme events is difficult. Improved governance is central to better water management. Integrated national sustainable development plans have been promulgated as a necessary improvement, but their relevance to island countries has been questioned. Tonga's national planning instrument is the Tonga Strategic Development Framework, 2015–2025 (TSDFII). Local Community Development Plans (CDPs), developed by rural villages throughout Tonga's five Island Divisions, are also available. Analyses are presented of island water sources from available census and limited hydrological data, and of the water supply priorities in TSDFII and in 117 accessible village CDPs. Census and hydrological data showed large water supply differences between islands. Nationally, TDSFII did not identify water supply as a priority. In CDPs, 84% of villages across all Island Divisions ranked water supply as a priority. Reasons for the mismatch are advanced. It is recommended that improved governance in water in Pacific Island countries should build on available census and hydrological data and increased investment in local island planning processes.

Keywords: groundwater; rainwater harvesting; climate variability; small island developing states; improved water governance; national sustainable development plans; SDG6; community participation

1. Introduction

The United Nations (UN) 2030 Sustainable Development Goal (2030) for water and sanitation, i.e., "Ensure availability and sustainable management of water and sanitation for all" (SDG 6 [1]), presents significant challenges for small island development states (SIDS), particularly in terms of securing universal and equitable access to safe and affordable water for all. Limited resources and institutions, dispersed island communities, increasing demands, scarce fresh groundwater resources vulnerable to salinization and pollution, variable and changing climates driven by large-scale ocean-atmosphere interactions and frequent, devastating, extreme events such as tropical cyclones, droughts and floods compound the difficulties of ensuring that communities have access to adequate and safe freshwater supplies [2,3], which is fundamental for sustainable development.

Faced with these difficulties, Oceania was one of the few regions in the world that did not meet the 2015 Millennium Development Goals for water and sanitation [4]. Better governance and increased water information have been claimed to be key factors in improving water security in Pacific island countries (PICs), particularly when facing impacts of climate change [5,6]. National strategic development strategies (NSDS) for SIDS have been promoted as a key mechanism for improved governance and for fulfilling government commitments to local, regional and international agendas on sustainable development [7], particularly SDGs, the 2005 Mauritius Strategy for the Further Implementation of the Programme of Action for the Sustainable Development of SIDS [8] and the 2014 SIDS Accelerated Modalities of Action (SAMOA) Pathway resolution [6].

NSDSs are used to identify national priorities, to allocate resources to government agencies and to monitor outcomes. Initially, creating an NSDS involved a two-phase approach, with national assessments in phase one, followed by selected interventions in phase two [6]. Inexperience, limited resources and institutions in some SIDS meant that donor and funding agency-supported external consultants played central roles in driving NSDS processes assisted by senior national bureaucrats. These planning processes tend to be top-down prescriptions, with the planning priorities and expected outcomes being predominantly economically focused. The applicability of similar, transplanted governance mechanisms to PICs has been questioned [9].

One of the key characteristics of SIDS, and particularly those with dispersed island communities, is their well-developed local institutions which are particularly suited to bottom-up, priority-setting processes [10]. Their advantage is inclusiveness, but a draw-back of bottom-up processes is the time involved to reach agreement or consensus [11]. Planners are faced with a dilemma: are efficient top-down national planning processes able to address the priorities of local communities, or are lengthy, expensive community processes required? The South Pacific Kingdom of Tonga, reliant predominantly on fresh groundwater lenses and rainwater harvesting for water supply, presents a unique opportunity to compare priorities in water supply identified by both top-down and bottom-up planning processes.

The government of Tonga, with support from the Asian Development Bank, developed the Tonga Strategic Development Framework 2015–2015 (TSDFII) through a high-level consultation process over three months in late 2014 [12]. TSDFII incorporated lessons learnt in the preceding Tonga Strategic Development Framework 2011–2014. TSDFII identified 29 highest priority issues as key planned "Organisational Outcomes" (OOs) for Ministries, Departments and agencies, while 153 more issues raised during consultations are listed as strategic concepts (SCs) under the OOs. These are meant to be planning aids to inform Ministry corporate plans and decisions regarding budgeting, staffing and reports.

From 2007 to 2016, a very extensive, nation-wide Community Development Plan (CDP) process, supported by a range of agencies, involved most rural villages throughout Tonga's five Island Divisions. Communities identified and ranked priority development issues in their village and then built and endorsed their CDP. The plans prioritized the most urgent issues in each village in terms of women's, youths' and men's perspectives, and 136 CDPs were presented in 2016 [13].

One of the difficulties in assessing water security and priorities in PICs is the limited information on water availability and its use, particularly in rural areas [3]. Census data usually provides relatively coarse information on household water sources. In this work, four questions are addressed:

1. In the absence of detailed water data, can census data, together with available hydrological data, be used to asses planning priorities for water supply in dispersed island countries?
2. Do variations in rainfall due to climate change need to be addressed for water supply in medium-term (10 years) national development plans?
3. Do top-down governance templates, such as national sustainable development strategies, capture the priorities of rural island communities in water supply?
4. Can national development plans be improved for water supply in multi-island countries?

In this work, we summarize census demographic and water source data, and limited information on water use, groundwater and rainfall characteristics of the Kingdom of Tonga, to identify Island Divisions with special needs. We then analyze the priority given to water supply in the national planning instrument, TSDFII. This is compared with Island Division-level priorities found by analyzing water supply priorities in the 117 available CDPs. These are analyzed in terms of women's, youths' and men's priorities. Island Division priorities are related to the analysis of census and hydrological data. The bottom-up CDP priorities in water supply are contrasted with those in the top-down TSDFII. The discrepancy between national and local priorities in water supply are discussed, and suggestions for improving processes are made.

2. Materials and Methods

2.1. Study Location

The Kingdom of Tonga's population of 102,000 people is dispersed over 169 islands in five Island Divisions covering a land area of about 750 km^2 scattered across 700,000 km^2 of the southwestern Pacific Ocean (Figure 1). The Kingdom adjoins the seismically very-active Tonga trench. Tonga's western islands are of volcanic origin, while the eastern islands are uplifted coral limestone and sand islands. Many of the eastern islands, such as Tongatapu, the largest island which contains the capital Nuku'alofa, have a mantle of fertile volcanic soil from past volcanic eruptions of the western islands. Volcanic eruptions, earthquakes and subsequent tsunamis, tropical cyclones (TCs), storm surges and El Niño Southern Oscillation- (ENSO) and Pacific Decadal Oscillation (PDO)-related droughts and floods are frequent natural hazards faced by the Kingdom's island communities. Recent TCs that have devastated parts of Tonga include TC Ian (2014), TC Gita (2018) and TC Harold (2020).

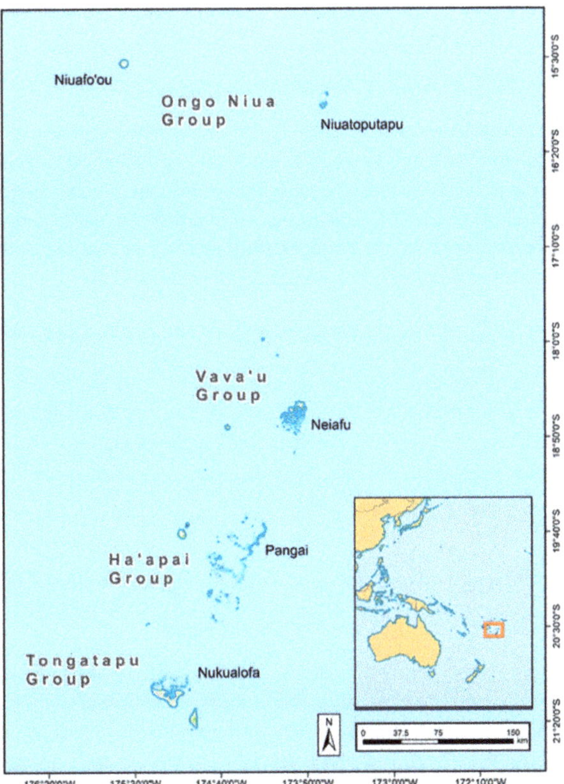

Figure 1. Map of the Kingdom of Tonga, showing main island Divisions and population centers [14].

Annual rainfall in Tonga increases from south to north, influenced by proximity to the South Pacific Convergence Zone (SPCZ). Annual rainfall variability is moderate and smaller in the north than in the south. Tonga has a wetter season from November to April, followed by a drier season from May to October, also influenced by the position of the SPCZ. The predominant sources of water are rainwater harvesting, which Tongans prefer for drinking, and groundwater from groundwater lenses or springs in the carbonate and sand islands, used for all other purposes including washing, bathing and sanitation. Groundwater salinity varies from fresh to brackish and even saline, depending on island geology, size, and ENSO and PDO conditions. In general, many of the volcanic islands do not appear to have useable groundwater [14].

In 2016, nearly three quarters of Tongans lived on the main island Tongatapu, with Greater Nuku'alofa, the capital region, having over a third of the country's population. Just over a quarter of the population are spread over the Kingdom's other four Island Divisions. About 55% of the population are under the age of 25, with youths aged 14 to 24 making up nearly 19% of the total population. Gross domestic product (GDP) in 2017 was estimated to be US$5900 per person, with an annual growth rate around 2.5% [15]. Per capita GDP varies widely between island groups, with Tongatapu being 15% above the national average in 2013, while Ha'apai was 40% below the national average. Natural disasters have been estimated to cost on average 4.4% of GDP and TC Ian in 2014, which affected 70% of the population of Ha'apai, cost 11% of GDP [10].

Community piped water supplies are sourced from groundwater, springs or, less commonly, from community rain tanks [14]. In Nuku'alofa, 'Eua and urban centers on Vava'u and Ha'apai, piped water is supplied by the Tonga Water Board. In villages throughout Tonga, piped water supplies are the responsibility of Village Water Committees, overseen by the Ministry of Health. Piped water supply in villages is often intermittent.

2.2. Demographics, Water Sources, Water Use and Water Demand

We use the results of the latest census in 2016 [16] to examine the urban/rural composition and the distribution of population and their trends across Tonga's five Island Divisions (Figure 1). Census results are also used to compare water sources used by households for different water uses, again between urban and rural communities and across Island Divisions. As a measure of the ease of accessing water supply for the greatest volumetric use of water for nondrinking purposes [17], we use the water source ratio (WSR), defined for the number of households (HH) at the Island Division level as:

$$WSR = (HH \text{ with rain water supply})/(HH \text{ with piped water supply}) \tag{1}$$

Outside urban areas and population centres, there is very little available information on water demand and unaccounted for water in Tonga. Here, we make use of recent estimates of water supply, Q (ML/day), for urban areas [14], and the small number of estimates of daily per capital water use, W (L/pers/day), for a handful of villages in outer islands [18–20]. For the urban areas, W was estimated from:

$$W = 10^6 \times Q \times (1 - UAW/100)/N, \tag{2}$$

where UAW is the percentage of unaccounted for water, and N the number of urban inhabitatnts.

2.3. Climate, Rainfall, Variability and Rainwater Harvesting Failures

A recent draft report on national water resources in Tonga [14] is used to extract data on average annual and seasonal rainfall and their variability and trends across Tonga's Island Divisions. Trends in annual rainfall are examined using linear regression and their statistical significance is tested. Trends are compared with climate model projections of the impact of climate change on rainfall in Tonga [21].

The frequencies of failure of rainwater harvesting systems over the six month wet (Novemeber to April) and dry (May to October) seasons are estimated from the percentile of 360 mm of rainfall over the full seasonal record for that location. The daily per capita water availability, S (L/person/day), for a household rainwater harvesting is given by:

$$S = CE \times P_S \times A/(d_S \times N_{HH}) \tag{3}$$

where P_S (mm) is the seasonal rainfall, CE is the capture efficiency of rainwater harvesting, A (m^2) is the roof catchment area, d_S is the number of days in the season and N_{HH} is the number of people in a household.

When P_S = 360 mm, and with a typical roof area of 100 m^2, a capture efficiency of 0.55 [18–20] and the average household size in Tonga (5.5 persons), Equation (3) gives S of less than 20 L/person/day, i.e., below the World Health Organization recommended minimum quantity of water required to satisfy essential health and hygiene needs in emergency situations [17].

2.4. Groundwater

A summary of the limited available information on groundwater resources is provided in [14], and the few available village integrated water management plans in [18–20]. Other information is sourced from reports and publications [22–28]. Estimates of per capita groundwater use in population centers are based on water supply data provided by the Tonga Water Board [29].

2.5. Tonga Strategic Development Framework 2025–2015

The TSDF II is: "the overarching framework of the planning system in Tonga. It provides an integrated vision of the direction that Tonga seeks to pursue." [12]. TSDFII is conceived as the top of a cascading system of planning and budgeting in Tonga which is intended to guide:

- medium-term sector and district/island master plans
- three year rolling Corporate Plans and Budgets for all Ministries, Departments and Agencies
- annual Divisional and Staff Plans and job descriptions.
- consultation, monitoring, and evaluation.

TSDF II identifies Government priorities, assigns Ministerial responsibilities and aims to focus resources. It is arranged in a hierarchy where 29 Organizational Outcomes (OO), grouped under three institutional pillars and two input pillars, feed into seven desired National Outcomes (NOs) which, in turn, feed into the single planned National Impact of the TSDF II: "A more progressive Tonga supporting a higher quality of life for all", which supports the Motto of TSDF II, given by the reformer monarch Tupou I: "God and Tonga are my inheritance." TSDFII also lists 153 Strategic Concepts (SCs) which were issues raise during the consultation process which lie outside TDSFII, but are intended as aids to sector, district and Ministry, Department and Agency planning and budgeting.

The Ministry of Finance and National Planning (MFNP) led the creation of TSDF II, which was based on a wide, but fairly rapid consultative process. In the period from October 2014 to December 2014, consultation meetings were held throughout Tongatapu and the Island Divisions of 'Eua, Ha'apai and Vava'u. The Ongo Niua Division was not visited.

The TDSFII was scanned for references to water, freshwater, rainwater, groundwater and water supply, and the planned National Outcomes, OOs and SCs were examined for their relevance to water supply. In NSDS, water supply is usually identified as an infrastructure service. The weight given to water supply in TDSFII is assessed relative to the weight given to infrastructure and other services identified in TSDFII, such as energy, transport, information and communications technology, building and structures, and research and development, in the listed OOs and SCs.

2.6. Community Development Plans, 2016

Work on CDPs began in 2007, under the Local Government Division of the Ministry of Internal Affairs. The CPDs were a response to the then National Vision, i.e., "a Progressive Tonga Supporting

Higher Life for All." Consultations throughout rural villages in Tonga's five Island Divisions were implemented by the nongovernment organization, Mainstreaming of Rural Development Innovation Trust Tonga (MORDI TT). The CDP process was supported by the International Fund for Agricultural Development, United Nations Development Programme, Australian Agency for International Development and the Tonga Government. One of the requirements of the project was the participation of 80% of the population of each rural village in the development, ranking of priorities and endorsement of the village CDP. This was a lengthy process which culminated in District Officers and Town Officers of 136 village communities presenting their CDPs to the then Prime Minister, the late Hon. 'Akilisi Pohiva on 4 October, 2016 [13].

During the planning process, the Department for Local Government was transferred from the Prime Minister's Office to the Ministry of Training, Employment, Youth, and Sports, and then to the Ministry of Internal Affairs. Analysis of and action on the Community Development Plans appears to have been deferred by these moves. We have been unable to find any analysis of the valuable information on island priorities contained in these CDPs.

Of the 136 CDPs presented, 117 are available online [30]. These represent over 77% of all rural villages in Tonga. CDPs were downloaded and the priority rankings of each village that mentioned water or water supply were examined and their ranking recorded (Table S1). Particular note was made when water or water supply ranked as the top priority or in the top three priorities for women, men and youths separately. Village level results were aggregated to Island Division level, and the percentage of villages in each Island Division that ranked water as the highest priority, that ranked it within the top three priorities or that ranked it anywhere within their list of priorities were recorded. A comparison was then made of the weight given to water supply as an infrastructure service in TSDFII and to indications of Island Division-level water supply limitations from the 2016 census and hydrological data.

3. Results

3.1. Demographics

The distribution of Tonga's population across its island Divisions in 2016 is given in Table 1 [16]. Also shown are the total rural, urban and the population of Greater Nuku'alofa, as well as the annual growth rate between 2011 and 2016, the population density and the average household size.

Table 1. Summary of 2016 population statistics for Tonga as a whole, for island Divisions, for urban, rural areas and for greater Nuku'alofa populations [16].

Item	TONGA	'Eua	Tongatapu	Vava'u	Ha'apai	Ongo Niua [1]	Urban [2]	Rural [3]	Greater [4] Nuku'alofa
Total Population	100,651	4945	74,611	13,738	6125	1232	23,221	77,430	35,184
Male	50,255	2486	37,135	6866	3118	650	11,529	38,726	17,490
Female	50,396	2459	37,476	6872	3007	582	11,692	38,704	17,694
Population change 2011–2016 (%)	−2.5	−1.4	−1.1	−7.9	−7.4	−3.9	−4.2	−2	−2.4
Av. Annual Growth (%)	−0.5	−0.3	−0.2	−1.7	−1.5	−0.8	−0.9	−0.4	−0.5
Population Density (pers/km^2)	155	57	286	114	56	17	2035	121	1010
Number of Households	18,198	889	13,096	2745	1193	275	4175	14,023	6240
Average Household Size (pers)	5.5	5.6	5.7	5	5.2	4.5	5.6	5.5	5.6
Number of Villages	165	15	67	44	27	12	3	162	14

[1] Niuafo'ou and Niuatoputapu combined. [2] Urban area comprises the villages of Kolofo'ou, Kolomotu'a and Ma'ufanga in Tongatapu. [3] Rural area consists of all villages in Tonga except Kolofo'ou, Kolomotu'a and Ma'ufanga in Tongatapu. [4] Greater Nuku'alofa is made up of the districts of Kolofo'ou and Kolomotu'a in Tongatapu.

The demographic data in Table 1 show a concentration of population in the main island, Tongatapu, with 35% of the population of just over 100,000 living in Greater Nuku'alofa, and a further 40% of the population living in 53 rural villages across Tongatapu Island Division. Just over 35% of all rural villages are in Tongatapu. The remaining 26% of Tonga's population is scattered over 98 villages in Tonga's other four Island Divisions.

Between 2011 and 2016, the population of Tonga shrank by an annual rate of 0.5%, and urban populations decreased at over twice the annual rate of rural populations. The rate of loss of population was highest in Vava'u and Ha'apai Island Divisions and lowest in Tongatapu. In 2015, 70% of the housing stock in Ha'apai was devasted by TC Ian [12]. Table 1 suggests an inward migration from outer islands, particularly to rural villages in Tongatapu, at an annual rate of about +0.9%. This suggests that water demand due to the number of people should be decreasing in outer islands and increasing in Tongatapu.

Population density was over 16 times higher in urban areas that in rural areas. In Tongatapu, the population density was nearly 17 times higher than that of the far northern Ongo Niua Division. Average household size in urban areas, however, was very close to that in rural areas, with Tongatapu having the highest average household size (5.7 persons) and Ongo Niua the lowest (4.5 persons).

3.2. Water Sources

Table 2 lists the percentage of households using water from different sources for drinking and for all other uses for Tonga as a whole and for each of the island Divisions from the 2016 census [16].

Table 2. Percentages of households in Tonga as a whole, in each island Division, in urban and rural areas and Greater Nuku'alofa using water from different sources for (a) drinking water and (b) all other water uses given by the 2016 census [16].

Water Source	TONGA	'Eua	Tongatapu	Vava'u	Ha'apai	Ongo Niua	Urban	Rural	Greater Nuku'alofa
(a) Drinking Water (%)									
Piped Supply	10.0	2.5	11.3	10.3	3.0	1.1	12.2	9.3	11.7
Rain Tank	60.5	73.1	54.8	69.2	86.2	92.3	53.5	62.5	53.8
Neighbour/Community [1]	19.6	23.4	20.7	18.8	10.0	6.6	14.3	21.2	15.8
Bottled water	9.5	0.9	12.8	1.4	0.6	0.0	19.3	6.6	18.1
Boiled Well Water	0.2	0.1	0.2	0.1	0.1	0.0	0.4	0.1	0.3
Other	0.2	0.0	0.2	0.2	0.2	0.0	0.3	0.1	0.2
(b) All Other Water Uses (%)									
Piped Supply	88.3	95.6	93.3	80.7	52.0	59.0	92.3	87.1	92.0
Rain Tank	10.9	4.0	6.0	18.9	44.6	40.3	7.1	12.0	7.1
Own Well	0.6	0.0	0.5	0.0	3.1	0.4	0.4	0.6	0.6
Other	0.3	0.5	0.3	0.3	0.3	0.4	0.2	0.3	0.3
Total Number Households	18,005	885	12,953	2715	1179	273	4089	13,916	6139

[1] Probably neighbour or community rain tanks.

Table 2 reveals the variety of water sources used for different purposes in Island Divisions. Households throughout Tonga prefer rainwater (over 60%) for drinking over piped supply (10%). Most piped water supply is sourced from groundwater. Nationally, nearly 20% of drinking water is sourced from neighbors' or community rain tanks.

Urban households in Table 2a have a 31% higher use of piped water for drinking than rural users, who access rainwater for drinking 17% more than urban users. Bottled water is increasingly being used for drinking, with urban use of bottled water being four times that of rural households. Boiled groundwater from household wells has very low use for drinking, i.e., 0.2% nationally. Local groundwater has the potential to be polluted from household sanitation systems, such as pit latrines or leaking septic tanks. Ha'apai and Ongo Niua stand out in the island Districts, with much less drinking water supplied from piped water than the other island Divisions, and correspondingly, much more from rainwater harvesting.

All other uses of water, i.e., bathing, washing, sanitation, make up the bulk of volumetric demand for water [17]. When all other uses of water are considered, Table 2b shows that nationally, 88% of household water is sourced from piped groundwater systems, with rain tanks supplying only 11% of other uses, since it is reserved for drinking. Only 0.6% access water from private wells. Rural households access piped water 5.7% less than urban households for nondrinking water uses; therefore, rural households rely on rain tanks nearly 1.7 times more than those in urban districts.

Again, Ha'apai and Ongo Niua are quite distinct from the other Island Divisions, with less than 60% of other household water use coming from the piped system and more than 40% coming from rainwater harvesting. Figure 2 shows the water source ratio, Equation (1) for other uses in each Island Division and Tonga as a whole. There is a large difference in WSR between, Ha'apai, Ongo Niua, Vava'u and Tongatapu and 'Eua. Piped water systems are less available in the Ha'apai and Ongo Niua Divisions than in the other Island Divisions [14]. This suggests that access to reliable water sources is more difficult in those islands.

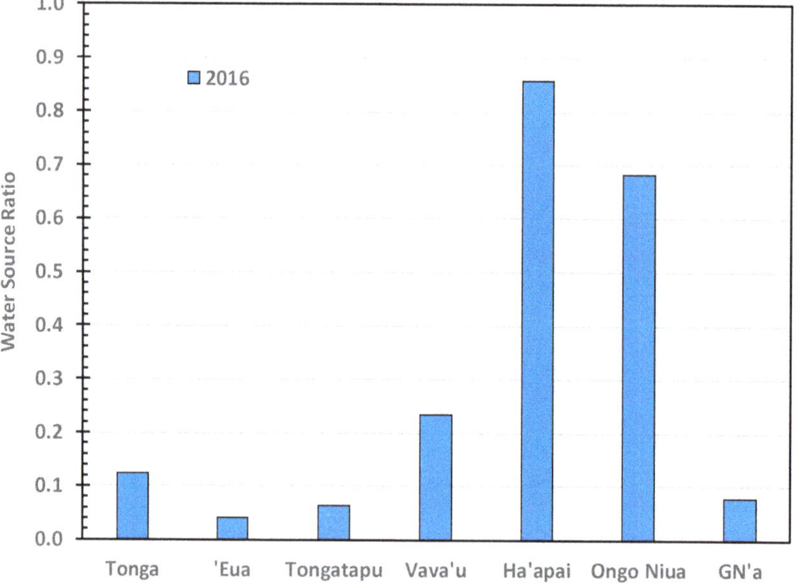

Figure 2. Average water source ratio for nondrinking uses in households for Tonga as a whole, for Tonga's Island Division, and for Greater Nuku'alofa (GN'a).

3.3. Fresh Groundwater

Fresh groundwater in Tonga's islands mostly consists of freshwater lenses overlying seawater in carbonate (limestone and sand) islands [14,22–26]. The salinity gradient increases with depth through the lens, going from low salinity water at the groundwater surface through a diffuse saline transition zone to underlying seawater [24]. Thin freshwater lenses have higher salinity than thicker lenses, but salinity varies with rainfall recharge, pumping and island geomorphology [5]. Groundwater may also exist in fractured rock aquifers in the volcanic islands, but these are yet to be assessed [14]. There are also a number of springs emanating from fresh perched groundwater on 'Eua, an island with mixed volcanic and carbonate geology, and small freshwater lakes on Niuafo'ou [14].

There is useful information about the freshwater lens in Tongatapu, used to supply the capital Nuku'alofa from a nearby wellfield location at Mataki'eua-Tongamai, but less about village pumping from local wells and boreholes throughout Tongatapu, in [22–25,27,28,31]. Local wells and boreholes are vulnerable to contamination from leaking septic tanks and pit latrines.

The mean maximum height of the surface of the freshwater lens in Tongatapu has been estimated to be about 0.6 m above mean sea level, and the lens thins toward the coastal margins. Its elevation varies slightly with the oceanic tide. In the period from 1997 to 2018, the maximum thickness of the freshwater lens was around 16 m. Maximum thickness varied with rainfall between 9.5 and 16 m.

During the same period, the salinity of water, as measured by the electrical conductivity (EC) of the water supplied to Nuku'alofa, varied between almost 1600 µS/cm, following the dry period in 1998, to about 800 µS/cm, following the wet period in 1999 [14].

There is no evidence of any increasing trend in groundwater salinity in Tongatapu over the period from 1997 to 2018 [14]. Increases in EC of individual bores due to progressive increases in wellfield pumping rates over time have been observed at Makahi'eua-Tongamai since 1971 (Figure 3 [27]). The current rate of fresh groundwater extraction across all Tongatapu has been estimated to be about 24,000 m³/day [29].

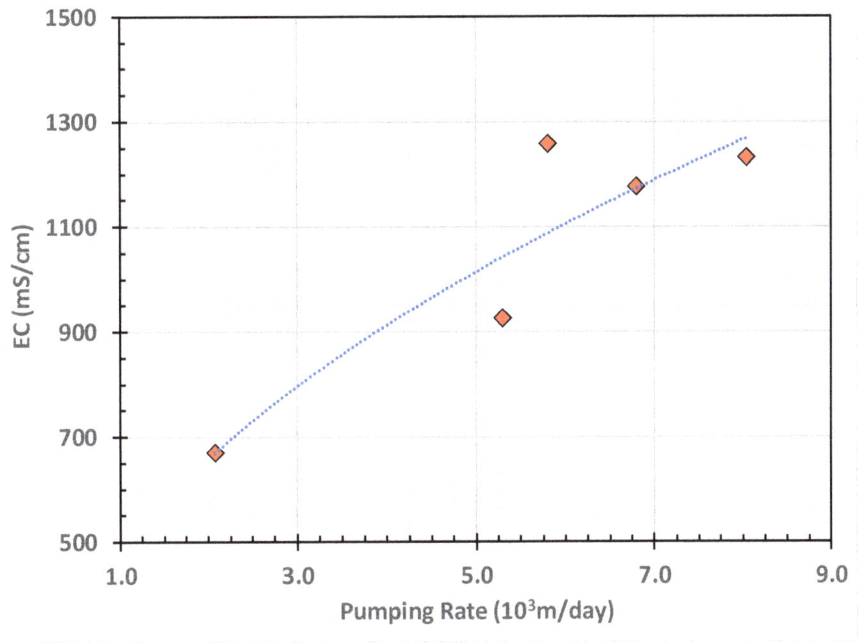

Figure 3. Impact of pumping rate from the Mataki'eua-Tongamai wellfield on the EC of water pumping from an individual well (PS106) between 1971 and 2007 [27].

There is limited information about groundwater used to supply population centers in Ha'apai or Vava'u. On Lifuka Island in the Ha'apai Division, the spatially limited fresh groundwater lens is much thinner than in Tongatapu, with maximum thickness ranging from 2 m in 1998 to 4 m at the end of 2011. Because of the thin freshwater lens, the salinity in the water supply system is higher than in Nuku'alofa, varying from a brackish EC of about 10,000 µS/cm to a low of around 1000 µS/cm. The salinity of the water there depends on the method of extraction, with groundwater pumped from vertical wells having a higher salinity, due to upconing of the underlying seawater, than that pumped from horizontal infiltration galleries or skimming wells. The approximate groundwater extraction rate for water supply in Lifuka is a about 500 m³/day [29].

The population center of Vava'u, Neiafu, in the main island Vava'u 'Utu, a raised limestone island, is supplied water from a fresh groundwater lens with maximum thickness between 5 to 8.5 m over the period from 1999 to 2018. The thickness of the lens varies across the island, with some village wells only producing brackish groundwater. The salinity of water supply to Neiafu in this period varied, from an EC of about 500 µS/cm in September 2000 to a brackish 4500 µS/cm in January 2016. The approximate rate of groundwater extraction to supply Neiafu is about 1000 m³/day [29].

In 'Eua, water is supplied from groundwater springs and wells. Salinity of the supply water in 'Eua is low, typically less than 500 μS/cm, and the supply of water is about 1100 m^3/day [24,27].

3.4. Water Demand

With the diversity of water sources used (Table 2) and the differences in use between rural and urban and Island Divisions, there is little information on the ranges of water demand throughout Tonga. For Greater Nuku'alofa, about 12.6 ML/day are pumped from the groundwater Mataki'eua-Tongamai wellfield [13]. Unaccounted for water amounts to between 30 to 40% [29]. It has been estimated that about 11 ML/day of groundwater is extracted to supply the rural villages throughout Tongatapu [28]. Information is also available on groundwater extraction for the urban centers in Vava'u and Ha'apai [29]. There is limited information on unaccounted for water in these or in Tongatapu's village water supply systems. Inspections of village groundwater supply systems suggest that 40 to 50% losses are conservative. With this estimate and the population numbers in the 2016 census [16], the estimated average per capita groundwater use, Equation (2), in Greater Nuku'alofa, in rural villages in Tongatapu and in the urban centers in Vava'u and Ha'apai, as well as spring and groundwater demand in 'Eua, are given in Table 3.

Table 3. Estimated average per capita water use, Equation (2), in population centers and measured use in some rural villages in Tonga's Island Divisions supplied by spring water (SW), groundwater(GW) and rainwater (RW) [14,18–20,29,31].

Island Division	Location	Water Source	Water Supply Rate (10^3 m^3/Day)	Unaccounted for Water (%)	Average Water Use (L/Pers/Day)
'Eua		SW, GW	1.1	40–50	110–130
Tongatapu	Greater Nuku'alofa	GW	12.6	30–40	230–270
	Rural [1] Villages	GW	11	40–50	130–160
Vava'u	Neiafu	GW	1.0	40–50	140–160
	Koloa	RW	-	-	18
Ha'apai	Lifuka	GW	0.5	40–50	125–150
	Nomuka	RW	-	-	22
Ongo Niua	Niuafo'ou	RW	-	-	14

[1] All villages not within Greater Nuku'alofa.

Rainwater is also used extensively throughout Tonga (Table 2). There is very little information on water use from rainwater harvesting systems. Measurements at villages on three outer islands, i.e., Koloa, Nomuka and Niuafo'ou, in the Vava'u, Ha'apai and Ongo Niua Divisions provide details on the average per capita rainwater and are listed in Table 3 [18–20]. In Koloa, the thin groundwater lens is saline and is only used for toilet flushing, cleaning and bathing [18]. In Nomuka, the village piped water supply, which was sourced from beneath a shallow ephemeral freshwater lake, is no longer operational, so household and community rain tanks are the only sources of freshwater [19].

Niuafo'ou island is a basalt shield volcano surmounted by an andesitic cone. Communities there rely on rainwater harvesting. It is not known if there is any viable fresh groundwater in Niuafo'ou [19]. The World Health Organization recommends that the minimum quantity of water required to satisfy essential health and hygiene needs in emergency situations is 20 L/person/day [17], a value similar to the regular average rainwater use for most purposes in these three outer islands.

3.5. Rainfall

Tonga's climate is tropical, with a warm period from December to April, when temperatures can reach 32 °C, and a cooler season from May to November, with temperatures infrequently rising above 27 °C [32]. Tonga's reliance of water sourced from rainfall harvesting and from groundwater (Table 2) means that rainfall and subsequent groundwater recharge are key determinants of water availability [22,

24,27,28]. Table 4 summarizes the rainfall characteristics at the six long-term meteorological stations throughout Tonga.

Table 4. Annual, wet (November to April) and dry season (May to October) rainfall characteristics at meteorology stations in Tonga [14].

Met Station	Island Division	Daily Average Temperature Range (°C) [1]	Average Annual Rainfall (mm)	CV of Annual Rainfall	Mean Nov–Apr Rainfall (mm)	Mean May–Oct Rainfall (mm)	Period
Niuafo'ou	Ongo Niua	25.9–27.9	2534	0.22	886	1648	1971–2019
Niuatoputapu		25.7–27.5	2315	0.21	803	1512	1947–2019
Lupepau'u	Vava'u	22.9–26.9	2290	0.22	793	1497	1947–2019
Ha'apai	Ha'apai	23.1–27.4	1754	0.24	599	1155	1947–2019
Fua'amotu	Tongatapu	21.4–26.6	1765	0.25	664	1101	1980–2019
Nuku'alofa		21.8–27.2	1863	0.26	735	1128	1945–2019

[1] Temperatures shown for the period from 1980 to 2017.

Average annual rainfall varies from about 1750 mm in the south to about 2500 mm in the northern islands closer to the Equator. Tonga has a wetter season from November to April and a drier season from May to October. Its rainfall is influenced by the position and strength of the South Pacific Convergence Zone (SPCZ) which is influenced both by season, ENSO events [21,33] and by the Pacific Decadal Oscillation (PDO) [24]. Tonga is periodically impacted by TCs whose intensity appears to be increasing [34], in line with climate change projections [21,35].

During the wet season, the SPCZ lies between Samoa and Tonga, while during the dry season, the SPCZ is normally positioned to the north-east of Samoa, where it is often weak, inactive and sometimes nonexistent [32]. In the northern islands, about 35% of annual rainfall occurs in the November to April period, while in the south, the percentage is slightly higher, i.e., 38–39%.

Estimates of average annual potential evaporation in Tonga range from about 1460 mm in Tongatapu to about 1670 mm in Niuatoputapu [32]. These high annual potential evaporation rates mean that open water storages lose large volumes of water. The losses from groundwater systems due to evaporation are much lower; therefore, groundwater storage has an advantage over surface storage.

3.6. Variability of Rainfall

The coefficient of variability (CV) of annual rainfall in Table 4 is moderate, at around 0.21 to 0.26, and is less in the wetter, northern islands than in the south. ENSO and the Pacific Decadal Oscillation (PDO) are key drivers of this variability. Two indicators of ENSO, the Niño Index (sea surface temperature anomaly in the Niño 3.4 region) and the Southern Oscillation Index (based on sea level pressure difference between Darwin and Tahiti), are strongly negatively and positively correlated, respectively, to wet season but not dry season rainfall. The PDO index is also strongly negatively correlated to wet season rainfall. In La Niña phases of ENSO (negative Niño Index), the SPCZ tends to move further south, and Tonga gets more rain in the wet season, while in El Niño phases (positive Niño Index), it moves further north and causes lower wet season rainfall [21]. Droughts tend to occur during El Niño events. During severe El Niños, the SPCZ can spread azonally along the equator [36], leading to widespread impacts across the Pacific. In negative phases of the longer-period PDO, the south-western Pacific is warmer than in positive phases, leading to higher rainfalls.

3.7. Changing Climate and Rainfall Trends

Projections for future climate in the 21st century based on the Coupled Model Intercomparison Project Phase 5 using global climate models have been made for Tonga under various Representative Concentration Pathways estimating possible future trends in greenhouse gas emissions [21]. These projections suggest that there will be more extreme rainfall events, continuing sea level rise, increasing ocean acidification and higher risk of coral bleaching, and that El Niño and La Niña events will continue (all very high confidence). It is projected that the frequency of tropical cyclones will decrease by the end of the 21st century (medium confidence) but their maximum intensity may increase [34,35]. There is no consensus on whether average annual rainfall will increase or decrease (low confidence), and it is projected that drought frequency may slightly decrease (low confidence) [21].

The CSIRO and Australian Bureau of Meteorology [21] have provided no projections of trends in potential evaporation or actual evapotranspiration. Earlier projections based on increasing temperatures [33] are erroneous. The available climate records for Tonga show average atmospheric temperatures increasing by about 1 °C per century [14], similar to increases in sea surface temperatures in the surrounding ocean. Table 5 lists the historic linear regression trends in annual and seasonal rainfall at the meteorological stations in Table 1.

There are no significant trends in Table 5 in annual or wet season rainfall over the period from 1947 to 2019. This is consistent with climate change projections [21]. Out of the four stations with long-term rainfall records, only Lupepau'u in Vava'u showed a significant increasing trend in dry season rainfall over this period. In contrast, the two stations with shorter rainfall records, i.e., 1980 to 2019, in the northern and southern island Divisions had significant and identical—within the margin of error—increasing trends in both annual rainfall and wet season rainfall. While it is tempting to attribute these more recent increasing trends over 39-year periods to climate change, they are very strongly negatively correlated with trends over 31 years in the PDO, unrelated to changing climate (Figure 4).

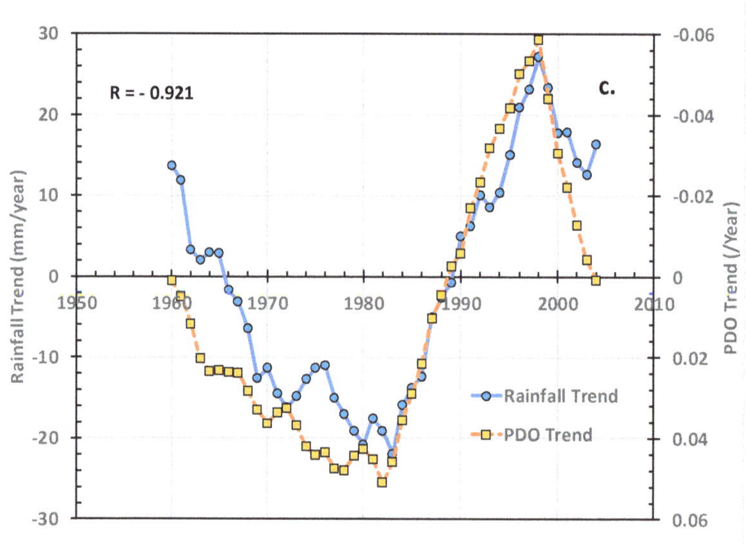

Figure 4. Comparison between the trends in rainfall at Nuku'alofa over 31 year consecutive periods and the consecutive trends in the PDO over the same period. The correlation, −0.921, is highly significant. The year shown is the midyear of the 31-year period.

Table 5. Historic linear regression trends in annual and seasonal rainfalls with standard errors and significance level (Signif) for the meteorological stations in Table 4. Long- and short-term records are compared over the same periods, i.e., 1947–2019 and 1980 to 2019, respectively. Bold values are statistically significant.

Met Station	Trend Annual Rainfall (mm/Decade)	Signif	Trend Nov-Apr Rainfall (mm/Decade)	Signif	Trend May-Oct Rainfall (mm/Decade)	Signif	Period
Niuafo'ou	**204 ± 71** [1]	**$p < 0.05$**	**82 ± 36**	**$p < 0.05$**	**141 ± 68** [1]	**$p < 0.05$**	1980–2019
Niuatoputapu	−3 ± 30	NS	−9 ± 23	NS	6 ± 19	NS	1947–2019
Lupepau'u	34 ± 28	NS	2 ± 24	NS	**34 ± 14**	**$p < 0.05$**	1947–2019
Ha'apai	14 ± 25	NS	3 ± 22	NS	14 ± 12	NS	1947–2019
Fua'amotu	**212 ± 57**	**$p < 0.001$**	**162 ± 58**	**$p < 0.01$**	60 ± 35	NS	1980–2019
Nuku'alofa	0 ± 24	NS	−7 ± 21	NS	9 ± 12	NS	1947–2019

[1] Period 1981–2019, data missing for 1980.

3.8. Droughts

Because of the nation-wide reliance on rainwater harvesting or relatively shallow groundwater, droughts are particularly consequential in Tonga. They are projected to slightly decrease in frequency under climate change (low confidence) [21]. Projections, however [37–40], suggest that the frequency of extreme ENSO events will increase from one in 20 years to one in 10 years within the 21st century, and the PDO will become less predictable [41]. This implies that severe droughts in Tonga may increase in frequency. Resident household sizes and their water demands mean that rain tanks supplying households often fail during protracted dry seasons or droughts. Table 6 lists the historic frequency of severe, i.e., 6-month, dry periods in which the average rainfall per month was less than or equal to 60 mm and when household rain tanks are predicted to run dry.

Table 6. Frequency of severe dry periods in both wet and dry season at meteorological stations in Tonga.

Met Station	Frequency Severe Dry Periods		Period
	Nov.–Apr.	May–Oct.	
Niuafo'ou	-	1/46	1980–2019
Niuatoputapu	-	1/43	1947–2019
Lupepau'u	-	1/28	1947–2019
Ha'apai	-	1/9	1947–2019
Fua'amotu	1/40	1/26	1980–2019
Nuku'alofa	1/39	1/28	1947–2019

In terms of dry season rainwater shortages in the island Divisions, the frequency in Table 6 increases in the order of Ongo Niua < Vava'u = Tongatapu < Ha'apai. The frequency of expected raintank failures in Ha'apai is three times that in Tongatapu/Vava'u and five times that in Ongo Niua. The frequency of expected rain tank failures in the periods from 1980 to 2019 and 1947 to 2019 at Fua'amotu and Nuku'alofa on Tongatapu are similar. This is also the case for the lower failure frequencies over these two periods at Niuafo'ou and Niuatoputapu in Ongo Niua. These suggest that, at present, there is no discernable impact of climate change on drought frequency. Surprisingly, Table 6 also indicates that, unlike other Island Divisions, the wet season in Tongatapu will also fail to supply adequate rainfall for household supply about once in forty years. Failure of the wet season is a serious concern. These results suggest that both Hapai and Tongatapu have higher risks of rainwater harvesting supply failures than the other island Divisions.

The larger water storages in groundwater systems mean that they are more robust supply sources during droughts, provided groundwater extraction is carefully monitored and managed [14].

Census and hydrological data have revealed major differences in access to acceptable and reliable sources of water across Tonga, with no overall long-term trends in rainfall, despite shorter-term trends driven by the PDO variations.

3.9. The Tonga Strategic Development Framework 2015–2025

The seven planned National Outcomes (NOs) of TSDFII are listed in Table 7. The five institutional and input pillars, together with their accompanying OO and the number of associated SCs, are listed in Appendix A. Infrastructure services are incorporated within NO E in Table 7. In total, 14% of the planned NOs concern infrastructure. The OO under the Infrastructure and Technology Inputs Pillar, associated with NO E, focus on more reliable, safe and affordable energy, transport and information and communications technology (ICT), building and structures and research and development services, but they do not include water supply services. Table 8 compares the number of OOs and SCs devoted to infrastructure services and compares them with those targeting water supply.

Table 7. The seven planned National Outcomes of TSDFII [12].

Code	National Outcome
A.	A more inclusive, sustainable and dynamic knowledge-based economy
B.	A more inclusive, sustainable and balanced urban and rural development across island groups
C.	A more inclusive, sustainable and empowering human development with gender equality
D.	A more inclusive, sustainable and responsive good governance with law and order
E.	A more inclusive, sustainable and successful provision and maintenance of infrastructure and technology
F.	A more inclusive, sustainable and effective land administration, environment management, and resilience to climate and risk
G.	A more inclusive, sustainable and consistent advancement of our external interests, security and sovereignty

Table 8. The number of Organizational Outcomes and Strategic Concepts assigned to services in TSDFII. The total number of each are in parentheses.

Service	Organizational Outcome (29)	Strategic Concepts (153)
Energy	1	4
Transport	1	9
Information & Communications	1	9
Building & Structures	1	5
Research & Development	1	8
Water Supply	0	1

Table 8 shows that over 17% of all OO in TSDFII are assigned to infrastructure services, an important sector of the government's agenda, but water supply is not included in any OO. Nearly 24% of all SCs highlight infrastructure services, with transport and ICT each having nearly 6% of SCs. Water has nearly an order of magnitude less, i.e., 0.7%, with only one mention under the environment and natural resources pillar: "improve the management and delivery of safe water supply for business and households". Under the Health component of the Social Institutions Pillar, "Percentage of population with safe water supply" is listed as a key performance indicator (KPI), but there is no OO or SC associated with this KPI), nor any associated with environment and natural resources. This means that government performance in managing and delivering water supply at the Island Division level cannot be measured.

In mapping the planned National Outcomes against the UN SDGs, the TSDFII identifies National Outcomes F, E, and B (Table 7) as contributing to the UN's SDG6, yet the Infrastructure and Technology Inputs Pillar associated with National Outcome E does not mention water supply services or sanitation. In summary, not one of the 29 OOs identifies water supply as a nationally important infrastructure service. Only one of the SC raised in consultations mentions safe water supply, and nowhere is the adequacy of water supply raised, despite the large differences between island Divisions found in Table 2 and shown in Figure 2.

3.10. Village Community Development Plans, 2016

Table 9 provides details of the number of village CDPs that were available for analysis for each of the Island Divisions and Districts in Tonga. In total, 117 CDPs were available of the original 136 that were presented in 2016. CDPs for the main island Tongatapu excluded the Districts of Kolofo'ou and Kolomotu'a, that comprise the capital area, Greater Nuku'alofa, and so represent rural areas of Tongatapu. Each village identified a different number of priority issues to be addressed. The number varied among Island Divisions and among women, youths and men.

Table 9. The number of accessed village community development plans for Districts and Island Divisions in Tonga and the medium number of identified priorities identified by women, youths and men within each Island Division [27].

Island Division	District	No. of Village CDPs	No. CDPs/Island Division	Median Number of Priorities in Island Division		
				Women	Youth	Men
'Eua	'Eua Motu'a	13	13	8	7.5	7.5
Tongatapu	Nukunuku	9	48	6	5	6
	Tatakamatonga	7				
	Vaini	11				
	Lapaha	10				
	Kolovai	11				
Vava'u	Hahake	8	39	7	6	7
	Hihifo	5				
	Leimatu'a	4				
	Pangaimotu	4				
	Motu	9				
	Neiafu	9				
Ha'apai		5	5	9	7	10
Ongo Niua	Niuatoputapu	4	12	5	5	5.5
	Niuafo'ou [1]	8				
Total		117	Country Median	6	6	6

[1] One village in Niuafo'ou gave aggregated priorities rather than separate women's, youths' and men's priorities.

For the whole country, Table 9 shows that the median numbers of development priority issues identified in villages by women, youths and men were the same, i.e., six each. Ongo Niua Island Division identified the fewest priority issues, while Ha'apai Island Division identified the most. This may be a result of the impact of TC Ian on Ha'apai in 2014. Youths in 'Eua identified slightly more priority issues than youths elsewhere. The maximum numbers of priorities identified by women, youths and men were 12, 11 and 12, respectively, while the minimum numbers were 4, 3 and 3.

Table 10 lists the percentage of villages in each Island Division that identified water supply issues as their number one priority. Over 56% of the available village CDPs throughout Tonga and over 53% in all individual Island Divisions identified water supply as the number one priority. Women in Tonga ranked water supply as a higher priority issue than did men or youths. Only in 'Eua did youths and men view water supply as a higher priority than did women. On 'Eua, youths had the highest concern about water supply among the youths in any Island Division. Women in the Ha'apai Division had the greatest concerns (80%) about water supply of any gender-age cohort and Island Division.

Table 10. The percentage of villages in each Island Division that identified water supply issues as first priority in terms of women's, youths' and men's perspectives.

Island Division	Percentage of Villages with Water Supply as Highest Priority			
	Women	Youth	Men	Average
'Eua	46%	62%	54%	54%
Tongatapu	67%	42%	54%	54%
Vava'u	67%	49%	62%	59%
Ha'apai	80%	60%	20%	53%
Ongo Niua	50%	58%	50%	53%
Tonga	**63%**	**49%**	**55%**	**56%**

Figure 5a shows the distribution of the percentage of villages throughout the Island Divisions where concerns over water supply were ranked within the top three priorities. The percentage of villages in Tonga that ranked water supply in the top three priorities was 76%, with women comprising a slightly higher percentage (83%) than men (80%), and considerably higher than youths (66%). Concerns over village water supply were highest in Ha'apai (93%), where both 100% of villages in the women and youths cohort ranked water within the top three priorities compared with 80% for men. Villages in the wetter, northern Ongo Niua Division had the lowest average percentage, but still 72% ranked water supply issues within the top three village priorities. Youth in Ongo Niua villages had the lowest percentage (58%) of any group who ranked water within the top three priorities.

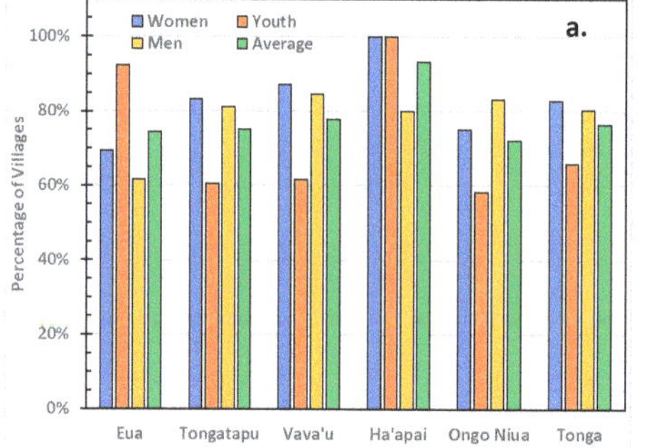

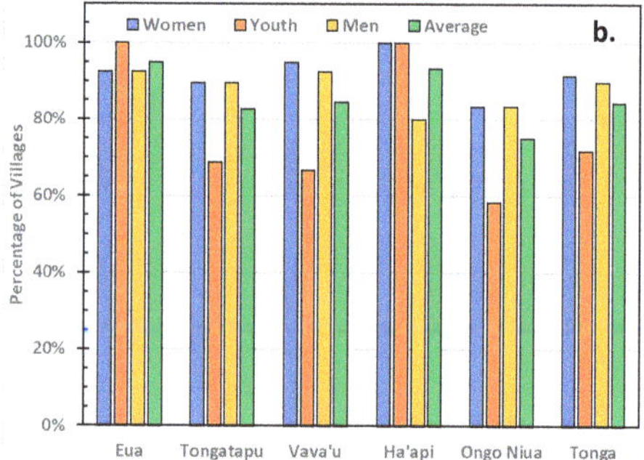

Figure 5. The percentage of villages throughout Tonga's Island Divisions, as well as Tonga collectively, which ranked water supply: (**a**) within the top three priorities; and (**b**) as a priority in terms of women's, youths' and men's perspective and the village average of all three.

Figure 5b shows the number of villages that ranked water supply as one of their development priorities. Overall, 84% of villages in Tonga listed water supply as a priority. The least concern among Island Groups over water supply was observed in the main island of Tongatapu (83%). Youth in Ongo Niua were the group with the lowest level of concern (58%) among all groups about water supply. The greatest concerns were in Ha'apai, where 100% of women and youths ranked water supply as a priority, as did youths in 'Eua. Nationally, women in villages were slightly more concerned, with 91% listing water as a priority, in contrast to 90% of men; both sexes had more concerns than youths (72%).

The correlations between the percentage of villages in Island Divisions that ranked water supply as their top priority, within their top three priorities or as a priority and the WSR of Equation (1) were not significant ($p < 0.1$). This means that, although the census data show wide variation in the water sources available to different Divisions, they are not reflected in differences in their priority concerns over water supply.

4. Discussion

4.1. Census, Hydrological Data and Water Supply

The census data in Table 1 show the concentration of population of Tonga in both the urban and rural areas of the main Tongatapu Island Division, which includes the capital, Nuku'alofa. Just over a quarter of the 26% the population is dispersed across villages in the other four Island Divisions. The overall population has declined although inward migration to the main island and the capital is occurring, some of which may have been the result of TC Ian which devastated Ha'apai in 2014. This inward migration has been a continuing feature of multi-island PICs for several decades [42].

Table 2 reveals that Tongans switch water sources between public piped groundwater supply and private household rainwater harvesting depending on water use. Rainwater is preferred for drinking. The greater volumetric use of water for nondrinking purposes, such as toilet flushing washing, and bathing means that the piped water supplies are important. We used here the ratio of the number of households accessing rainwater supply to those accessing piped water supply, Equation (1) as a coarse measure of the reliability of supply. The results show urban areas and population centers have greater access to piped groundwater supply than in rural areas with very limited use of private local groundwater wells which are prone to contamination from household sanitation systems.

The WSR of Island Divisions (Figure 2) identified that Ha'apai, Ongo Niua and Vava'u Island Divisions are vulnerable because of their larger reliance on drought-sensitive household rainwater harvesting water supply systems. Outer islands in these three Island Divisions are more reliant on rainwater harvesting, and from the few examples in Table 3 have extremely limited average daily per capita water use about equal to the World Health Organization's recommended minimum quantity of water required to satisfy essential health and hygiene needs in emergency situations [17]. Estimations of the frequency of failure of rain tank systems based on available rainfall data (Table 4) reveal that even well-managed rainwater harvesting systems will fail periodically especially in the May to October drier season. Ha'apai is more vulnerable to rainwater system failure than southern or northern islands (Table 5). The diversity in access to water sources of different reliability throughout Tonga's islands, as evidenced by WSR (Equation (1)), means that Target 6.1 of UN SDG6 of achieving universal and equitable access to safe and affordable water for all remains a major challenge.

The greatest volumetric use of water in Tonga is sourced from groundwater. In the southern Island Divisions of 'Eua and Tongatapu, water is sourced from springs and well-studied relatively thick freshwater lenses with comparatively low salinity and significant water yields. In the northern Island Divisions of Vava'u and Ha'apai, fresh groundwater lenses are much thinner, and the salinities of water supplied to population centers are more saline than in Tongatapu and can be even brackish in droughts with limited groundwater yields. In other islands, the availability of groundwater for water supply remains to be determined.

This work has sought to address three questions. The first was: In the absence of detailed water data, can census data, together with available hydrological data, be used to asses planning priorities for water supply in dispersed island countries? We have sought to answer this question at the Island Division level. It was found that census data and the limited hydrological data available identified significant differences in water availability that can be used to better inform planning processes.

Rainfall is centrally important to groundwater recharge, piped water and rainwater supply systems in Tonga. The distribution and variability of rainfall in Tonga (Table 4) differs across Tonga's five Island Division. Both depend on proximity to the SPCZ (20, 33) and on major ocean-atmosphere interactions, ENSO events and the PDO. Projections on the impact of climate change on annual rainfall and drought frequency and intensity are equivocal and of low confidence [21,33]. While air and ocean temperatures have increased by about 1 °C over the period from 1947 to 2019, the historic record shows no statistically significant long-term trends in annual rainfall (Table 5). There are shorter-term trends in rainfall (Table 5) over 31-year periods which are highly significantly negatively correlated with trends in the PDO over the same period (Figure 4).

The second question this work has sought to address is: Do variations in rainfall due to climate change need to be addressed for water supply in medium-term (10 years) national development plans? We have shown here in Tonga that the historic increase in temperatures over the period from 1947 to 2019 has not be associated with any historic increase in annual rainfall or its variability. Since both are key factors in water supply in Tonga, the answer here, compared with the variability imposed by ENSO and PDO, is that climate change is of secondary importance in decade-long development plans addressing water supply. The conclusion is consistent with findings that other factors, including governance, increasing demand and urbanization pose much greater shorter-term risks to water supply in PICs [43]. Continued impacts of tropical cyclones on water supply infrastructure, especially supply and rain tanks, can be devastating. The change in TC frequency and intensity with climate change [21,33–35] could pose increased risks to water supply systems as would climate change induced alterations to the frequency and intensity of ENSO and PDO events [36–41].

4.2. Top-Down versus Bottom-Up Planning Priorities

The largely top-down TSDFII [12] involved discussions with key sectors of the economy including District and Town officers, church leader forums, nongovernment organizations and all main sectors of private business forums over a three-month period. TDSFII claims that the planned National Outcomes, F, E and B (Table 7) are contributions to the UN's SDG6 on water and sanitation. Water and sanitation do not appear in any of the Organizational Outcomes associated with National Outcomes F, E or B.

Investment in infrastructure is a major driver of economic productivity and development [44]. Our analysis of the weight given to infrastructure services in Table 8 shows that while services are assigned to nearly 17% of the total 29 OO, water supply was not mentioned. Of the 153 Strategic Concepts raised in consultations leading to TSDFII, infrastructure services were raised in nearly 24% of SCs, with transport and ICT services each identified in nearly 6% of SCs. In contrast, water supply services ranked nearly an order of magnitude lower in SCs, i.e., at 0.7%. TSDFII conveys the strong impression that water supply is a low national priority in Tonga compared to other services.

The rural village, bottom-up, Community Development Plans in Tonga evolved over a 9-year period, partly because each stage of the planning process required participation of 80% of the population of each village, and partly because of the dispersal of villages over Tonga's many villages. The available 117 CDPs, analyzed for the first time here, represents 77.5% of the total number of rural villages in Tonga and 86% of the 136 CDPs presented in 2016. The greatest number of available CDPs come from the main island Tongatapu, followed by the Vava'u Division, reflecting the distribution of the rural population (Table 1). The least number were from Ha'apai, which may reflect the devastation caused by TC Ian in 2014.

The median number of priority issues ranked in the CDPs for Tonga was six, the same across gender and age groups. The highest number of priorities identified was in the Ha'apai Division, and the lowest number was in the wetter, northern Ongo Niua Division (Table 9). Over 56% of all rural villages ranked water supply as their top priority, with women ranking it as the first priority more frequently than youths or men (Table 10). This reflects the fact that in rural villages, household water supply is largely the responsibility of women. In Ha'apai, women in 80% of the villages ranked water supply as the first priority, in contrast to men in Ha'apai villages, where only 20% ranked it as the highest priority.

Around 76% of all villages ranked water supply within the top three priority issues, with women giving a higher ranking than men, followed by youths (Figure 5a). The highest rankings were in villages in Ha'apai, with women and youths in 100% of villages ranking water supply within the top three priorities. The lowest top three ranking was in the Ongo Niua Division, due to low rankings by youths. Even in rural Tongatapu, 75% of villages ranked water supply within the top three priorities. For Tonga as a whole, 84% of villages included water supply within their village priorities with the lowest Island Division being Ongo Niua, 75%, and the highest being 'Eua, 95%. In Ha'apai, 100% of women and youths as well a 100% of youths in 'Eua, in all villages ranked water supply as a development priority

(Figure 5b). These overwhelming local village development priorities for improved water supply are in sharp contrast to very low concern over water supply in TSDFII.

One hypothesis was that village priorities in water supply may be linked to the WSR from census data as a coarse measure of the accessibility of water for nondrinking purposes. There was no statistically significant ($p > 0.1$) correlation between village water supply priorities and WSR from the census data. There is an inbuilt assumption in WSR that all Island Division piped water supplies are equivalent in adequacy and quality. While the capital area, Greater Nuku'alofa, has access to continuous piped water supply, village piped water supplies are almost all intermittent, even on the main island, Tongatapu. Moreover, the groundwater supply in Tongatapu is relatively fresh, while those in Vava'u and Ha'apai have higher salinity and can even be brackish during droughts. The wide-spread concerns over water supply expressed in the village CDPs involve supply from rain tanks, and village piped water supply.

The third question we have attempted to answer is: Do top-down governance templates, such as national sustainable development strategies, capture the priorities of rural island communities in water supply? From our analysis of TSDFII and CDPs, the answer in Tonga is an emphatic "NO!".

4.3. Possible Reasons for the Mismatch in National and Local Planning Priorities

There is an enormous difference in the priorities given to water supply in the top down TSDSII compared with the bottom up CDPs. Traditionally in many PICs, water supply was a responsibility of the family or extended family. In Tonga, water supply is still a shared responsibility between households with their private rainwater harvesting systems and authorities who supply piped water. It may be that the ministries that developed TSDFII do not see water as a government responsibility, compared to other infrastructure services. The recent passage by Parliament of the Tonga Water Resources ACT 2020 may change that view.

Another factor could be that the ministries, departments and agencies involved in developing TSDFII are mostly based in Greater Nuku'alofa, which has a continuous, very adequate per capita piped supply (Table 3) of lower salinity groundwater, so water supply for them is not a priority.

A final factor could be that prior to 2020, there was no ministry with overall legal responsibility for managing, conserving and protecting the nation's water resources. In effect, water resources did not have a voice in the national planning process.

4.4. Limitations of This Work

Comparison between national and local development plans involved assumptions about priorities given to different sectors. We have only analyzed water supply and not all sector priorities in the CDPs. In an attempt to remove bias in our analysis of TSDFII, we analyzed the national priority given to water supply services relative to other infrastructure services (Table 8). The omission of water supply services is thus put into the context of other infrastructure services which are necessary for development.

This work only analyzed census and CDP data at the Island Division level to facilitate a comparison with the national level TSDFII. Both contain a wealth of valuable information at the District and village levels which could better inform planning processes. We compared the lack of emphasis on water supply in TSDFII with water supply data in the 2016 census and water priorities in village CDPs. TSDFII was produced in 2015, while CDPs were presented in 2016, and the 2016 census results were not available until late 2017. Examination of the 2006 census results, which were available in 2008 in plenty of time to be incorporated in TSDFII, show even larger challenges in water supply than in the 2016 census. Work on the CDPs commenced in 2007, and widespread concerns about water supply have been evident for decades [24].

WSR, Equation (1), was used as a coarse measure of the ease of accessing appropriate quantities of reliable water supplies throughout Tonga's Island Divisions. This assumes that piped water supplies are more reliable, are of larger volume than rainwater harvesting supplies and that piped supplies are similar in all Island Divisions. The lack of correlation between WSR and Island Division level water supply priorities indicates that intermittent piped water supplies, a feature of village water supply systems, salinity of water supply and the cost of local water supply may influence village priority rankings.

5. Conclusions

Better governance has been proposed as a key strategy for improving water security in PICs. National Sustainable Development Strategies have been promulgated as an improved overarching governance instrument to identify national priorities, plan their solutions, identify responsibilities, allocate resources, monitor and evaluate outcomes, as well as fulfil international and regional commitments, especially the UN SDGs. The relevance of transplanting governance instruments in the Pacific has been questioned, as has the applicability of time-efficient, top-down planning processes to local village priorities, particularly in terms of water supply. Many SIDs, with dispersed island communities, have a wide range of local conditions that need to be accommodated by planning instruments, as well as the differing concerns of women, youths and men. The Kingdom of Tonga presents the unique opportunity to contrast priorities given to water supply in the top-down Tonga Strategic Development Framework 2015–2025, developed after three months of high-level consultations, with water supply priorities identified in 9 years of community consultations that led to Community Development Plans for rural villages through Tonga's five Island Divisions. Priorities in these CDPs appear not to have been analyzed previously. They contain a wealth of information which could better inform priorities in revising TSDFII.

The TSDFII analysis revealed that, although improved infrastructure services were a significant proportion of planned organizational outcomes, water supply was not included. This implies that water supply is not a national priority, and that TSDFII, therefore, does not address SDG6. In contrast, 84% of villages ranked water supply as a priority, and 56% of villages ranked water supply as their highest priority. Island Divisions with highest concerns were clearly identified, as were the contrasting priorities of women, youths and men. Since NSDS have been widely promulgated throughout PICs, the mismatch found here in national and local water supply priorities warrants investigation in other island countries. It also points to the importance in multi-island countries of investing in bottom-up planning processes, building on the strengths of island communities.

In many PICs, information on water sources and their uses is limited. In this work, census results and the available hydrological data were used to contrast differences in Island Division water supplies. There were large differences between the capital area, main island and outer islands. Census and hydrological data, no matter how limited, are valuable for better informing priorities in NSDS processes. The analysis also showed that in Tonga, over the 10-year planning horizon, climate change was not a significant factor compared with the variability imposed by ENSO and PDO events. The impact of climate change on the frequency and intensity of extreme events, such as droughts and tropical cyclones, which impact water supply, remains a pressing research question.

The lack of correlation between the census-derived water source ratio for nondrinking purposes and Island Division CDP water supply priorities identified in all Island Division CDPs is interesting, since all water sources have been deemed safe for drinking. The overwhelming concern in village communities appears to be the adequacy and reliability of water supply for higher volume usage, i.e., bathing, washing, sanitation and hygiene. Village per capita water supplies from rainwater harvesting systems of around 20 L/pers/day, found here in the limited number of outer islands with data available, are not adequate for these purposes.

A final question raised in this work was: Can national development plans be improved for water supply in multi-island countries? Improved governance and increased water information may be keys to better water management, but governance instruments such as overarching NSDS need to be adapted for use in PICs. They need to be informed by available census and hydrological data and by the results of continuing local planning processes, for which island communities have considerable strengths and long experience. Local processes may be time-consuming, but as shown here, more efficient top-down processes can fail completely to capture widespread local development priorities.

The Fale Alea 'o Tonga (National Parliament) has very recently passed the Water Resources Act 2020 to conserve, protect and manage the Kingdom's water resources. This requires, amongst other objectives, the "implementation of urban and rural planning regimes that take account of water management". This has the potential to address gaps in knowledge of the Kingdom's water resources and their use, and to focus on improving water supplies in Island Divisions and villages identified in the CDPs. It also provides a stimulus to revise TDSFII to better reflect island community concerns.

Supplementary Materials: The following are available online at http://www.mdpi.com/2306-5338/7/4/81/s1, Table S1: Summary Water Priorities in Community Development Plans. Rural Villages in Tonga.

Author Contributions: Conceptualization, I.W., T.F., and T.K.; methodology, I.W., T.F., and T.K.; formal analysis, T.F. and I.W.; writing—original draft preparation, I.W.; writing—review and editing, I.W., T.F., and T.K.; project administration, T.K. All authors have read and agreed to the published version of the manuscript.

Funding: This research received no external funding.

Acknowledgments: This work arose as a result of a strategic planning project supported by the World Meteorology Organization under its Climate Risk and Early Warning Systems Initiative for Pacific Island Countries focusing on improved governance. WMO Pacific are thanked for that support. We are grateful to Quddus Fidelia, Tonga Water Board, for providing helpful information on urban water supplies in Tonga, to Lopeti Tufui, Victoria University of Wellington, for information on rural groundwater pumping in Tongatapu and Karin Nagorcka for editorial assistance. We thank an unnamed reviewer for helpful comments which improved this paper.

Conflicts of Interest: The authors declare no conflict of interest.

Appendix A

Table A1. The Pillars and associated Organisational Outcomes and number of Strategic Concepts connected with each OO in the TSDFII [12].

Pillars	Organisational Outcomes	No. Strategic Concepts
1. Economic Institutions	1.1 Improved macroeconomic management & stability with deeper financial markets	6
	1.2 Closer private/public partnerships for economic growth	4
	1.3 Strengthened business enabling environment	4
	1.4 Improved public enterprise performance	5
	1.5 Better access to, & improved use of overseas trade and employment, & foreign investment	5
2. Social Institutions	2.1 Improved collaboration with and support to civil society organizations & community Divisions	7
	2.2 Closer partnerships between government & churches & other stakeholders for community development	4
	2.3 More appropriate social & cultural practices	7

	2.4 Improved education & training providing lifetime learning	11
	2.5 Improved health care & delivery systems (universal healthcare coverage)	5
	2.6 Stronger integrated approaches to address both to address both communicable and noncommunicable diseases	4
	2.7 Better care & support for vulnerable people, in particular the disabled	6
	2.8 Improved collaboration with the Tongan diaspora	3
3. Political Institutions	3.1 More efficient, effective, affordable, honest, transparent & apolitical public service focussed on clear priorities	7
	3.2 Improved Law & order & domestic security appropriately applied	7
	3.3 Appropriate decentralisation of government admin. With better scope for engagement with the public	3
	3.4 Modern & appropriate Constitution with laws & regulations reflecting international standards of democratic processes	4
	3.5 Improved working relations and collaboration between Privy Council, executive, legislative & judiciary	2
	3.6 Improved collaboration with development partners ensuring programs better aligned behind govt. priorities	4
	3.7 Improved political & defence engagement within the Pacific & the rest of the World	4
4. Infrastructure & Technology Inputs	4.1 More reliable, safe & affordable energy services	4
	4.2 More reliable, safe & affordable transport services	7
	4.3 More reliable, safe & affordable information & communication technology (ICT) used in more innovative ways	4
	4.4 More reliable, safe & affordable buildings & other structures	5
	4.5 Improved use of research & development focussing on priority needs based on stronger foresight	3
5. Natural Resources & Environment Inputs	5.1 Improved land use planning, administration & management for private and public use	7
	5.2 Improved use of natural resources for long-term flow of benefits	7
	5.3 Cleaner environment with improved waste recycling	5
	5.4 Improved resilience to extreme natural events and climate change	9

References

1. Clean Water and Sanitation. Available online: https://www.un.org/sustainabledevelopment/water-and-sanitation/ (accessed on 21 July 2020).
2. UN. *Report of the Global Conference on the Sustainable Development of Small Island Developing States (SIDS), Bridgetown, Barbados, 25 April–6 May 1994*; United Nations: New York, NY, USA, 1994; p. 77. Available online: https://undocs.org/pdf?symbol=en/A/CONF.167/9 (accessed on 9 September 2020).
3. Falkland, T.; White, I. Freshwater availability under climate change, Chapter 11. In *Climate Change and Impacts in the Pacific*; Kumar, L., Ed.; Springer Nature: Cham, Switzerland, 2020; pp. 403–448. [CrossRef]

4. UNICEF; WHO. *Progress on Sanitation and Drinking-Water—2015 Update and MDG Assessment*; United Nations International Children's Emergency Fund, World Health Organization: New York, NY, USA, 2016; 90p, Available online: https://apps.who.int/iris/bitstream/handle/10665/177752/9789241509145_eng.pdf; jsessionid=677BFB8F2BA902C267BE57C04203E1E9?sequence=1 (accessed on 30 September 2020).

5. White, I.; Falkland, A. Practical responses to climate change: Developing national water policy and implementation plans for Pacific small island countries. In *Water and Climate: Policy Implementation Challenges, Proceedings of the 2nd Practical Responses to Climate Change Conference, Canberra, Australia, 1–3 May 2012*; Daniell, K.A., Barton, A.C.T., Eds.; Engineers Australia: Barton, ACT, Australia, 2012; pp. 439–449. ISBN 9780858259119. Available online: https://search.informit.com.au/documentSummary;dn=889974847944755; res=IELENG (accessed on 2 September 2020).

6. UN. SIDS Accelerated Modalities of Action (SAMOA) Pathway, Resolution Adopted by the United Nations General Assembly, 51st Plenary Meeting, 14 November 2014. Available online: https: //sustainabledevelopment.un.org/sids2014/samoapathway (accessed on 30 September 2020).

7. National Sustainable Development Strategies. Available online: https://www.un.org/esa/sustdev/natlinfo/ nsds/pacific_sids/pacific_sids.htm (accessed on 21 July 2020).

8. UN. *Report of the International Meeting to Review the Implementation of the Programme of Action for the Sustainable Development of Small Island Developing States, Port Louis, Mauritius, 10–14 January 2005*; A/61/277; United Nations: New York, NY, USA, 2005; Available online: https://www.un.org/ga/search/view_doc.asp? symbol=A/CONF.207/11&Lang=E (accessed on 24 July 2020).

9. Lamour, P. *Foreign Flowers: Institutional Transfer and Good Governance in the Pacific Islands*; University of Hawaii Press: Honolulu, HI, USA, 2005; 220p.

10. Barnett, J. Titanic states? Impacts and responses to climate change in the Pacific Islands. *J. Int. Aff.* **2005**, *59*, 203–219.

11. Daniell, K.A. *Co-Engineering and Participatory Water Management, Organisational Challenges for Water Governance*; UNESCO IHP International Hydrology Series; Cambridge University Press: Cambridge, UK, 2012; 215p.

12. Government of Tonga. *Tonga Strategic Development Framework, 2015–2025. A More Progressive Tonga: Enhancing Our Inheritance*; Ministry of Finance and National Planning: Nuku'alofa, Tonga, 2015; 130p. Available online: http://www.finance.gov.to/tonga-strategic-development-framework (accessed on 4 September 2020).

13. Tonga Launches First Community Development Plans, *Pacific Islands Report* 13 October 2016. Available online: http://www.pireport.org/articles/2016/10/13/tonga-launches-first-community-development-plans (accessed on 25 July 2020).

14. *National Water Resources Report (Draft)*; Water Resources Management Section, Natural Resources Division, Ministry of Lands and Natural Resources; Prepared for Climate Resilience Sector Project; ADB Grant (03-78 TON); WRMS: Nuku'alofa, Tonga, 2019; 97p.

15. CIA. *The World Factbook Australia-Oceania: Tonga*; US Central Intelligence Agency: Washington, DC, USA, 2020. Available online: https://www.cia.gov/library/publications/the-world-factbook/geos/tn.html (accessed on 30 September 2020).

16. Basic Tables and Administrative Report. In *Tonga 2016, Census of Population and Housing*; Tonga Statistics Department: Nuku'alofa, Tonga, 2017; Volume 1, 259p. Available online: https://tongastats.gov.to/census/ population-statistics/ (accessed on 9 September 2020).

17. World Health Organization. Chapter 9. In *Technical Notes on Drinking Water, Sanitation and Hygiene in Emergencies*, 2nd ed.; Prepared by Water, Engineering and Development Centre; Loughborough University: Leicestershire, UK, 2013; p. 4.

18. *Integrated Water Resources Management Plan for Koloa, Vava'u*; Water Resources Section, Natural Resources Division, Ministry of Lands and Natural Resources; Prepared for Climate Resilience Sector Project; ADB Grant (03-78 TON); WRMS: Nuku'alofa, Tonga, 2018; 43p.

19. *Integrated Water Resources Management Plan for Nomuka, Ha'apai*; Water Resources Section, Natural Resources Division, Ministry of Lands and Natural Resources; Prepared for Climate Resilience Sector Project; ADB Grant (03-78 TON); WRMS: Nuku'alofa, Tonga, 2018; 68p.

20. *Integrated Water Resources Management Plan for Niuafo'ou, Ongo Niu*; Water Resources Section, Natural Resources Division, Ministry of Lands and Natural Resources; Prepared for Climate Resilience Sector Project; ADB Grant (03-78 TON); WRMS: Nuku'alofa, Tonga, 2018; 77p.

21. ABoM; CSIRO. *Climate Variability, Extremes and Change in the Western Tropical Pacific: New Science and Updated Country Reports*; Pacific-Australia Climate Change Science and Adaptation Planning Program Technical Report; Australian Bureau of Meteorology and Commonwealth Scientific and Industrial Research Organisation: Melbourne, Australia, 2014; 372p.

22. Hunt, B. An analysis of the groundwater resources of Tongatapu island, Kingdom of Tonga. *J. Hydrol.* **1979**, *40*, 185–196. [CrossRef]

23. Furness, L.; Helu, S.P. *The Hydrogeology and Water Supply of the Kingdom of Tonga*; Ministry of Lands, Survey and Natural Resources: Nuku'alofa, Tonga, 1993.

24. Falkland, A. *Water Resources Report. Tonga Water Supply Master Plan Project*; PPK Consultants Pty Ltd.: Brisbane, QLD, Australia; Australian International Development Assistance Bureau: Canberra, ACT, Australia, 1992.

25. Furness, L.J. Hydrogeology of Carbonate Islands of the Kingdom of Tonga, Chapter 18. In *Geology and Hydrogeology of Carbonate Islands Geology and Hydrogeology of Carbonate Islands*; Vacher, H.L., Quinn, T., Eds.; Developments in Sedimentology 54; Elsevier Science: Amsterdam, The Netherlands, 1997; pp. 565–576.

26. Falkland, A. *An Outline of Recent Water Supply Improvements for Pangai-Hihifo, Lifuka, Kingdom of Tonga*; Prepared for Tonga Water Board; Ecowise Environmental: Canberra, ACT, Australia, 2000.

27. White, I.; Falkland, A.; Fatai, T. *Groundwater Evaluation and Monitoring Assessment, Vulnerability of Groundwater in Tongatapu, Kingdom of Tonga*; Report to Pacific Islands Applied Geoscience Commission (SOPAC); Canberra Australia European Union EDF8 project; Reducing the Vulnerability of Pacific APC States; Australian National University: Canberra, ACT, Australia, 2009; 361p.

28. Waterloo, M.J.; Ijzermans, S. Groundwater availability in relation to water demands in Tongatapu. In *Dutch Risk Reduction Team—Reducing the Risk of Water Related Disasters*; SPREP Inform project: Apia, Samoa, 2017; 81p.

29. Fielea, Q.V.; (Tonga Water Board, Nuku'alofa, Kingdom of Tonga). Personal Communication, 2020.

30. Tonga Local Government, Ministry of Internal Affairs. Tonga Community Development Plans. Launched 4 October 2016. Available online: https://www.tongalocal.gov.to/community-development-plans (accessed on 2 September 2020).

31. Tufui, L.; (Victoria University of Wellington, WLG, New Zealand). Personal Communication, 2020.

32. Thompson, C.S. *The Climate and Weather of Tonga*; NZ Met. Service, Misc. Publication 188; New Zealand Meteorological Service, Te Ratonga Tirorangi: Wellington, New Zealand, 1986; 60p.

33. ABoM; CSIRO. *Climate Change in the Pacific: Scientific Assessment and New Research. Volume 1: Regional Overview*; Hennessy, K., Power, S., Cambers, G., Eds.; Pacific Climate Change Science Program; Australian Bureau of Meteorology and Commonwealth Scientific and Industrial Research Organisation: Melbourne, VIC, Australia, 2011; 257p.

34. Diamond, H.J.; Lorrey, A.M.; Renwick, J.A. A Southwest Pacific tropical cyclone climatology and linkages to the El Niño—Southern Oscillation. *J. Clim.* **2013**, *26*, 3–25. [CrossRef]

35. Lafale, P.F.; Diamond, H.J.; Anderson, C.L. Effects of climate change on extreme events relevant to the Pacific Islands. In *Pacific Marine Climate Change Report Card*; Science Review 2018; Cefas: Suffolk, UK; UK Hydrographic Office: Taunton, UK; National Oceanography Centre: Southampton, UK, 2018; pp. 50–73. Available online: https://assets.publishing.service.gov.uk/government/uploads/system/uploads/attachment_data/file/714530/5_Extreme_Events.pdf (accessed on 30 July 2020).

36. Cai, W.; Lengaigne, M.; Borlace, S.; Collins, M.; Cowan, T.; McPhaden, M.J.; Timmermann, A.; Power, S.; Brown, J.; Menkes, C.; et al. More extreme swings of the South Pacific convergence zone due to greenhouse warming. *Nature* **2012**, *488*, 365–369. [CrossRef] [PubMed]

37. Pörtner, H.-O.; Roberts, D.C.; Masson-Delmotte, V.; Zhai, P.; Tignor, M.; Poloczanska, E.; Mintenbeck, K.; Alegría, A.; Nicolai, M.; Okem, A.; et al. *IPCC Special Report on the Ocean and Cryosphere in a Changing Climate. Approved at the Second Joint Session of Working Groups I and II of the IPCC and accepted by the 51th Session of the IPCC, Principality of Monaco, 24th September 2019*; Intergovernmental Panel on Climate Change: Geneva, Switzerland, . 2019; 1170p; (in press). Available online: https://www.ipcc.ch/site/assets/uploads/sites/3/2019/12/SROCC_FullReport_FINAL.pdf (accessed on 2 September 2020).

38. Cai, W.; Borlace, S.; Lengaigne, M.; van Rensch, P.; Collins, M.; Vecchi, G.; Timmermann, A.; Santoso, A.; McPhaden, M.J.; Wu, L.; et al. Increasing frequency of extreme El Niño events due to greenhouse warming. *Nat. Clim. Change* **2014**, *4*, 111–116. [CrossRef]

39. Cai, W.; Wang, G.; Santoso, A.; McPhaden, M.J.; Wu, L.; Jin, F.-F.; Timmermann, A.; Collins, M.; Vecchi, G.; Lengaigne, M.; et al. Increased frequency of extreme La Niña events under greenhouse warming. *Nat. Clim. Change* **2015**, *5*, 132–137. [CrossRef]

40. Cai, W.; Wang, G.; Dewitte, B.; Wu, L.; Santoso, A.; Takahashi, K.; Yang, Y.; Carréric, A.; McPhaden, M.J. Increased variability of eastern Pacific El Niño under greenhouse warming. *Nature* **2018**, *564*, 201–206. [CrossRef] [PubMed]

41. Li, S.; Wu, L.; Yang, Y.; Geng, T.; Cai, W.; Gan, B.; Chen, Z.; Jing, Z.; Wang, G.; Ma, X. The Pacific Decadal Oscillation less predictable under greenhouse warming. *Nat. Clim. Chang.* **2020**, *10*, 30–34. [CrossRef]

42. Ward, R.G. *Widening Worlds, Shrinking Worlds, the Reshaping of Oceania*; Pacific Distinguished Lecture 1999; Centre for the Contemporary Pacific, Australian National University: Canberra, ACT, Australia, 1999; 38p.

43. Falkland, A. *Report on Water Security and Vulnerability to Climate Change and Other Impacts in Pacific Island Countries and East Timor (Draft). 2011 For Pacific Adaptation Strategy Assistance Program, Department of Climate Change and Energy Efficiency*; GHD Pty Ltd.: Canberra, ACT, Australia, 2011; 133p.

44. *Nauru Economic Infrastructure Strategy and Investment Plan*; Government of Nauru: Yaren, Nauru, 2011; 137p. Available online: https://www.theprif.org/documents/nauru/other/nauru-economic-infrastructure-strategy-and-investment-plan-2011 (accessed on 30 September 2020).

Publisher's Note: MDPI stays neutral with regard to jurisdictional claims in published maps and institutional affiliations.

 *hydrology*

Article

A New Physically-Based Spatially-Distributed Groundwater Flow Module for SWAT+

Ryan T. Bailey [1,*], Katrin Bieger [2], Jeffrey G. Arnold [3] and David D. Bosch [4]

[1] Dept. of Civil and Environmental Engineering, Colorado State University, Fort Collins, CO 80523, USA
[2] Blackland Research & Extension Center, Texas A&M AgriLife, Temple, TX 76502, USA;
    kbieger@brc.tamus.edu
[3] Grassland Soil and Water Research Laboratory, USDA-ARS, Temple, TX 76502, USA; jeff.arnold@usda.gov
[4] Southeast Watershed Research, USDA-ARS, Tifton, GA 31794, USA; david.bosch@usda.gov
* Correspondence: rtbailey@colostate.edu

Received: 17 September 2020; Accepted: 6 October 2020; Published: 9 October 2020

Abstract: Watershed models are used worldwide to assist with water and nutrient management under conditions of changing climate, land use, and population. Of these models, the Soil and Water Assessment Tool (SWAT) and SWAT+ are the most widely used, although their performance in groundwater-driven watersheds can sometimes be poor due to a simplistic representation of groundwater processes. The purpose of this paper is to introduce a new physically-based spatially-distributed groundwater flow module called *gwflow* for the SWAT+ watershed model. The module is embedded in the SWAT+ modeling code and is intended to replace the current SWAT+ aquifer module. The model accounts for recharge from SWAT+ Hydrologic Response Units (HRUs), lateral flow within the aquifer, Evapotranspiration (ET) from shallow groundwater, groundwater pumping, groundwater–surface water interactions through the streambed, and saturation excess flow. Groundwater head and groundwater storage are solved throughout the watershed domain using a water balance equation for each grid cell. The modified SWAT+ modeling code is applied to the Little River Experimental Watershed (LREW) (327 km^2) in southern Georgia, USA for demonstration purposes. Using the *gwflow* module for the LREW increased run-time by 20% compared to the original SWAT+ modeling code. Results from an uncalibrated model are compared against streamflow discharge and groundwater head time series. Although further calibration is required if the LREW model is to be used for scenario analysis, results highlight the capabilities of the new SWAT+ code to simulate both land surface and subsurface hydrological processes and represent the watershed-wide water balance. Using the modified SWAT+ model can provide physically realistic groundwater flow gradients, fluxes, and interactions with streams for modeling studies that assess water supply and conservation practices. This paper also serves as a tutorial on modeling groundwater flow for general watershed modelers.

Keywords: SWAT+; groundwater; modeling; groundwater–surface water interactions

1. Introduction

The SWAT (Soil and Water Assessment Tool) watershed model [1] is used frequently worldwide to assess water supply, nutrient dynamics, and sediment dynamics under scenarios of climate change, water management practices, and conservation practices. More recently, the SWAT+ model [2] has been presented as an alternative modeling tool with an emphasis on connectivity between spatial features (hydrologic response units, aquifers, channels, reservoirs, ponds, canals, pumps, etc.) within the watershed. Both models, however, often are limited in groundwater-driven watersheds due to the use of simplified, steady-state flow equations to represent water table fluctuation, groundwater storage, and groundwater discharge to streams [3–5]. Simulating groundwater head and flow using equations for unsteady flow in a heterogeneous aquifer system is important for estimating groundwater supply, groundwater–surface water interaction location and rates, and nutrient (nitrogen, phosphorus) loading to streams via subsurface flow for such watersheds and is essential for the correct assessment of scenarios.

To provide for more accurate simulation groundwater flow in watershed models, SWAT and SWAT+ have been linked to physically-based, spatially-distributed groundwater flow models, most notably MODFLOW [6], which simulates flow in heterogeneous three-dimensional (3D) aquifer systems. Examples for linked SWAT-MODFLOW models include [5,7–10], with [10] being used recently for many watersheds worldwide, e.g., [11–16]. To the authors' knowledge, only one study [17] has linked SWAT+ with MODFLOW.

Although these linked models have provided improved simulation capabilities for coupled stream-aquifer systems within watersheds, their general use is hampered by three principal limitations:

i. The development of a MODFLOW model can be time-intensive [18–20] and require the use of software that may not be available to the user;

ii. SWAT-MODFLOW simulations can have long run-times, often many times the duration of a stand-alone SWAT model; this can impede the use of the linked model in applications that involve ensembles of simulations (automated calibration, uncertainty analysis, sensitivity analysis, e.g., [21–25]); and

iii. Both SWAT-MODFLOW and SWAT+MODFLOW require extensive modifications to the SWAT and SWAT+ modeling codes and the inclusion of numerous additional source files, resulting in a cumbersome modeling code and inhibiting model developers from making modifications.

The first limitation has been overcome largely by the development of the graphical user interface QSWATMOD [26], a QGIS plugin tool that facilitates the linkage of SWAT and MODFLOW models and the running and visualization of SWAT-MODFLOW simulations. However, the second and third limitations are ongoing and unsolved.

The objective of this paper is to present a new physically-based spatially-distributed groundwater flow module for SWAT+ that addresses the three limitations listed above. Called *gwflow*, the module: (1) does not require the use of MODFLOW or other groundwater modeling codes, many of which are commercialized; (2) does not add significant run time to SWAT+ simulations; and (3) consists of only three new code subroutines, in keeping with the modular nature of the SWAT+ code. In addition, all data for the *gwflow* module can be prepared using a GIS (ArcMap or QGIS). Due to the detailed description of the *gwflow* module and the underlying solution algorithm presented in Section 2, this paper also serves as a primer for watershed modelers that desire to better understand groundwater flow modeling and how it relates to watershed hydrologic processes.

The capabilities of the *gwflow* module for SWAT+ are demonstrated through an application to the Little River Experimental Watershed (LREW) (327 km^2) in southern Georgia, wherein groundwater flow has been observed to be a significant portion of streamflow. In addition, global data sets for aquifer thickness and aquifer permeability are used to populate necessary data for the module to demonstrate how the model could be applied to data-scarce watersheds. Model results are compared against observed streamflow and groundwater levels at eight stream gages and eight groundwater monitoring sites, respectively, to demonstrate the general correctness of the module in simulating hydrological processes. As the purpose of this paper is to present the new *gwflow* module, detailed calibration is not performed. Hence, further calibration is needed for the LREW model if it is to be used for scenario analysis.

2. Materials and Methods

This section presents an overview of the SWAT+ model, the theoretical foundation for the new *gwflow* module and its implementation into the SWAT+ modeling code, and an application of the new SWAT+ modeling code to the Little River Experimental Watershed.

2.1. SWAT+

SWAT+ [2,27,28] is a restructured version of the SWAT watershed modeling code in which watershed hydrologic entities (hydrologic response units, aquifers, channels, reservoirs, ponds, point sources) are designated as spatial objects that can exchange water in any user-specified system. The watershed is divided into subbasins, which are then each divided into one or more landscape units, which lump hydrographs and route them to any spatial object. Within SWAT+, groundwater flow is simulated in the same manner as in the original SWAT model, with the following assumptions and limitations:

i.   Groundwater head (i.e., water table elevation) changes only according to soil recharge and groundwater discharge to streams; in reality, there are many other sources and sinks of groundwater;

ii.   a single groundwater head is computed for each Hydrologic Response Unit (HRU);

iii.   groundwater flow to streams is based on the presence of a groundwater divide and an assumption of steady-state conditions;

iv.   groundwater flow to streams is simulated only if groundwater storage exceeds a user-specified threshold, rather than due to hydraulic gradients;

v.   seepage from streams to the aquifer due to hydraulic gradients is not simulated;

vi.   each aquifer is treated as a homogeneous system in which aquifer properties (hydraulic conductivity K, specific yield S_y) do not vary in space.

2.2. Groundwater Flow Module (gwflow) for SWAT+

2.2.1. Overview of the gwflow Module

The *gwflow* module for SWAT+ presented in this section is based on a physically-based, spatially distributed groundwater flow model for unconfined aquifers that are hydraulically connected to streams. The aquifer is discretized laterally into a collection of square grid cells, with groundwater volume and groundwater head calculated for each cell on a daily time step. The module uses a single layer to represent the aquifer from the ground surface to the bedrock and is based on the Dupuit–Forchheimer assumption of horizontal flow in unconfined aquifers, and therefore ignores any vertical gradients in groundwater head. Groundwater stresses include recharge, lateral flow, groundwater evapotranspiration (ET), discharge to streams, stream seepage to the aquifer, pumping, and saturation excess flow. These stresses are listed in Table 1, along with whether each stress is a source (groundwater entering the aquifer) or a sink (groundwater leaving the aquifer) if the stress flux (i.e., volumetric flow rate m³/day) depends on groundwater head and/or groundwater storage, the required aquifer and streambed properties to compute the stress flux, and the general method for computation. Recharge is provided to cells by connecting SWAT+ HRUs, and cells that intersect SWAT+ channels can exchange water with these channels. The module is embedded into the SWAT+ code, as described later in this section, and replaces the current aquifer module of SWAT+. Methods for calculating groundwater volume and groundwater head are presented in Section 2.2.4.

Table 1. Groundwater stresses included in the *gwflow* module. Specific details regarding the method for computing each stress are presented in Section 2.2.4.

Groundwater Stress	Source/Sink	Dependent on Groundwater Head	Dependent on Groundwater Storage	Required Aquifer/Stream Properties	Method for Computing Flow Rate
Soil Recharge	Source	No	No	-	SWAT+ HRUs
Lateral Flow	Mixed	Yes	Yes	K	Darcy's Law
Groundwater ET	Sink	Yes	Yes	-	Linear Equation
Discharge to streams	Sink	Yes	Yes	K_{bed}, d_{bed}	Darcy's Law
Stream seepage	Source	Yes	No	K_{bed}, d_{bed}	Darcy's Law
Pumping	Sink	No	Yes	-	User Specified
Saturation excess flow	Sink	Yes	No	S_y	Storage equation

K = aquifer hydraulic conductivity (m/day). S_y = aquifer specific yield (-). K_{bed} = streambed hydraulic conductivity (m/day). d_{bed} = streambed thickness (m).

2.2.2. Solution Strategy for the *gwflow* Module

This section describes the method to calculate groundwater volume and groundwater head throughout a heterogeneous unconfined aquifer system within a watershed domain. The hydrologic fluxes in a typical watershed domain are shown in Figure 1a: land surface and soil fluxes include rainfall, runoff, lateral flow, ET, and percolation; groundwater fluxes include lateral flow from each direction (Q_{north}, Q_{south}, Q_{west}, Q_{east}); flow across the watershed boundary, which is a component of lateral flow; recharge Q_{rech}; groundwater ET Q_{gwet}; pumping Q_{pump}; groundwater discharge to streams Q_{gwsw}; stream seepage to the aquifer Q_{swgw}; and saturation excess flow Q_{satex}, which occurs typically during runoff events related to a rapid rise in the water table and subsequent runoff to streams [29–31]. For SWAT+, the land surface and soil hydrologic fluxes are simulated by original SWAT+ routines, whereas the groundwater fluxes are simulated by the new *gwflow* module. Figure 1a also shows the saturated thickness s of the aquifer (water table to the bedrock) and the head h of groundwater, measured from a datum, typically mean sea level. Within an unconfined flow system, the head h is generally equivalent to water table elevation.

As with all groundwater flow models (e.g., MODFLOW [6], SUTRA [32], CATHY [33], HydroGeoSphere [34]), a control volume approach based on mass conservation is used to establish a water balance equation for each grid cell in the model domain. These equations are then solved for groundwater volume and corresponding groundwater head for each time step of the simulation. Most models use an implicit approach to solve the system of equations, in which head values for all grid cells are updated concurrently (i.e., each head value depends on the head values of surrounding cells at the same time step), and therefore requires linear algebra algorithms or iteration methods. In contrast, the *gwflow* module uses an explicit approach, in which volume and head values for each grid cell are calculated using information (head values of surrounding grid cells; source and sink fluxes) only from the previous time step. Therefore, each grid cell water balance equation is independent of the equations for surrounding cells and can be solved using a simple loop within a computer program, negating the need for computationally intensive solution algorithms. The explicit solution method was chosen for the *gwflow* module due to (1) simplicity in coding; and (2) facility in explaining groundwater modeling within the context of watershed systems for watershed modelers. Readers interested in the implicit approach for groundwater modeling are referred to [35].

Although explicit methods have been available for modeling groundwater heads for many years (e.g., [36]), they often are not used due to the requirement of relatively small time steps for solution stability [36]. However, for linking with watershed models that often use daily time steps, the required time step is reasonable, as will be shown in the model application in Section 2.3.

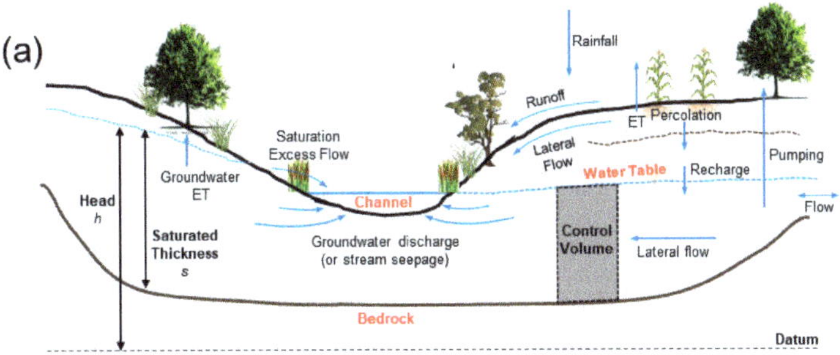

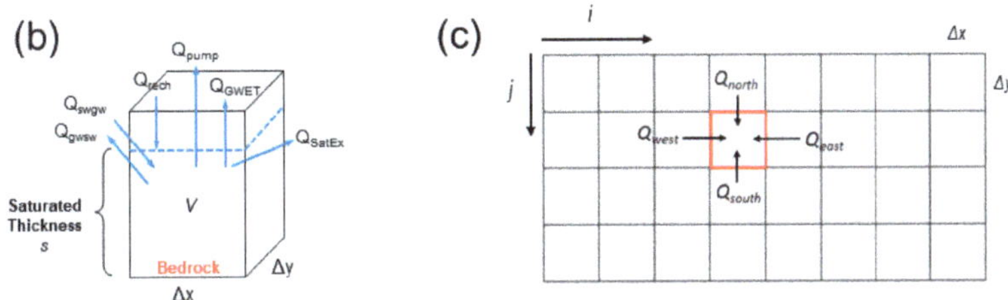

Figure 1. Schematics showing the hydrologic fluxes in a typical watershed stream-aquifer system:
(**a**) cross-section of stream-aquifer system showing main hydrologic elements and hydrologic fluxes;
(**b**) general control volume from (**a**) showing all groundwater inputs/outputs and the groundwater
volume V; (**c**) plan view of collection of connected control volumes (i.e., grid cells), showing lateral
groundwater flow into a grid cell from the four surrounding grid cells.

The derivation of the generic aquifer water balance equation is now presented. The aquifer
domain within the watershed is discretized into individual control volume cells (i.e., 3D cubes), which
in the *gwflow* module are required to be square on top, e.g., 100 m × 100 m or 200 m × 200, with the
thickness of the cell equal to the thickness of the aquifer (ground surface to the bedrock). An example
control volume cell and corresponding groundwater fluxes are shown in Figure 1b, with a plan view of
the cells shown in Figure 1c. The collection of cells shown in Figure 1c is referred to as a "grid" and
covers the area of the watershed. In Figure 1b, the water table is denoted by a dashed blue line, and
the saturated thickness s of the cell is the vertical distance from the water table to the bedrock. As
the cell is composed of both aquifer material and groundwater, total groundwater volume (m³) in the
cell is equal to ($\Delta x \cdot \Delta y \cdot s \cdot$ porosity). However, as we are concerned only with groundwater that can be
added to/removed from storage, and not with that which is retained due to suction forces between
aquifer sediment and groundwater, the groundwater volume V (m³) of concern is:

$$V = (\Delta x \cdot \Delta y \cdot s) \cdot S_y \tag{1}$$

where S_y is specific yield (m³ water / m³ of bulk material), i.e., the volume of groundwater released
from storage per unit surface area of aquifer per unit decline in groundwater head (i.e., water table)
due to gravity.

A volumetric water balance is performed daily for each cell by tracking all groundwater fluxes
into/out of the cell:

$$\frac{\Delta V}{\Delta t} = \sum Q_{in} - \sum Q_{out} \tag{2}$$

where t is time (day), Q_{in} represents fluxes (m^3/day) that add groundwater to the cell, and Q_{out} (m^3/day) represent fluxes that remove groundwater from the cell. Groundwater inputs and outputs can be further categorized into source fluxes, sink fluxes, and lateral fluxes:

$$\frac{\Delta V}{\Delta t} = \sum Q_{source} - \sum Q_{sink} \pm Q_{lateral} \tag{3}$$

where source fluxes include recharge (Q_{rech}) and stream seepage ($Q_{sw \rightarrow gw}$) and sink fluxes include groundwater discharge to streams ($Q_{gw \rightarrow sw}$), groundwater ET (Q_{gwet}), pumping (Q_{pump}), and saturation excess flow (Q_{satex}), as shown in Figure 1b. Lateral fluxes are gradient-driven fluxes that flow across the four sides of the cell. The grid in the *gwflow* module is required to be oriented along north-south and west-east axes, resulting in lateral fluxes in the north, south, west, and east directions (Q_{north}, Q_{south}, Q_{west}, Q_{east}), as shown in Figure 1c. The "±" sign indicates that flow can either enter the cell from the cell in that direction, or leave the cell towards the cell in that direction. Including these individual flux terms into Equation (2) yields:

$$\frac{\Delta V}{\Delta t} = \left(Q_{rech} + Q_{sw \rightarrow gw}\right) - \left(Q_{gwet} + Q_{gw \rightarrow sw} + Q_{satex} + Q_{pump}\right) \pm Q_{north} \pm Q_{south} \pm Q_{west} \pm Q_{east} \tag{4}$$

where lateral fluxes are assumed to add groundwater to the cell. The actual direction of lateral fluxes, i.e., into or out of the cell, depends on groundwater head patterns, as described in Section 2.2.3.

Saturated thickness can be included in the water balance equation of (4) by inserting Equation (1):

$$\frac{\Delta\left[(\Delta x \cdot \Delta y \cdot s) \cdot S_y\right]}{\Delta t} = \left(Q_{rech} + Q_{sw \rightarrow gw}\right) - \left(Q_{gwet} + Q_{gw \rightarrow sw} + Q_{satex} + Q_{pump}\right) \pm Q_{north} \pm Q_{south} \pm Q_{west} \pm Q_{east} \tag{5}$$

Since a change in saturated thickness is equal to a change in groundwater head, h can be substituted for s in Equation (5). Assuming S_y of the aquifer does not change in time yields:

$$\tfrac{\Delta h}{\Delta t}\left(\Delta x \Delta y S_y\right) = \left(Q_{rech} + Q_{sw \rightarrow gw}\right) - \left(Q_{gwet} + Q_{gw \rightarrow sw} + Q_{satex} + Q_{pump}\right) \pm Q_{north} \pm Q_{south} \pm Q_{west} \pm Q_{east} \tag{6}$$

Using Equation (3) to simplify notation, the change in head Δh for the grid cell is computed by:

$$\Delta h = \left(\sum Q_{source} - \sum Q_{sink} \pm Q_{lateral}\right)\left(\frac{\Delta t}{S_y \Delta x \Delta y}\right) \tag{7}$$

which, if assessing a change in head between two points in time n and $n+1$, h at the next time $n+1$ can be computed by:

$$h^{n+1} = h^n + \left(\sum Q_{source} - \sum Q_{sink} \pm Q_{lateral}\right)\left(\frac{\Delta t}{S_y \Delta x \Delta y}\right) \tag{8}$$

Equation (8) is solved for each cell in the grid using cell information (flux terms, h) from the previous time n. Each cell is identified using a row and column index using the i and j notation shown in Figure 1c, leading to:

$$h_{i,j}^{n+1} = h_{i,j}^n + \left(\sum Q_{source_{i,j}} - \sum Q_{sink_{i,j}} \pm Q_{lateral_{i,j}}\right)\left(\frac{\Delta t}{S_{y_{i,j}} \Delta x \Delta y}\right) \tag{9}$$

and now including all flux terms for the *gwflow* module:

$$h_{i,j}^{n+1} = h_{i,j}^n + \left[\left(Q_{rech} + Q_{sw \rightarrow gw}\right) - \left(Q_{gwet} + Q_{gw \rightarrow sw} + Q_{satex} + Q_{pump}\right) \pm Q_{north} \pm Q_{south} \pm Q_{west} \pm Q_{east}\right]\left(\tfrac{\Delta t}{S_{y_{i,j}} \Delta x \Delta y}\right) \tag{10}$$

As the computation of h for each cell depends only on information from the previous time n, this solution strategy is an *explicit* numerical method. Inside the SWAT+ code (see Section 2.2.4 for an explanation of the code structure), the *gwflow* subroutine loops over the grid cells in the watershed domain, solving for head in each cell using Equation (10). As the head value h^n must be known to solve for the head value at the next time step h^{n+1}, the model user must specify the initial head value for each grid cell so that the head values can be calculated on the first day of the SWAT+ simulation. The calculation of each source and sink flux is described in Section 2.2.3. The requirement for solution stability for this explicit approach has been reported [36] as:

$$\frac{\Delta t \cdot K \cdot s}{\Delta x \cdot \Delta y \cdot S_y} \leq \frac{1}{4} \tag{11}$$

Note that Equation (11) does not account for source and sink flux terms. Therefore, the largest Δt that can be used without causing solution instabilities should be found by trial and error when running the model.

2.2.3. Calculating Groundwater Stress Volumetric Fluxes

This section provides details for calculating the individual flux terms in Equation (10) within the context of the SWAT+ modeling code.

Soil Recharge

Recharge to the water table is assumed equivalent to the water percolating from the bottom of the soil profile in each HRU of the SWAT+ model. Similar to SWAT-MODFLOW [10], HRU soil percolation is mapped to grid cell recharge using intersected areas of the HRUs and grid cells. During each daily time step, the depth of recharge from an HRU is multiplied by the area of the HRU to provide a volumetric flow rate in m³/day. This volume is then distributed to grid cells that overlap the HRU, multiplying the volume by the fraction of the HRU that overlaps the grid cell. These fractions are calculated during model construction by intersecting the HRU shapefile and the grid cell shapefile within a GIS, and then placed in an input file *gwflow.hrucell*.

Lateral Flow

The lateral rate of groundwater flow across the north, south, west, and east sides of each grid cell (see Figure 1c) is calculated using Darcy's Law:

$$Q = A \cdot K \cdot \frac{\Delta h}{\Delta x} \tag{12}$$

where A is the cross-sectional area of groundwater flow (m²), K is the hydraulic conductivity of the aquifer material (m/day), and $\Delta h/\Delta x$ is the hydraulic gradient, with groundwater flowing from areas of high head to areas of low head. Using the i and j notation from Figure 1c, lateral flux across the west side of a cell (i,j) is written as:

$$Q_{west} = \left(\frac{\Delta x}{\frac{\Delta x/2}{K_{i-1,j}} + \frac{\Delta x/2}{K_{i,j}}} \right) \left[\left(\frac{s_{i-1,j} + s_{i,j}}{2} \right) \Delta y \right] \left(\frac{h_{i-1,j} - h_{i,j}}{\Delta x} \right) \tag{13}$$

where the first term is the harmonic mean K value of the cell (i,j) and the cell to the west $(i-1,j)$, often used to simulate flow perpendicular to aquifer units; the second term uses the average saturated thickness s (head h–bedrock elevation) from the two cells to estimate A, and the third term uses the h values from the two cells to compute the hydraulic gradient. The third term is written so that higher head in the cell to the west results in a positive gradient, indicating an input of groundwater into the cell (i,j); conversely, lower head in the cell to the west results in a negative gradient, indicating removal of groundwater from the cell (i,j). Similar equations are written for the other three sides:

$$
\begin{aligned}
Q_{east} &= \left(\frac{\Delta x}{\frac{\Delta x/2}{K_{i+1,j}} + \frac{\Delta x/2}{K_{i,j}}} \right) \left[\left(\frac{s_{i+1,j} + s_{i,j}}{2} \right) \Delta y \right] \left(\frac{h_{i+1,j} - h_{i,j}}{\Delta x} \right) \\
Q_{north} &= \left(\frac{\Delta x}{\frac{\Delta x/2}{K_{i,j-1}} + \frac{\Delta x/2}{K_{i,j}}} \right) \left[\left(\frac{s_{i,j-1} + s_{i,j}}{2} \right) \Delta x \right] \left(\frac{h_{i,j-1} - h_{i,j}}{\Delta y} \right) \\
Q_{south} &= \left(\frac{\Delta x}{\frac{\Delta x/2}{K_{i,j+1}} + \frac{\Delta x/2}{K_{i,j}}} \right) \left[\left(\frac{s_{i,j+1} + s_{i,j}}{2} \right) \Delta x \right] \left(\frac{h_{i,j+1} - h_{i,j}}{\Delta y} \right)
\end{aligned}
\tag{14}
$$

Note that for homogeneous aquifer systems, the harmonic mean is simplified to the arithmetic mean of hydraulic conductivity.

Groundwater ET

ET from the saturated zone is simulated only if the water table in a cell is above a specified elevation z_{bot} (m), calculated by subtracting a specified ET extinction depth $EXDP$ (m) (i.e., the depth below which ET cannot occur) from the ground surface z_{surf} (Figure 2a). The maximum depth of ET that can be removed from the saturated zone is equal to the unsatisfied ET ET_{remain} (mm), set equal to the difference between the potential ET (mm) and the actual ET (mm) simulated for each HRU for the day. The connection between HRUs and grid cells for mapping unsatisfied ET is the same as for mapping soil percolation to recharge. The depth of groundwater ET_{GW} (mm) removed from the cell is calculated using the linear relationship from MODFLOW's EVT package [6]:

$$
\begin{aligned}
h_{i,j} &< z_{bot} \rightarrow ET_{GW} = 0 \\
h_{i,j} &> z_{bot} \rightarrow ET_{GW} = ET_{remain} \cdot \left(\frac{h_{i,j} - z_{bot}}{z_{surf} - z_{bot}} \right)
\end{aligned}
\tag{15}
$$

(a)   (b)

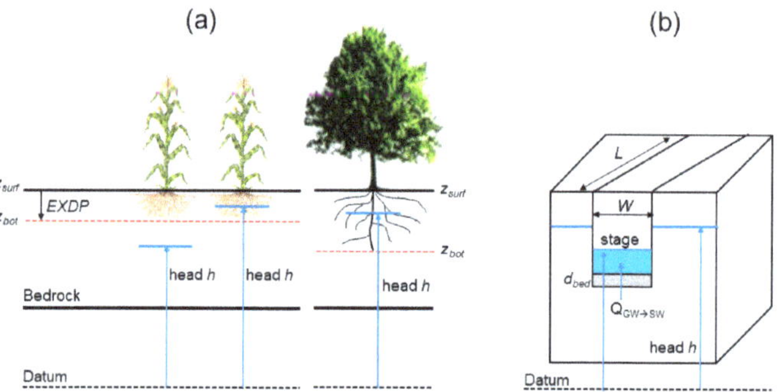

Figure 2. Variable depictions and cross-section schematics for (**a**) groundwater Evapotranspiration (ET) calculations and (**b**) groundwater–surface water exchange.

The depth of ET_{GW} is multiplied by the horizontal spatial area of the HRU to provide a volumetric flow rate in m^3/day, and then divided amongst the cells connected to the HRU, resulting in a Q_{gwet} value for each cell. This groundwater volume is removed from the cell (see Figure 1b). Figure 2a shows the scenario of Equation (15) within the context of the variables. For the row crop (corn as an example), the condition on the left results in no groundwater ET, whereas the condition on the right would result in ET_{GW} equal to approximately half of ET_{remain}. For the tree, the condition would result in ET_{GW} equal to more than half of ET_{remain}. Groundwater ET can be significant for areas with high water tables and deep-reaching vegetation roots, e.g., in riparian areas of streams [37,38].

Discharge to Streams and Stream Seepage to the Aquifer

For grid cells that contain streams, Darcy's Law is used to estimate flow rates between the aquifer and the stream. The direction of flow depends on the relationship between the groundwater head h and the stream stage h_{stream}. Figure 2b shows the scenario of $h > h_{stream}$, resulting in groundwater flow from the aquifer to the stream channel through the streambed with thickness d_{bed} (m). Flow occurs along the entire length L (m) of the stream within the cell and along the entire bottom width W (m) of the channel. $Q_{GW \to SW}$ (m^3/day) for this scenario is estimated by:

$$Q_{GW \to SW} = K_{bed}(WL)\left(\frac{h - h_{stream}}{d_{bed}}\right) \tag{16}$$

where K_{bed} (m/day) is the hydraulic conductivity of the streambed material and (WL) (m^2) is the cross-section area of flow. If groundwater head $h < h_{stream}$, then stream water recharges the aquifer and $Q_{SW \to GW}$ is calculated. If h is greater than the streambed elevation z_{bed}, the gradient uses the head difference between h_{stream} and h; if h is lower than z_{bed}, the gradient uses the channel depth. These scenarios are summarized in the following set of equations, similar to those in MODFLOW's River package:

$$
\begin{array}{ll}
h > h_{stream} & \left\{
\begin{array}{l}
Q_{GW \to SW} = K_{bed}(WL)\left(\frac{h - h_{stream}}{d_{bed}}\right) \\
Q_{SW \to GW} = 0
\end{array}\right. \\[2em]
z_{bed} < h < h_{stream} & \left\{
\begin{array}{l}
Q_{GW \to SW} = 0 \\
Q_{SW \to GW} = K_{bed}(WL)\left(\frac{h_{stream} - h}{d_{bed}}\right)
\end{array}\right. \\[2em]
h < z_{bed} & \left\{
\begin{array}{l}
Q_{GW \to SW} = 0 \\
Q_{SW \to GW} = K_{bed}(WL)\left(\frac{h_{stream} - z_{bed}}{d_{bed}}\right)
\end{array}\right.
\end{array}
\tag{17}
$$

Pumping

Pumping rates Q_{pump} (m^3/day) are specified by the model user. These can be specified for any grid cell, for any day of the simulation in the *gwflow.aqu* input file.

Saturation Excess Flow

Saturation excess flow occurs when groundwater head h rises above the ground surface, typically during rainfall events that rapidly increase the water table. This condition is tested for each cell during each daily time step. If $h >$ ground surface elevation, the volumetric flux (m^3/day) of groundwater excess flow is given by:

$$Q_{satex} = \left(h_{i,j} - z_{surf_{i,j}}\right)(\Delta x \Delta y) S_{y_{i,j}} \tag{18}$$

The water is removed from the grid cell and added to the closest stream channel on that same day.

2.2.4. SWAT+ Code Structure with the *gwflow* Module

The structure of the SWAT+ modeling code with the embedded *gwflow* module is presented in Figure 3. The *gwflow* module adds only two subroutines to the SWAT+ code: *gwflow_read*, which reads data for all grid cells from the input file *gwflow.aqu*, and *gwflow_simulate*, which performs the computations of groundwater fluxes and changes in groundwater head and groundwater storage. The subroutine *hyd_read_connect*, which reads connections between all spatial objects in the SWAT+ model, also reads the input file *gwflow.con*, which contains a list of connections between grid cells and SWAT+ channels to enable groundwater–surface water interactions (see Section 2.2.4). The inputs required in the *gwflow.aqu* and *gwflow.con* files will be discussed in Section 2.2.5.

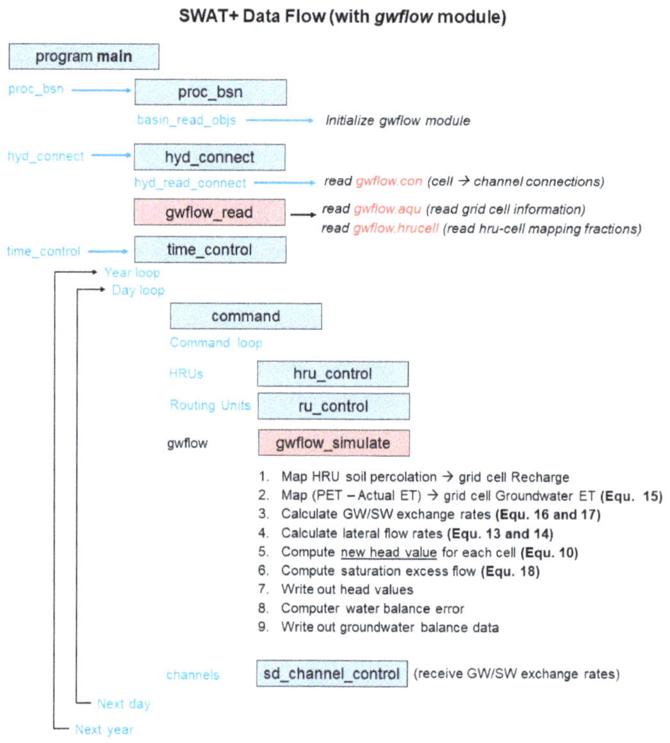

Figure 3. Data flow of the Soil and Water Assessment Tool+ (SWAT+) modeling code. Subroutines for the new *gwflow* module are shaded in light red, input files for the *gwflow* module are highlighted in red text, and the calculations performed in the *gwflow* module are listed in bullet points for the *gwflow_simulate* subroutine.

Following the reading of all input data, the *time_control* subroutine is called, which controls the yearly and daily computation loops. For each daily time step of the simulation, the *command* subroutine is called, which loops through all objects (HRUs, Routing Units, channels) in the model, including the *gwflow* grid cells. When the *gwflow_simulate* subroutine is called, all groundwater fluxes are calculated, whereupon the change in groundwater storage and groundwater head is calculated for each grid cell. Equations (see Sections 2.2.2 and 2.2.3) for each calculation are indicated in Figure 3. Head values, groundwater flux values for each grid cell, and a groundwater balance for the entire watershed model domain are then written to files. The groundwater balance is calculated and output for each day, each year, and then for the entire simulation (average annual).

2.2.5. Required Inputs for the *gwflow* Module

All data for the *gwflow* module are contained in three input files: *gwflow.con*, *gwflow.aqu*, and gwflow.*hrucell* (see Figure 3). *gwflow.con* contains a list of all grid cells connected to stream channels, with the ID number listed for each, so that $Q_{SW \rightarrow GW}$ and $Q_{GW \rightarrow SW}$ can be calculated for the correct cells. These cells are called "River Cells".

The file *gwflow.aqu* contains general information for the module and a list of data values for each grid cell. A complete list of information in this file is shown in Table 2. Basic information includes the number of rows and columns in the grid, the cell size (m), flags for water table initiation (at the beginning of the simulation) and the inclusion of saturation excess flow and groundwater ET, a recharge delay term (to transfer recharge from the bottom of the soil profile to the water table), and the time step. The water table can be initiated by (1) ground surface minus a specified depth for each grid cell, (2) a fraction of the distance between the ground surface and the bedrock, or (3) a value specified for each cell. For K and S_y, the grid cells are divided into zones, with K and S_y specified for each zone. This facilitates calibration by changing only values for each zone rather than values for each grid cell. Groundwater head h for each cell can be output for any day of the simulation. Certain cells can also be designated as "observation cells", with h values for these cells output for each day of the simulation, resulting in time series that can be compared to data from groundwater monitoring wells. Many of these data can be obtained from national or global datasets. Sources for these data will be discussed in Section 2.3.

Table 2. Information required in the *gwflow.aqu* file for the *gwflow* module. Many of the parameters can be found from national or global datasets, as described in the model application in Section 2.3.

Basic Information	Units
Cell size	m
Number of Rows and Columns in the grid	-
Water table initiation flag (1, 2, 3)	-
Saturation excess flow flag (yes/no)	-
Groundwater ET flag (yes / no)	-
Recharge delay	*day*
Time step Δt	*day*
Grid Cell Information	
Row Index	-
Column Index	-
Status (active = 1, inactive = 0, boundary = 2)	-
Longitude	*degree*
Latitude	*degree*
Ground surface elevation	m
Aquifer thickness	m
Hydraulic conductivity zone	-
Specific yield zone	-
Groundwater ET extinction depth *EXDP*	m
Output Control	
Days for groundwater head output	*day*
Cells for daily groundwater head output	-
River Cell Information	
Depth of streambed below DEM value	m
Row Index	-
Column Index	-
Channel ID	-
Stream length in cell	m
Streambed elevation z_{bed}	m
Streambed hydraulic conductivity zone K_{bed}	-
Streambed thickness zone d_{bed}	-

2.3. Application to the Little River Experimental Watershed, Georgia

2.3.1. Overview of Study Region

The SWAT+ model using the new *gwflow* module is applied in this study to the 327 km^2 Little River Experimental Watershed (LREW), located within the Upper Suwannee River Basin in south-central Georgia (Figure 4). This watershed is ideal for testing the *gwflow* module due to the strong connection between the shallow unconfined aquifer and the many streams that make up the river system [31]. In addition, there are a number of streamflow gages and groundwater monitoring wells located throughout the watershed for model testing.

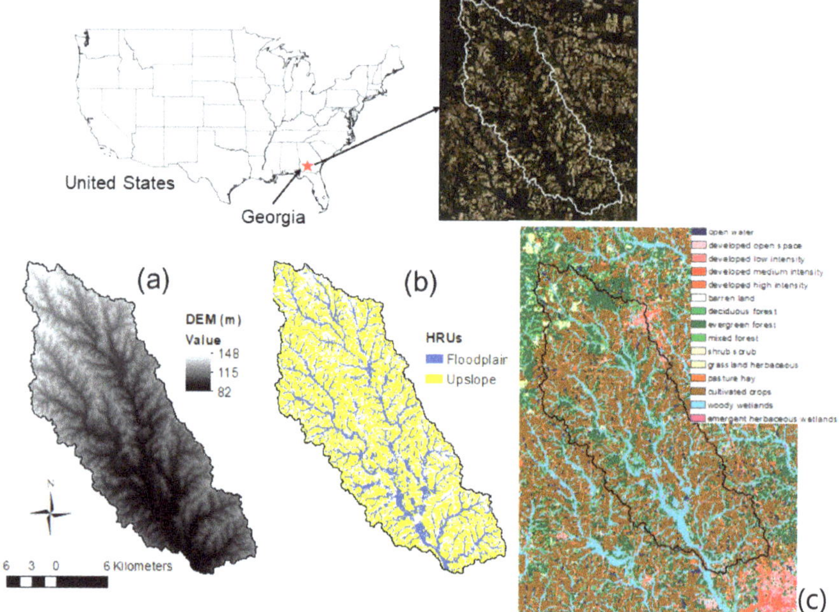

Figure 4. Geographic location and SWAT+ datasets and computational units, showing (**a**) digital elevation model (DEM) (m); (**b**) Hydrologic Response Units (HRUs), showing HRUs located in the upland areas and the floodplain areas; and (**c**) land use.

The climate of the region is humid subtropical, with hot and humid summers and mild winters. Summers are characterized by high-intensity thunderstorms, leading to sharp increases in streamflow. The average annual rainfall is approximately 1200 mm, and the annual mean temperature is 19 °C. The average annual ET has been estimated to be approximately 70% of annual precipitation [37]. The watershed elevation ranges between 82 m and 148 m, with broad floodplains (Figure 4a). Land use (Figure 4c) consists of forest (50%), primarily pine trees in upland areas and hardwoods and evergreens in riparian areas; agriculture, primary cotton and peanut row-crops (41%); urban areas (7%), and open water (2%). Soils are loamy sands and sandy loams [39], which are underlain by the relatively impermeable Hawthorne formation which limits groundwater flow to deeper geologic layers [40]. The presence of the Hawthorne formation, referred to as the bedrock through the remainder of this paper, leads to significant lateral groundwater flow and associated groundwater discharge to streams [31,37,41]. Saturation excess flow conditions occur within the riparian areas [42]. The thickness of the alluvial aquifer is presented in Section 2.3.3.

Many variations and applications of the SWAT and SWAT+ models have been applied to the LREW, including for model evaluation, calibration, and parameter sensitivity [43–46], analyzing the effect of conservation practices on water quality [47], landscape routing [48] and connectivity between upland areas and floodplains [28]. Indeed, the SWAT+ model was first published and introduced using an application to this watershed [2]. These applications, however, treated groundwater flow and groundwater–surface water interactions simplistically, based on steady flow assumptions and unconnected aquifer regions. The inclusion of the *gwflow* module is expected to simulate more realistic groundwater hydrologic fluxes, which can lead to improved assessments of riparian-stream hydrology and conservation practices within the study region.

2.3.2. SWAT+ Model Construction

The base SWAT+ model was constructed using QSWAT+, a QGIS interface for SWAT+ (https: //swat.tamu.edu/software/plus/). The interface guides the user through loading datasets (DEM, land use and crop data, soil type, and climate station data) and creating hydrologic response units (HRUs), routing units, and channels. The resolution and source for each data set are shown in Table 3. The interface then uses the SWAT+ Editor to write all necessary input files to run the SWAT+ modeling code. In this study, SWAT+ version 59.3 is used. The process resulted in 10796 HRUs and 343 channels. Figure 4b shows the HRUs divided into upland areas (yellow) and floodplains areas (blue). Channel delineation is shown in Section 2.3.3.

Before including the *gwflow* module, an initial calibration was performed for SWAT+ to provide hydrologic fluxes (surface runoff, lateral flow, groundwater discharge, ET) that are similar to those observed from field data assessment. This calibration was performed manually as well as using IPEAT+ (Integrated Parameter Estimation and Uncertainty Analysis Tool Plus) [49], an automated calibration tool developed for SWAT+, based on the Dynamically Dimensioned Search (DDS) algorithm [50], using the streamflow at the watershed outlet as testing data.

Table 3. Datasets and corresponding sources for the construction of the Little River Watershed SWAT+ model.

Data	Resolution	Source
Topo-graphy	30 m	U.S. Geological Survey, National Elevation Data *Accessed: 15 October 2019, https://viewer.nationalmap.gov*
Land use	30 m	U.S. Geological Survey, National Land Cover Data (2016) *Accessed: 15 October 2019, https://www.mrlc.gov/data*
Crop Data	30 m	U.S. Dept. of Agriculture, CropScape *Accessed: 15 October 2019, https://nassgeodata.gmu.edu/CropScape/*
Soil	30 m	USDA-NRCS, Soil Survey Geographic (SSURGO) *Accessed: 15 October 2019, -https://datagateway.nrcs.usda.gov/*
Climate	Multiple stations	Precipitation: daily watershed weighted average [51]. Other climate variables: SWAT+ global database.
Aquifer thickness (cm)	250 m	[52] *Accessed: 10 December 2019, https://soilgrids.org/*
Hydraulic conduct. (m/day)	Vector Polygons	[53] *Accessed: 10 December 2019,* *https://dataverse.scholarsportal.info/dataset.xhtml?persistentId=doi:* *10.5683/SP2/TTJNIU*

2.3.3. Preparing the gwflow Module

Preparing the *gwflow* module for SWAT+ consists of populating the *gwflow.aqu*, *gwflow.hrucell*, and *gwflow.con* files. The following list provides details regarding the preparation of all required data for the *gwflow.aqu* file (see Table 2). All data can be prepared in a GIS (e.g., ArcMap, QGIS).

i.   <u>Cell size</u>: defined by the user, typically between 100 and 500 m on a side based on previous applications of coupled surface water–groundwater models such as SWAT-MODFLOW (e.g., [14,16]). In the current edition of the code, all cells must be the same size.

ii.   <u>Number of rows and columns</u>: based on the extent of the watershed and the specified cell size.

iii.   <u>Time step Δt</u>: The maximum time step is 1 day, since the *gwflow* module is called each day within the SWAT+ code (see Figure 3). The minimum required time step must be found using a trial and error approach. We recommend starting with 1 day, checking results in the daily groundwater balance file to verify that the percent error in groundwater balance is equal to 0.

iv.   <u>Status</u> (active, inactive, boundary): as the grid is rectangular in shape, many cells will be located outside the watershed boundary; these cells are "inactive" and assigned status = 0. All cells within the watershed boundary are "active" and assigned status = 1, except for cells that lie along the boundary and are designated as "boundary" cells and assigned status = 2. Boundary cells are simulated as constant-head cells, i.e., the groundwater head assigned to these cells at the beginning of the simulation remains constant throughout the simulation period.

v. <u>Groundwater surface elevation</u>: obtained from the DEM of the watershed.

vi. <u>Aquifer thickness</u> (m): obtained from [52], who provide a global map of unconsolidated sediment thickness (cm) from the ground surface to bedrock (see Table 2). This dataset is particularly useful for watersheds with limited borehole and drilling information. These values must be converted to meters for the *gwflow* module.

vii. <u>Hydraulic conductivity K</u> (m/day): obtained from [53], who provide a global map of permeability (see Table 2). Permeability must be converted to K in m/day.

viii. <u>Specific Yield S_y</u>: specific yield values (typically 0.10 to 0.30 for alluvial aquifer sediments) are estimated based on the values of K, i.e., if values of K coincide with a certain material (e.g., sand, silt), specific yield values for this material type are also used.

ix. <u>River Cell information</u>: River Cells are identified by intersecting the grid cells with the SWAT+ channels; streambed elevation is calculated using the elevation of the cell (from the DEM) minus a specific depth (Depth of streambed below DEM value), recognizing that the actual streambed elevation is likely much lower than the average elevation of a DEM raster cell. Initial K_{bed} and d_{bed} values can be based on literature.

This study uses 200 m grid cells for the *gwflow* module, resulting in 8647 active cells. The grid and associated K values are shown in Figure 5a. These K values (3 m/day, 5 m/day, 20 m/day) are the final values based on manual calibration (see Section 2.3.4), and were increased from the values (0.001 m/dy, 0.086 m/day, 1.369 m/day) provided by the global data set. Corresponding S_y values are 0.10, 0.15, and 0.11. As shown in Figure 5a, there are only 3 zones of aquifer materials. K_{bed} and d_{bed} were set to 0.001 m/day and 2 m, respectively, for all River Cells. The aquifer thickness (cm) (Figure 5b) ranged from 0.13 m to 88.2 m. The identified River Cells are shown in Figure 5c.

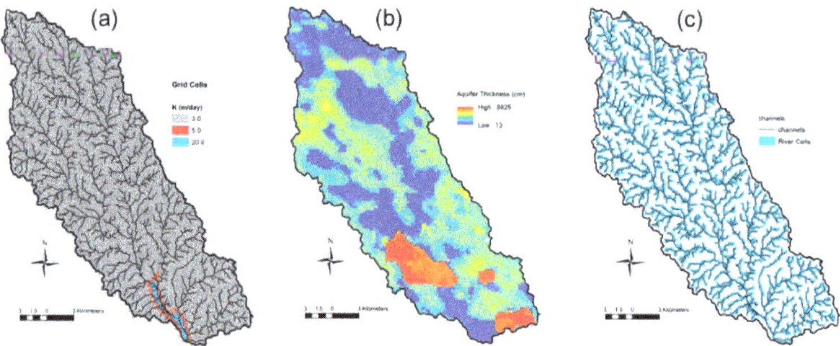

Figure 5. (**a**) gwflow grid cells showing SWAT+ channels and final hydraulic conductivity (m/day) values; (**b**) aquifer thickness (cm); and (**c**) grid cells designated as River Cells (i.e., cells that can exchange water with SWAT+ channels).

2.3.4. Overall Simulation

The simulation is run for the 1980–2013 period, with the first two years used as a warm-up (i.e., the results from 1980 and 1981 are not compared to observed watershed data). Based on initial simulations, a time step Δt of 0.25 days was required for the stability of the groundwater balance equation (Equation (10)). The run time of the model is only 20% higher than the original SWAT+ simulation due to the additional equations added to SWAT+ by the *gwflow* module. Simulated daily streamflow was compared to average daily streamflow at 8 gage sites, provided by the USDA-ARS

(Agricultural Research Service) Southeast Watershed Research Laboratory, and simulated daily groundwater head was compared to periodic groundwater levels measured at 8 monitoring wells, provided by the U.S. Geological Survey (USGS) (https://maps.waterdata.usgs.gov/mapper/index.html, accessed 15 February 2020). Figure 6 shows the location of the streamflow gages (green dots) and monitoring wells (red dots).

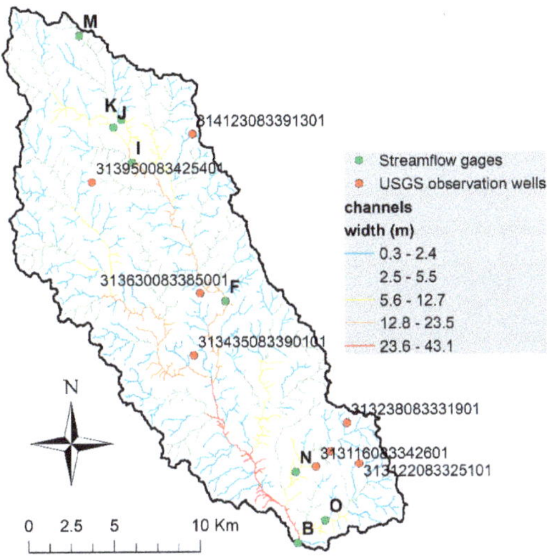

Figure 6. Map of the Little River Experimental Watershed (LREW) showing the location of the streamflow gages (B, O, N, F, I, J, K, M) and the U.S. Geological Survey (USGS) observation wells. The SWAT+ delineated channels also are shown, using color to indicate approximate channel width.

Only manual calibration was performed in this study, to provide reasonable matches between simulated and observed streamflow and groundwater head. The Nash–Sutcliffe Efficiency (NSE) coefficient is used to quantify the performance of the model regarding streamflow simulation. For aquifer properties K and S_y, values were changed only according to the three identified zones (see Figure 5a) to remain consistent with the global data set, and therefore point-by-point calibration was not pursued to yield optimal matches between simulated and observed groundwater head at the 8 monitoring wells. Automated calibration could be performed to yield optimal matches for both groundwater head and streamflow. However, the intent of this paper is to present the *gwflow* module and its basic workings rather than prepare a model for scenario analysis.

3. Results and Discussion

3.1. Water Balance

The overall water balance of the SWAT+ simulation is shown in Figure 7 using average annual flux values in mm (depth over the entire watershed). For simulations using the original SWAT+ code, fluxes would be limited to surface runoff (78 mm), lateral flow (57 mm), ET (843 mm), percolation and recharge (184 mm), and groundwater discharge. With the inclusion of the *gwflow* module, additional fluxes include boundary flow (0 mm), groundwater discharge (6 mm), stream seepage (1 mm), groundwater ET (11 mm), and saturation excess flow (161 mm), with state variables groundwater head (average of 107.2 m for the watershed) and saturated thickness (average of 19.2 m). Surface runoff (78 mm) and lateral flow (57 mm) account for 12% of annual precipitation, whereas groundwater accounts for 14% of annual precipitation. Water yield is 26% of annual precipitation, close to the observed value of

27% [31]. Groundwater entering streams occurs primarily via saturation excess flow (161 mm) rather than groundwater discharge (6 mm).

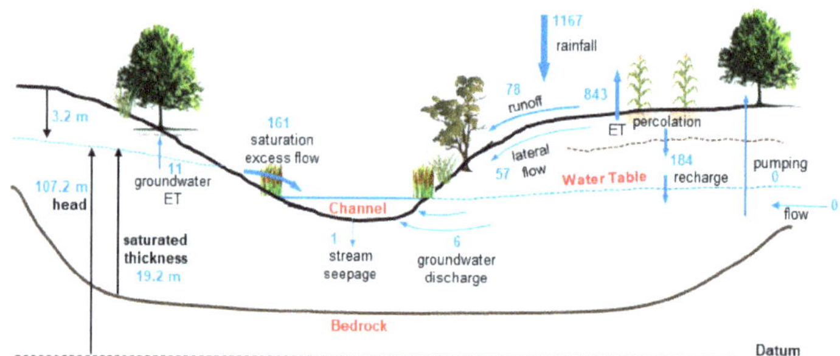

Figure 7. Average annual water balance for 1980–2013 for the LRW SWAT+ simulation, with fluxes shown in mm/yr. Fluxes groundwater ET, saturation excess flow, stream seepage, groundwater discharge, boundary flow, and recharge and state variables groundwater head (m) and saturated thickness (m) are simulated using the new *gwflow* module of SWAT+.

Total water inputs include rainfall (1167 mm), total outputs equal 1155 mm, and the total change in groundwater storage and soil water storage is 15 mm and 1 mm, respectively, resulting in a water balance error of 0.4%. A time series of volumetric amounts (m^3 × 10^6) for all hydrologic fluxes is shown in Figure 8a, with positive values indicating water entering the watershed, and negative values indicating water leaving the watershed. The year-by-year subsurface storage volume (groundwater + soil water) also is shown. A similar time series is shown in Figure 8b for the aquifer system, with positive values indicating water entering the aquifer (sources), and negative values indicating groundwater leaving the aquifer (sinks). The year-by-year groundwater storage volume also is shown. The percent error in the groundwater balance is < 0.00001%.

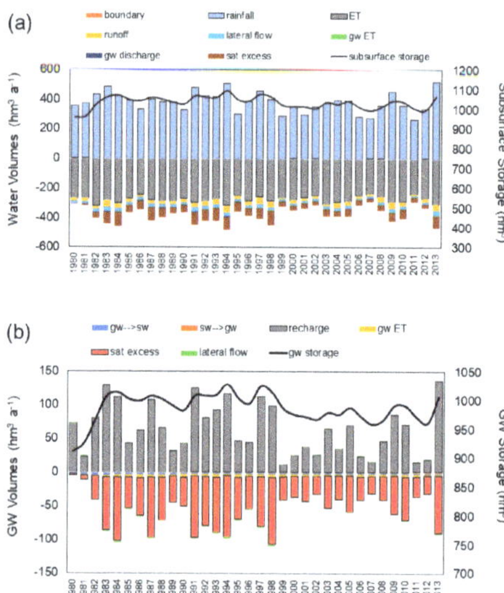

Figure 8. Time series of hydrologic flux volumes for (**a**) the entire watershed and (**b**) the aquifer system, for 1980-2013. For the legends, "gw" refers to groundwater, "ET" refers to evapotranspiration, "sw" refers to surface water, "gw→sw" refers to groundwater discharge to streams, "sw→gw" refers to stream seepage to the aquifer, "sat excess" refers to saturation excess flow, and "boundary" refers to groundwater flow across the watershed boundary.

The aquifer system time series (Figure 8b) shows that recharge is largely balanced by saturation excess flow, i.e., recharge events during large storm events produce saturation excess flow near the streams. This is demonstrated further in the maps (Figure 9) of spatial varying volumetric fluxes (m^3) over the 1980–2013 simulation period for recharge, saturation excess flow, groundwater–surface water exchange, and groundwater ET. Saturation excess flow occurs primarily in the riparian areas, as does groundwater ET, with the latter due to riparian trees extracting water from the water table [54].

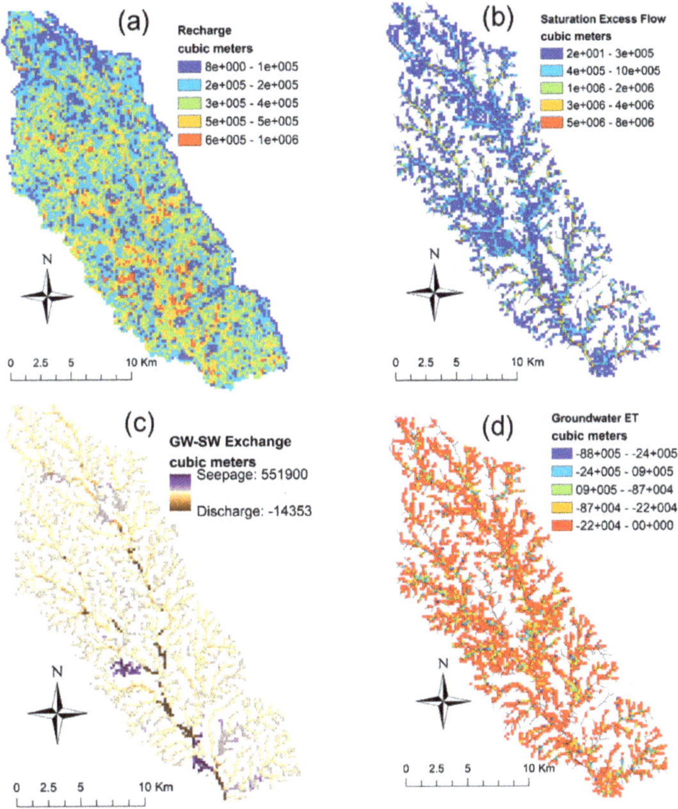

Figure 9. Maps of total volume (m^3) of groundwater stress flux for the 1980-2013 simulation period: (**a**) recharge, (**b**) lateral flow, (**c**) groundwater–surface water exchange, (**d**) groundwater ET.

3.2. Streamflow Results

Time series of daily simulated and observed streamflow (m^3/sec) are shown in Figure 10 for four of the stream gage sites from the upstream to downstream end of the watershed. The NSE value for each of the eight sites is shown in the inset table. Although several sites show adequate values (J, I, F, B: $\geq$ 0.30), others show poor matches (K, O: 0.22 and 0.27) and two are below 0 (M, N: -0.24, -0.37). Note that sites K, O, M and N are along small tributaries of the river system with moderate ($<$ 15 m^3/s) or low flows ($<$ 5 m^3/s). However, visual comparisons (see time series charts for M and O in Figure 10) indicate that the model does track the timing of measured flow. This is encouraging, as these flows are controlled to a large degree by subsurface flows calculated by the *gwflow* module. Detailed calibration of subbasin parameters for these tributary regions would be necessary if the model is to be used for scenario analysis. As this model application is intended solely for *gwflow* module demonstration, the results are adequate.

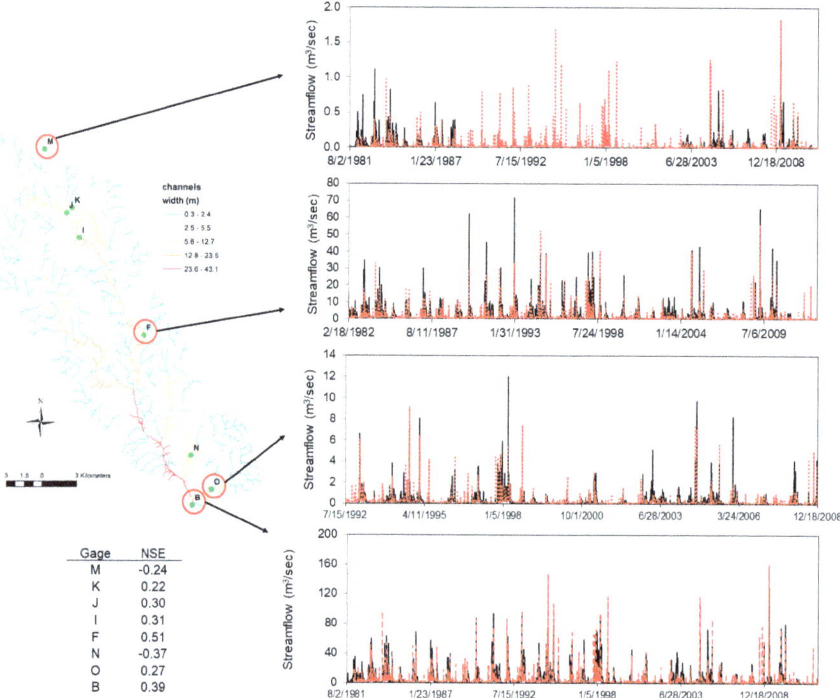

Gage	NSE
M	-0.24
K	0.22
J	0.30
I	0.31
F	0.51
N	-0.37
O	0.27
B	0.39

Figure 10. Time series of daily simulated and observed streamflow (m³/s) for the eight stream gage locations in the LREW study region (see Figure 6). The circled gage *B* is the outlet of the watershed. The Nash–Sutcliffe Efficiency (NSE) coefficient is shown on the chart for each stream gage site.

The differences in simulated streamflow between the original SWAT+ model and the SWAT+ model with the *gwflow* module are shown in Figure 11 for the outlet (site B) for several years of the simulation. As seen by the time series, SWAT+ simulates near-zero flow between rainfall events, consistent with the measured streamflow, whereas SWAT+ with *gwflow* simulates low flows (1–3 m³/s) during these times. Groundwater discharge in SWAT+ is simulated in the "pulse" form, based on the magnitude of rainfall events and the resulting recharge. Using the *gwflow* module, however, groundwater can enter the streams at any time based on groundwater gradients, leading to small flows between rainfall events. The results of SWAT+ are more accurate in this instance in this regard, as observed flows also approach zero between rainfall events during certain times of the year. As such, further calibration is required for the *gwflow* module if these low-flow periods are of importance for the intended use of the model. However, the inclusion of the *gwflow* module allows groundwater discharge to streams, either via the streambed or via saturation excess flow, to be simulated in a more physically realistic manner for scenario analysis (e.g., climate change, water conservation, nutrient management).

3.3. Groundwater Head Results

Groundwater head (m) at the end of the simulation (end of 2013) for each grid cell is shown in Figure 12. The groundwater head (i.e., water table elevation) mimics the ground surface elevation (see Figure 4a). Groundwater head fluctuations are also shown at the eight monitoring well locations, in comparison with the observed groundwater head values from the USGS data set. The dotted brown line on each time series chart represents the ground surface at that site. Observed rapid groundwater head fluctuations generally are captured by the model. The rapid changes occur due to recharge events and resulting lateral flow and/or saturation excess flow.

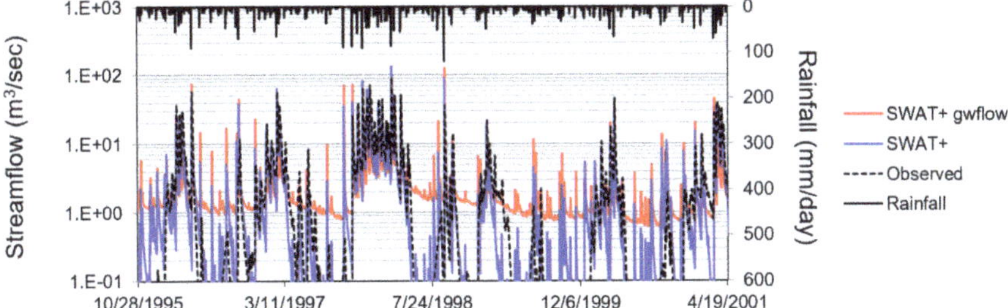

Figure 11. Comparison of daily streamflow at the outlet of the watershed (site B in Figure 6) simulated for the original SWAT+ model and the SWAT+ model with the new *gwflow* module. A semi-log plot is used to show the differences between the model results for low flows. The measured streamflow is also shown.

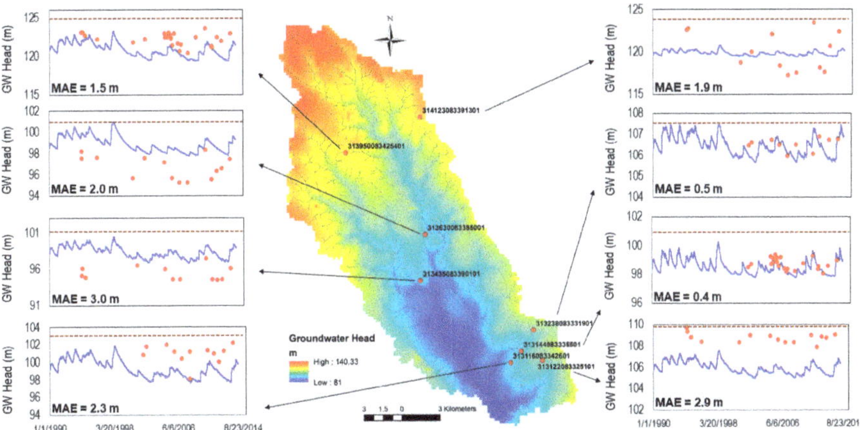

Figure 12. Map of groundwater head (m) for each grid cell at the end of the simulation period, and time series of daily simulated groundwater head (m) and periodically observed groundwater head (m) for the eight USGS groundwater monitoring well locations (see Figure 6). The dotted brown line on each time series chart represents the ground surface at that site. MAE = Mean Absolute Error (m), i.e., the average residual between the observed and simulated groundwater head values.

The mean absolute error (MAE) of simulated head values that correspond in location and timing to measured head values is 1.7 m. All sites have an MAE of 3.0 m or less. Two locations (313144083335501, 313238083331901) show excellent tracking of the measured head by the model, while others show moderate agreement. For several locations, the model underpredicts head levels, and at two locations (313435083390101, 313630083385001) the model overestimates head levels. The underestimation and overestimation could be due to the misrepresentation of aquifer properties (K, S_y) or underestimation of localized recharge by HRUs. Likely the aquifer properties are not represented correctly for these localized areas. However, as with the streamflow results shown in Section 3.2, in this study aquifer property values are constrained by the zones established by the global permeability map (see Figure 5a). Point-by-point calibration could be performed to provide better matches between the observed and simulated head values; however, the purpose of this paper is to present the *gwflow* module and to highlight its capabilities. If the model is to be used for scenario analysis, localized aquifer conditions can be better represented using borehole lithology data. Such an approach would also use automated parameter estimation methods and employ ensemble-based uncertainty analysis and sensitivity analysis methods (e.g., [21–23]) to explore the effect of parameter values (e.g., hydraulic conductivity: [24,25]) on system-response variables such as water table elevation, groundwater–surface water exchange rates, and streamflow.

Finally, depth to water table (m) and saturated thickness (m) for each grid cell at the end of the simulation are shown in Figure 13. The map of depth to water table shows depths near 0 or at 0 for much of the riparian and floodplain areas, indicating near-saturated conditions, leading to groundwater ET (see Figure 9d) and saturation excess flow (see Figure 9b). The map of saturated thickness shows spatial patterns similar to the map of aquifer thickness (see Figure 5b). Such maps can assist in determining groundwater storage throughout the aquifer system and the strong interplay between the land surface and subsurface hydrologic processes.

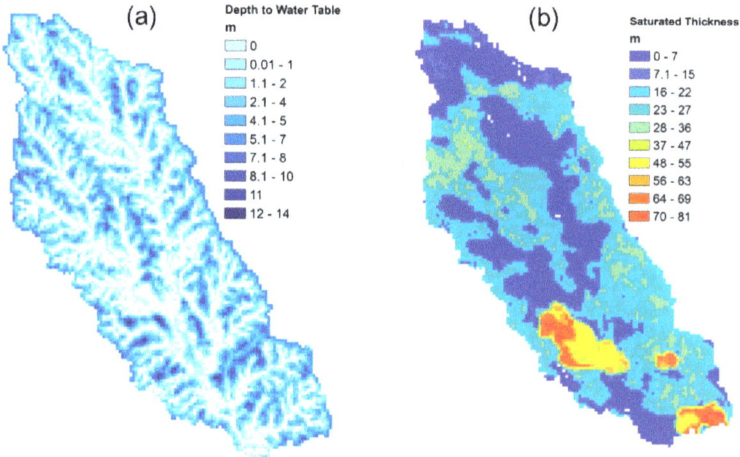

Figure 13. (**a**) Depth to Water Table (m) and (**b**) Saturated Thickness (m) for each grid cell at the end of the simulation period. Depth to Water Table was calculated by subtracting groundwater head (see Figure 12 for map) from ground surface elevation for each grid cell. Saturated Thickness was calculated by subtracting bedrock elevation from groundwater head.

4. Summary and Conclusions

This paper presents a new physically-based spatially-distributed groundwater flow module (*gwflow*) for the SWAT+ watershed model. The module is embedded in the SWAT+ modeling code as with other hydrologic modules and is intended to replace the current SWAT+ aquifer module [2] for watersheds with strong stream-aquifer connections. The model accounts for recharge from SWAT+ HRUs, lateral flow within the aquifer, ET from shallow groundwater, pumping, groundwater–surface water interactions through the streambed, and saturation excess flow. Daily groundwater head and groundwater storage are solved using an explicit numerical method approach for the groundwater balance equation, with head and flow values for the current day based on head and flow values from the previous day. The module outputs groundwater head and flux rates for each groundwater stress for analysis and mapping.

The modified SWAT+ model is applied to the 327 km² Little River Experimental Watershed in southern Georgia, USA, demonstrating its capabilities of simulating both surface and subsurface hydrological processes. Results from an uncalibrated model are compared against measured streamflow and groundwater levels at eight stream gage sites and eight groundwater monitoring locations, respectively. The model performed adequately regarding visual comparisons of time series and quantitative statistics but can be improved with additional calibration if it is to be used for scenario analysis. For this watershed, groundwater additions to the stream system are governed by saturation excess flow rather than discharge via the streambed, based on continual near-saturation conditions in the riparian areas and measured low baseflow between rainfall events. For this watershed, inclusion of the *gwflow* module increases SWAT+ model run time by approximately 20%.

The modified SWAT+ modeling code with the *gwflow* module can be an important tool for modeling hydrological fluxes in watersheds that have a strong connection between the shallow unconfined aquifer and the stream system. Using this code can provide physically realistic groundwater flow gradients, flux values, and interactions with streams, which are important for assessing the effects of conservation practices on hydrology and nutrient transport.

Author Contributions: Conceptualization, R.T.B. and J.G.A.; methodology, R.T.B.; software, R.T.B. and J.G.A.; validation, R.T.B., K.B., J.G.A., and D.D.B.; formal analysis, R.T.B.; investigation, R.T.B. and K.B.; resources, K.B. and D.D.B.; data curation, K.B. and D.D.B.; writing—original draft preparation, R.T.B.; writing—review and editing, K.B., J.G.A., and D.D.; visualization, R.T.B. and K.B.; supervision, R.T.B.; project administration, R.T.B. and J.G.A.; funding acquisition, R.T.B. and J.G.A. All authors have read and agreed to the published version of the manuscript.

Funding: This research was funded by the United State Department of Agriculture–Agricultural Research Service through Cooperative Agreement number 59-3098-8-002 and a grant from the Agriculture and Food Research Initiative of the USDA National Institute of Food and Agriculture, grant number #2017-67019-26332.

Acknowledgments: The authors wish to thank several anonymous reviewers for their suggestions which greatly improved the manuscript.

Conflicts of Interest: The authors declare no conflict of interest. The funders had no role in the design of the study; in the collection, analyses, or interpretation of data; in the writing of the manuscript, or in the decision to publish the results.

References

1. Arnold, J.G.; Srinivasan, R.; Muttiah, R.S.; Williams, J.R. Large area hydrologic modeling and assessment part I: Model development. *JAWRA J. Am. Water Resour. Assoc.* **1998**, *34*, 73–89. [CrossRef]

2. Bieger, K.; Arnold, J.G.; Rathjens, H.; White, M.J.; Bosch, D.D.; Allen, P.M.; Volk, M.; Srinivasan, R. Introduction to SWAT+, A Completely Restructured Version of the Soil and Water Assessment Tool. *JAWRA J. Am. Water Resour. Assoc.* **2017**, *53*, 115–130. [CrossRef]

3. Srivastava, P.; McNair, J.N.; Johnson, T.E. Comparison of process-based and artificial neural network approaches for streamflow modeling in an agricultural watershed. *JAWRA J. Am. Water Resour. Assoc.* **2006**, *42*, 545–563. [CrossRef]

4. Gassman, P.W.; Reyes, M.R.; Green, C.H.; Arnold, J.G. The Soil and Water Assessment Tool: Historical Development, Applications, and Future Research Directions. *Trans. ASABE* **2007**, *50*, 1211–1250. [CrossRef]

5. Deb, P.; Kiem, A.S.; Willgoose, G. A linked surface water-groundwater modelling approach to more realistically simulate rainfall-runoff non-stationarity in semi-arid regions. *J. Hydrol.* **2019**, *575*, 273–291. [CrossRef]

6. Harbaugh, A.W. *MODFLOW-2005, The U.S. Geological Survey Modular Ground-Water Model—The Ground-Water Flow Process*; USGS Techniques and Methods 2005 6-A16, US Geological Survey; US Department of the Interior: Reston, VA, USA, 2005.

7. Galbiati, L.; Bouraoui, F.; Elorza, F.J.; Bidoglio, G. Modeling diffuse pollution loading into a Mediterranean lagoon: Development and application of an integrated surface–subsurface model tool. *Ecol. Model.* **2006**, *193*, 4–18. [CrossRef]

8. Kim, N.W.; Chung, I.-M.; Won, Y.S.; Arnold, J.G. Development and application of the integrated SWAT-MODFLOW model. *J. Hydrol.* **2008**, *356*, 1–16. [CrossRef]

9. Guzman, J.A.; Moriasi, D.N.; Gowda, P.H.; Steiner, J.L.; Starks, P.J.; Arnold, J.G.; Srinivasan, R. A model integration framework for linking SWAT and MODFLOW. *Environ. Model. Softw.* **2015**, *73*, 103–116. [CrossRef]

10. Bailey, R.T.; Wible, T.C.; Arabi, M.; Records, R.M.; Ditty, J. Assessing regional-scale spatio-temporal patterns of groundwater-surface water interactions using a coupled SWAT-MODFLOW model. *Hydrol. Process.* **2016**, *30*, 4420–4433. [CrossRef]

11. Aliyari, F.; Bailey, R.T.; Tasdighi, A.; Dozier, A.; Arabi, M.; Zeiler, K. Coupled SWAT-MODFLOW model for large-scale mixed agro-urban river basins. *Environ. Model. Softw.* **2019**, *115*, 200–210. [CrossRef]

12. Chunn, D.; Faramarzi, M.; Smerdon, B.; Alessi, D.S. Application of an Integrated SWAT-MODFLOW Model to Evaluate Potential Impacts of Climate Change and Water Withdrawals on Groundwater–Surface Water Interactions in West-Central Alberta. *Water* **2019**, *11*, 110. [CrossRef]
13. Gao, F.; Feng, G.; Han, M.; Dash, P.; Jenkins, J.; Liu, C. Assessment of Surface Water Resources in the Big Sunflower River Watershed Using Coupled SWAT-MODFLOW Model. *Water* **2019**, *11*, 528. [CrossRef]
14. Molina-Navarro, E.; Bailey, R.T.; Andersen, H.E.; Thodsen, H.; Nielsen, A.; Park, S.; Jensen, J.S.; Jensen, J.B.; Trolle, D. Comparison of abstraction scenarios simulated by SWAT and SWAT-MODFLOW. *Hydrol. Sci. J.* **2019**, *64*, 434–454. [CrossRef]
15. Semiromi, M.T.; Koch, M. Analysis of spatio-temporal variability of surface-groundwater interactions in the Gharehsoo river basin, Iran, using a coupled SWAT-MODFLOW model. *Environ. Earth Sci.* **2019**, *78*, 201. [CrossRef]
16. Wei, X.; Bailey, R.T. Assessment of System Responses in Intensively Irrigated Stream–Aquifer Systems Using SWAT-MODFLOW. *Water* **2019**, *11*, 1576. [CrossRef]
17. Bailey, R.T.; Park, S.; Bieger, K.; Arnold, J.G.; Allen, P.M. Enhancing SWAT+ simulation of groundwater flow and groundwater-surface water interactions using MODFLOW routines. *Environ. Model. Softw.* **2020**, *126*, 104660. [CrossRef]
18. Raymond, H.A.; Bondoc, M.; McGinnis, J.; Metropulos, K.; Heider, P.; Reed, A.; Saines, S. Using Analytic Element Models to Delineate Drinking Water Source Protection Areas. *Ground Water* **2006**, *44*, 16–23. [CrossRef]
19. Brown, K.; Trott, S. Groundwater Flow Models in Open Pit Mining: Can We Do Better? *Mine Water Environ.* **2014**, *33*, 187–190. [CrossRef]
20. Jones, D.; Jones, N.L.; Greer, J.; Nelson, J. A cloud-based MODFLOW service for aquifer management decision support. *Comput. Geosci.* **2015**, *78*, 81–87. [CrossRef]
21. Yang, J.; Reichert, P.; Abbaspour, K.C.; Xia, J.; Yang, H. Comparing uncertainty analysis techniques for a SWAT application to the Chaohe Basin in China. *J. Hydrol.* **2008**, *358*, 1–23. [CrossRef]
22. Strauch, M.; Bernhofer, C.; Koide, S.; Volk, M.; Lorz, C.; Makeschin, F. Using precipitation data ensemble for uncertainty analysis in SWAT streamflow simulation. *J. Hydrol.* **2012**, *414*, 413–424. [CrossRef]
23. Muleta, M.K.; Nicklow, J.W. Sensitivity and uncertainty analysis coupled with automatic calibration for a distributed watershed model. *J. Hydrol.* **2005**, *306*, 127–145. [CrossRef]
24. Colombo, L.; Alberti, L.; Mazzon, P.; Formentin, G. Transient Flow and Transport Modelling of an Historical CHC Source in North-West Milano. *Water* **2019**, *11*, 1745. [CrossRef]
25. Moeck, C.; Molson, J.; Schirmer, M. Pathline Density Distributions in a Null-Space Monte Carlo Approach to Assess Groundwater Pathways. *Ground Water* **2020**, *58*, 189–207. [CrossRef] [PubMed]
26. Park, S.; Nielsen, A.; Bailey, R.T.; Trolle, D.; Bieger, K. A QGIS-based graphical user interface for application and evaluation of SWAT-MODFLOW models. *Environ. Model. Softw.* **2019**, *111*, 493–497. [CrossRef]
27. Arnold, J.G.; Bieger, K.; White, M.J.; Srinivasan, R.; Dunbar, J.; Allen, P. Use of Decision Tables to Simulate Management in SWAT+. *Water* **2018**, *10*, 713. [CrossRef]
28. Bieger, K.; Arnold, J.G.; Rathjens, H.; White, M.J.; Bosch, D.D.; Allen, P.M. Representing the Connectivity of Upland Areas to Floodplains and Streams in SWAT+. *JAWRA J. Am. Water Resour. Assoc.* **2019**, *55*, 578–590. [CrossRef]
29. Sklash, M.G.; Farvolden, R.N. The role of groundwater in storm runoff. *J. Hydrol.* **1979**, *43*, 45–65. [CrossRef]
30. Pearce, A.J. Streamflow Generation Processes: An Austral View. *Water Resour. Res.* **1990**, *26*, 3037–3047. [CrossRef]
31. Bosch, D.D.; Arnold, J.G.; Allen, P.G.; Lim, K.J.; Park, Y.S. Temporal variations in baseflow for the Little River experimental watershed in South Georgia, USA. *J. Hydrol. Reg. Stud.* **2017**, *10*, 110–121. [CrossRef]
32. Voss, C.I.; Provost, A.M. SUTRA: A model for 2D or 3D saturated-unsaturated, variable-density ground-water flow with solute or energy transport. *USGS* **2002**, *4231*. [CrossRef]
33. Camporese, M.; Paniconi, C.; Putti, M.; Orlandini, S. Surface-subsurface flow modeling with path-based runoff routing, boundary condition-based coupling, and assimilation of multisource observation data. *Water Resour. Res.* **2010**, *46*, W02512. [CrossRef]
34. Brunner, P.A.; Simmons, C.T. HydroGeoSphere: A Fully Integrated, Physically Based Hydrological Model. *Ground Water* **2012**, *50*, 170–176. [CrossRef]

35. Bedekar, V.; Niswonger, R.G.; Kipp, K.; Panday, S.; Tonkin, M. Approaches to the Simulation of Unconfined Flow and Perched Groundwater Flow in MODFLOW. *Ground Water* **2012**, *50*, 187–198. [CrossRef] [PubMed]

36. Chu, W.-S.; Willis, R. An Explicit Finite Difference Model for Unconfined Aquifers. *Ground Water* **1984**, *22*, 728–734. [CrossRef]

37. Sheridan, J.M. Rainfall-streamflow relations for coastal plain watersheds. *Appl. Eng. Agric.* **1997**, *13*, 333–344. [CrossRef]

38. Lupon, A.; Ledesma, J.L.J.; Bernal, S. Riparian evapotranspiration is essential to simulate streamflow dynamics and water budgets in a Mediterranean catchment. *Hydrol. Earth Syst. Sci.* **2018**, *22*, 4033–4045. [CrossRef]

39. Sullivan, D.G.; Batten, H.L.; Bosch, D.D.; Sheridan, J.; Strickland, T.C. Little River Experimental watershed, Tifton, Georgia, United States: A historical geographic database of conservation practice implementation. *Water Resour. Res.* **2007**, *43*. [CrossRef]

40. Stringfield, V.T. *Artesian Water in Tertiary Limestone in the Southeastern States*; US Geological Survey Professional Paper 517; US Government Printing Office: Denver, CO, USA, 1966; p. 226.

41. Bosch, D.D.; Sheridan, J.M.; Lowrance, R.R. Hydraulic Gradients and Flow Rates of a Shallow Coastal Plain Aquifer in a Forested Riparian Buffer. *Trans. ASAE* **1996**, *39*, 865–871. [CrossRef]

42. Bosch, D.D.; Arnold, J.G.; Volk, M.; Allen, P.M. Simulation of a Low-Gradient Coastal Plain Watershed Using the SWAT Landscape Model. *Trans. ASABE* **2010**, *53*, 1445–1456. [CrossRef]

43. Van Liew, M.W.; Arnold, J.G.; Bosch, D.D. Problems and potential of autocalibrating a hydrologic model. *Trans. ASAE* **2005**, *48*, 1025–1040. [CrossRef]

44. Zhang, X.; Srinivasan, R.; Bosch, D. Calibration and uncertainty analysis of the SWAT model using Genetic Algorithms and Bayesian Model Averaging. *J. Hydrol.* **2009**, *374*, 307–317. [CrossRef]

45. Veith, T.L.; Van Liew, M.W.; Bosch, D.D.; Arnold, J.G. Parameter Sensitivity and Uncertainty in SWAT: A Comparison Across Five USDA-ARS Watersheds. *Trans. ASABE* **2010**, *53*, 1477–1486. [CrossRef]

46. Arnold, J.G.; Moriasi, D.N.; Gassman, P.W.; Abbaspour, K.C.; White, M.J.; Srinivasan, R.; Santhi, C.; Harmel, R.D.; Van Griensven, A.; Van Liew, M.W.; et al. SWAT: Model Use, Calibration, and Validation. *Trans. ASABE* **2012**, *55*, 1491–1508. [CrossRef]

47. Cho, J.; Vellidis, G.; Bosch, D.D.; Lowrance, R.; Strickland, T.C. Water quality effects of simulated conservation practice scenarios in the Little River Experimental watershed. *J. Soil Water Conserv.* **2010**, *65*, 463–473. [CrossRef]

48. Rathjens, H.; Oppelt, N.; Bosch, D.D.; Arnold, J.G.; Volk, M. Development of a grid-based version of the SWAT landscape model. *Hydrol. Process.* **2015**, *29*, 900–914. [CrossRef]

49. Yen, H.; Park, S.; Arnold, J.G.; Srinivasan, R.; Chawanda, C.J.; Wang, R.; Feng, Q.; Wu, J.; Miao, C.; Bieger, K.; et al. IPEAT+: A Built-In Optimization and Automatic Calibration Tool of SWAT+. *Water* **2019**, *11*, 1681. [CrossRef]

50. Tolson, B.A.; Shoemaker, C.A. Dynamically dimensioned search algorithm for computationally efficient watershed model calibration. *Water Resour. Res.* **2007**, *43*. [CrossRef]

51. Bosch, D.D.; Sheridan, J.M.; Marshall, L.K. Precipitation, soil moisture, and climate database, Little River Experimental Watershed, Georgia, United States. *Water Resour. Res.* **2007**, *43*. [CrossRef]

52. Shangguan, W.; Hengl, T.; De Jesus, J.M.; Yuan, H.; Dai, Y. Mapping the global depth to bedrock for land surface modeling. *J. Adv. Model. Earth Syst.* **2017**, *9*, 65–88. [CrossRef]

53. Huscroft, J.; Gleeson, T.; Hartmann, J.; Börker, J. Compiling and Mapping Global Permeability of the Unconsolidated and Consolidated Earth: GLobal HYdrogeology MaPS 2.0 (GLHYMPS 2.0). *Geophys. Res. Lett.* **2018**, *45*, 1897–1904. [CrossRef]

54. Bosch, D.D.; Marshall, L.K.; Teskey, R. Forest transpiration from sap flux density measurements in a Southeastern Coastal Plain riparian buffer system. *Agric. For. Meteorol.* **2014**, *187*, 72–82. [CrossRef]

 *hydrology*

Article

Integrated Surface Water and Groundwater Analysis under the Effects of Climate Change, Hydraulic Fracturing and its Associated Activities: A Case Study from Northwestern Alberta, Canada

Gopal Chandra Saha [1,2,*] and Michael Quinn [2]

[1] Global Water Futures Program, Wilfrid Laurier University, Waterloo, ON N2L 3C5, Canada
[2] Institute for Environmental Sustainability, Mount Royal University, Calgary, AB T3E 6K6, Canada; mquinn@mtroyal.ca
[*] Correspondence: gsaha@wlu.ca; Tel.: +1-519-884-0710 (ext. 3918)

Received: 27 July 2020; Accepted: 21 September 2020; Published: 23 September 2020

Abstract: This study assessed how hydraulic fracturing (HF) (water withdrawals from nearby river water source) and its associated activities (construction of well pads) would affect surface water and groundwater in 2021–2036 under changing climate (RCP4.5 and RCP8.5 scenarios of the CanESM2) in a shale gas and oil play area (23,984.9 km^2) of northwestern Alberta, Canada. An integrated hydrologic model (MIKE-SHE and MIKE-11 models), and a cumulative effects landscape simulator (ALCES) were used for this assessment. The simulation results show an increase in stream flow and groundwater discharge in 2021–2036 under both RCP4.5 and RCP8.5 scenarios with respect to those under the base modeling period (2000–2012). This occurs because of the increased precipitation and temperature predicted in the study area under both RCP4.5 and RCP8.5 scenarios. The results found that HF has very small (less than 1%) subtractive impacts on stream flow in 2021–2036 because of the large size of the study area, although groundwater discharge would increase minimally (less than 1%) due to the increase in the gradient between groundwater and surface water systems. The simulation results also found that the construction of well pads related to HF have very small (less than 1%) additive impacts on stream flow and groundwater discharge due to the non-significant changes in land use. The obtained results from this study provide valuable information for effective long-term water resources decision making in terms of seasonal and annual water extractions from the river, and allocation of water to the oil and gas industries for HF in the study area to meet future energy demand considering future climate change.

Keywords: integrated surface water and groundwater analysis; climate change; hydraulic fracturing; construction of well pads; MIKE-SHE; MIKE-11; northwestern Alberta

1. Introduction

Surface water and groundwater are essential resources for the survival of human beings, livestock, wildlife, terrestrial and aquatic ecosystems. They are extensively used in agricultural, industrial, oil and gas exploration, household and recreation activities. Surface water and groundwater are closely connected components of the hydrologic system. Because of their close connectivity, the use of any one component can affect the other. As a result, it is necessary to conduct integrated surface water and groundwater analysis for developing sustainable water resources management. However, surface water (e.g., rivers, streams, lakes, wetlands, estuaries) and groundwater are extremely vulnerable to climate change [1]. The Intergovernmental Panel on Climate Change (IPCC) reported that the projection of global atmospheric concentrations of greenhouse gases (GHG) will continue to increase in the following decades, which will result in increased temperature and lead to continuing climate change [2]. Therefore, climate change might have significant effects on the temporal pattern of annual temperature as well as precipitation at the regional level, which in turn will affect the regional water resources (i.e., surface water and groundwater) and future water availability.

The extraction of oil and gas using hydraulic fracturing (HF) from vast shale reserves often requires large volumes of water from nearby water sources to be used as a fracturing fluid. The volume of water used by the oil and gas industries varies depending on geologic formations, type of well, number of hydraulic fracturing stages, length of the reach within the production zone and the type of hydraulic fracturing fluid (i.e., cross-linked gel or slick water) [3]. For example, in northeast British Columbia of the Western Canadian Sedimentary Basin, the water volume varies widely from less than 1000 m^3 to more than 70,000 m^3 per well [4]. In addition to water withdrawals, associated activities (i.e., construction of roads, well pads, pipelines, seismic lines, and power transmission lines) related to HF, change the natural soil lithology significantly by altering the upper soil layers. The alteration of soil layers results in different soil infiltration rates, which in turn affects groundwater recharge and discharge, surface runoff and stream flow significantly [5,6]. The use of HF has increased significantly in North America, and forecasts show continued growth and application of HF across the world [7]. For example, in the United States, natural gas production from shale resources increased from 0.1 to 3 Tcf (Trillion cubic feet) in the last decade [8]. By 2050, shale gas production is expected to account for 91% of the United States natural gas production [9]. The number of wells in North America that used HF for extracting oil and gas from shale reserves has changed over time to meet energy demand. For example, in the United States, about 278,000 wells were completed using HF from 2000 to 2010 [10]. However, there is considerable public concern regarding the sustainability of water withdrawals from nearby water sources for HF due to the potential negative impact on water resources (i.e., surface water and groundwater) especially during low flow period [11], as well as environmental (e.g., spills) [12] and health [13] related issues. Therefore, it is necessary to forecast climate change effects on water resources for developing future water resources management plan at regional level, so that HF and its associated activities meet future energy demand without causing significant negative effects to surface water and groundwater.

Due to the significance of water resources, the quantification of the effects of water withdrawal for HF on water resources has received increasing attention from a research point of view during the last 6 years. Although there are missing information (i.e., location of water withdrawal, type of water source, timing of water withdrawals, and whether any water was recycled), various assumptions were made related to these missing information for assessing the effects of water withdrawal for HF on water resources. Those research activities addressed daily stream flow [14–16], monthly stream flow [16,17], annual stream flow [18,19], stream low flow [20–22], environmental flow components (i.e., high flow, low flow and extremely low flow) of the stream [18,23], annual surface water and groundwater availability [24], and annual groundwater table [19,22]. In addition to water withdrawals, very few research activities have been conducted on associated activities related to HF on water resources. Those research activities highlighted annual stream flow [18,19], and annual groundwater table [19]. However, there is little knowledge regarding how HF and its associated activities would affect temporal patterns (i.e., monthly, seasonal and annual) of groundwater discharge under changing climate. This study attempted to fill up this gap.

The objective of this study was to quantify the effects of HF (i.e., water withdrawals from nearby river water source) and its associated activities on temporal patterns of stream flow and groundwater discharge under changing climate. The assessment of temporal variation of stream flow and groundwater discharge due to HF and its associated activities under changing climate will provide useful information for future planning of water uses to meet the industry's water demand, and the natural hydrologic system. In this study, the effects of HF and its associated activities on mean monthly, seasonal and annual stream flow and groundwater discharge were evaluated for 2021–2036 under the Representative Concentration Pathways (RCP) 4.5 and 8.5 of the CanESM2 (Second Generation Earth

System Model) from the Fifth Assessment Report (AR5) of the IPCC [25]. An Integrated Hydrologic Model (i.e., MIKE-SHE and MIKE-11 models [26]), and a cumulative effects landscape simulator (i.e., ALCES: A Landscape Cumulative Effects Simulator [27]) were used for the evaluation. A case study was used in a shale gas and oil play area of northwestern Alberta, Canada. Here, we define water withdrawal as the amount of water extracted from the river in a particular month to be used in HF.

2. Materials and Methods

2.1. Study Area

A study area (23,984.9 km^2) was selected in a rich shale gas and oil region of the Upper Peace Region of northwestern Alberta, Canada based on data availability for a significant number of hydraulically fractured wells, coupled with a number of active surface water monitoring stations and groundwater monitoring wells (Figure 1). It contains parts of the Montney, the Duvernay and the Muskwa formations. Among those, the Montney and the Duvernay are the most productive shale gas and oil reserves in Alberta. Forest (34.4%) and agriculture (34.1%) dominate land use in the study area. Other land uses are perineal crops (forage) and pasture (18.2%), water (i.e., river, lake, and wetland) (6.7%), grassland (4.9%), shrub land (1.2%), road (0.4%) and clear cut area (0.1%), based on the land use/land cover map of the study area for year 2000 collected from Natural Resources Canada (www.nrcan.gc.ca). Surface water is mostly used to meet forestry and agriculture needs [28]. Clay loam, loam, silty loam, silty clay, paved area and sand cover 31.74%, 29.2%, 24.46%, 14.1%, 0.45% and 0.05% of the study area, respectively, based on the soil map of the study area collected from Alberta Environment and Parks (https://soil.agric.gov.ab.ca/agrasidviewer/).

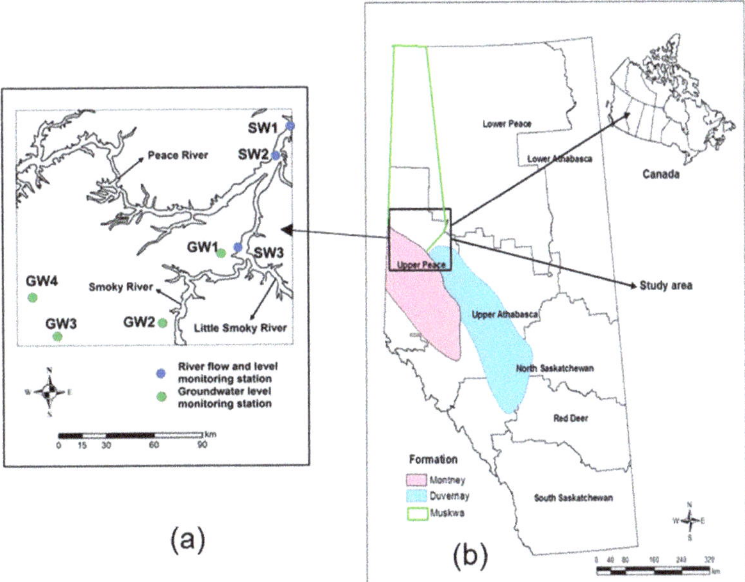

Figure 1. (a) Surface water monitoring stations and groundwater monitoring wells in the study area. Only those groundwater monitoring wells are shown here which were used for results and discussion section. (b) The location of the study area in Alberta, Canada.

The study area has an elevation ranging from 302 m to 1024 m, with an average slope of 2.1%. The hydrologic system in the study area is mainly rainfall dominated. The mean annual precipitation and temperature of the study area were 423 mm (312 mm of rain, and 111 mm of snow), and 1.9 °C, respectively, for the period of 1985–2014. The study area contains parts of three rivers: the Peace River, the Smoky River and the Little Smoky River. The Little Smoky River joins with the Smoky River, and then the Smoky River joins with the Peace River. There are three surface water monitoring stations in the study area maintained by Water Survey of Canada (https://wateroffice.ec.gc.ca/). One station is located in the Smoky River named as Smoky River at Watino (here it is named as SW3 for convenient results discussion). The others named as Peace River above Smoky River confluence (SW2) and Peace River at Peace River city (SW1) are situated in the Peace River. The SW1 station is the outlet of the study area. The SW1 and SW2 stations are approximately 76 and 54 km away from the SW3 station, respectively. The upstream parts of the Peace River, the Smoky River, and the Little Smoky River (which are outside of the study area) contributed 88%, 8.1% and 2.3% of the stream flow at the outlet (SW1) of the study area, respectively, based on the observed data at the monitoring stations of those areas in 2000–2012. There are 1235 active and inactive monitoring wells in the study area based on the information of Alberta provincial groundwater monitoring wells database (http://environment.alberta.ca/apps/GOWN/). In Figure 1a, only four groundwater wells (i.e., GW1 and GW2 are situated in lower elevation areas; GW3 and GW4 are located in higher elevation areas) are shown, which were used in the results and discussions section.

2.2. Integrated Hydrological Modeling

An integrated hydrological model was developed for the study area by using MIKE-SHE and MIKE-11 models. MIKE-SHE is a physically-based, distributed, and structured grid based hydrologic model. It simulates various hydrological processes for example, snowfall accumulation and melting, evapotranspiration, unsaturated flow, saturated groundwater flow, overland flow and infiltration in a watershed under given hydrometeorological inputs. In this study, snow melting was estimated by using the modified degree-day method [29]. Overland flow was computed by using the finite difference method by solving the diffusive wave approximation of the Saint Venant Equations [30]. Saturated groundwater flow was simulated with a finite difference representation of 3-D saturated groundwater flow equation. The saturated zone was divided into two layers: unconfined aquifer and bedrock (underlain by unconfined aquifer). Unsaturated flow was simulated by using the two-layer water balance method [31] because of the lack of detailed soil characteristics and geological layer data. On the other hand, MIKE-11 is a 1-D hydrodynamic model, and computes channel flow by using 1-D Saint Venant equations. In this study, channel flow was calculated by using the implicit finite difference scheme [32] to solve the dynamic wave version of the Saint Venant equations [33]. Then, the MIKE-11 model was coupled with the MIKE-SHE model to simulate stream flow as well stream water level along the channels, and groundwater level in the aquifer under given hydrometeorological inputs. Water flux between the stream and the saturated zone was estimated based on Darcy's law. The details of MIKE-SHE and MIKE-11 models can be found in MIKE-SHE user manual [30] and MIKE-11 reference manual [33], respectively. The coupled MIKE-SHE/MIKE-11 model needs a number of inputs. These are watershed specific data (i.e., elevation, land use/land cover, channel geometry, and soil type), vegetation characteristic (i.e., rooting depth and leaf area index) data, climatic (i.e., precipitation, temperature and reference evapotranspiration) data, and hydrological (i.e., stream flow and level, and groundwater level) data. Table 1 provides the details of these data for this study.

The MIKE-SHE model domain was discretized into 284 m by 284 m grid cells. The initial potential head (i.e., groundwater table) maps in the aquifer (unconfined) and bedrock were prepared by using observed groundwater table data collected from 1235 active and inactive monitoring wells in the study area from Alberta provincial groundwater monitoring wells database (http://environment.alberta.ca/ apps/GOWN/) and inverse distance weighted (IDW) interpolation method [34]. Aquifer and bedrock lower level maps were prepared by using bore log data of those wells in the study area and IDW interpolation method.

Table 1. Details of input data used for coupled MIKE-SHE/MIKE-11 model.

Type of Data	Data and Format	Source
Watershed	• Canadian Digital Elevation Data of 17.77 m grid	• Natural Resources Canada
	• Land use/land cover of 30 m by 30 m grid for year 2000	• Natural Resources Canada
	• Channel geometry	• Digitizing Digital Elevation Data and Google maps
	• Soil	• Alberta Environment and Parks
Vegetation Characteristics	• Rooting depths and Leaf area index of each land use type	• Published reports and articles [35–37]
Climate	• Observed precipitation, temperature, and reference evapotranspiration from 1985 to 2014	• 14 weather stations in the study area from Alberta Agroclimatic Information Service, and Alberta Environment and Sustainable Resource Development
Hydrological	• Daily stream flow and water level from 2000 to 2012	• Water Survey of Canada
	• Daily Groundwater level from 2000 to 2012	• Alberta provincial groundwater monitoring wells database

No-flow boundary condition was assumed around the perimeter of the study area for the developed MIKE-SHE model for model simplicity, and due to the lack of adequate information in the study area for setting appropriate boundary conditions (e.g., general head or specified head). Similarly, for MIKE-11 model, no-flow boundary condition was assumed for all unconnected ends (branches) of the river network except the Peace River, the Smoky River, and the Little Smoky River. Since the upstream parts of these large rivers (which are outside of the study area) contributed 98.4% of the flow at the outlet of the study area based on the observed stream flow data in 2000–2012, inflow boundary condition was chosen at the upstream parts of these three rivers. Water level boundary was selected at the downstream (the Peace River) of the model based on the relationship of stream flow vs. water level at the downstream location. A sensitivity analysis of the modeling parameters was performed before calibration to determine which parameters are sensitive to the model outputs (stream flow, river water and groundwater levels). The shuffled complex evolution method [38] was used for automated model calibration. The coupled model was calibrated and validated by using observed climate data (i.e., precipitation, temperature and reference evapotranspiration) at monitoring weather stations, and stream flow and stream water levels at three monitoring stations (SW1, SW2, and SW3), and groundwater levels at monitoring wells (only GW1 well is shown here). The coefficient of determination (R^2), and coefficient of efficiency (NSE: Nash-Sutcliffe efficiency) were used for evaluation of the goodness-of-fit of this integrated hydrologic model. The model calibration was conducted from 1 January 2000 to 31 December 2006, and the validation was conducted from 1 January 2007 to 31 December 2012. The average inflow (98.4% of the flow at the outlet of the study area) of the 2000–2012 period was used in numerical simulation for future climate change, HF and its associated activities impacts on water resources in this study due to the lack of future stream flow data at those upstream parts of those rivers.

2.3. Climate Scenarios

Statistically downscaled daily temperature and precipitation for the period of 2021–2036 under the RCP4.5 and RCP8.5 scenarios from the AR5 of the IPCC were directly obtained from the Pacific Climate Impacts Consortium (PCIC) data portal [39]. Those RCP4.5 and RCP8.5 outputs were generated by using the Second Generation Earth System Model (CanESM2) outputs of the CCCma (Canadian Centre for Climate Modeling and Analysis), and Bias Correction/Constructed Analogues with Quantile mapping reordering (BCCAQv2) method. The CanESM2 was used in this study area because the CanESM2 historical simulations on precipitation and temperature in 2000–2012 mimic well with the corresponding historical observations of precipitation and temperature, respectively. The RCP4.5 and RCP8.5 outputs are of roughly 10 km grid resolution. In RCP4.5, GHG emissions peak around 2040 and then decline [40]. The RCP4.5 scenario was chosen here because it is the pathway of stabilized GHG emission, whereas, in RCP8.5 GHG, emissions rise continuously over time [41]. The RCP8.5 scenario was chosen because it is a high-emission scenario, which is frequently referred to as "business as usual". This scenario will likely occur if the society does not make any efforts to cut GHG emissions. Future climate change scenarios assessed for two decades were used in a number of climate change impacts studies on water resources [42–44]. However, in order to be consistent with the forecast of future number of hydraulically fractured wells in Alberta until 2036, in the study conducted by Johnson et al. [45], the period of 2021–2036 (less than two decades) was used in this study.

2.4. Generation of Future HF Scenarios

HF data (i.e., number of wells and water use) for 2-year (2013–2014) were collected from the publicly available Frac Focus Chemical Disclosure Registry (www.fracfocus.ca). In Alberta, fracking data are publicly available according to the requirements of the Alberta provincial regulator since 19 December 2012 [46]. These data were collected for this study because HF activities occurred during this period when oil and gas prices were relatively high (e.g., oil at USD$100/barrel in 2013–2014) [47]. It represents the traditional HF scenario in Alberta, when oil price is good from a business point of view. The annual number of hydraulically fractured wells in 2013 and 2014 was 186 and 247, respectively. The monthly variation of those wells is presented in Table 2.

Table 2. Monthly variation of hydraulically fractured wells in 2013 and 2014.

Month	Number of Wells in 2013	Number of Wells in 2014	Total Wells in 2013 and 2014	Percentage to the Total Annual Wells in 2013 and 2014 (%)
January	18	20	38	8.78
February	29	19	48	11.09
March	10	35	45	10.39
April	3	9	12	2.77
May	7	13	20	4.62
June	9	25	34	7.85
July	20	13	33	7.62
August	19	22	41	9.47
September	15	25	40	9.24
October	22	25	47	10.85
November	18	20	38	8.78
December	16	21	37	8.55
Total	186	247	433	100

The future number of hydraulically fractured wells in the study area for 2021–2036 was projected based on the published report by Johnson et al. [45], which forecasted the number of hydraulically fractured wells in various provinces (i.e., Alberta, British Columbia, Saskatchewan, Manitoba, and Newfoundland and Labrador) of Canada from 2016 to 2036. In this study, the projection of wells in 2021–2036 was conducted based on the ratio of the total number of wells that was completed by HF in the study area in 2013 and 2014 to the total number of wells completed by HF in Alberta in 2013 and 2014 collected from Alberta Energy Regulator [48,49]. On average, that ratio was 5.3%. The annual variation in the number of hydraulically fractured wells from 2021 to 2036 is presented in Figure 2. The annual number of wells of each year from 2021 to 2036 was distributed monthly according to the monthly percentage of wells to the total annual number of wells in 2013 and 2014 (Table 2). This approach resulted in a prediction of 2014 wells being completed in the study area using HF.

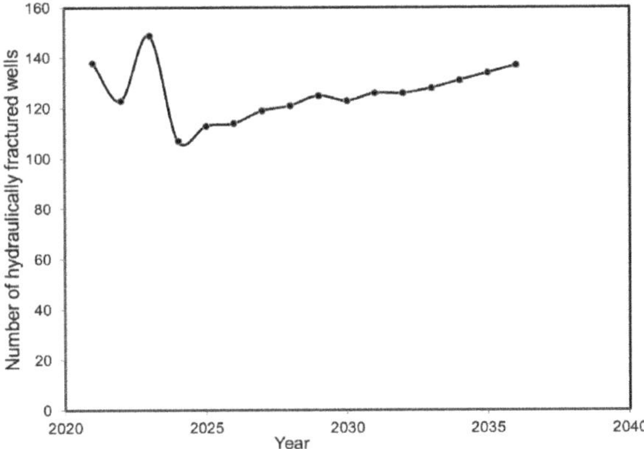

Figure 2. Annual variation of the number of hydraulically fractured wells from 2021 to 2036.

The annual water use in HF in 2013 and 2014 was 997,291 m^3 and 948,931 m^3, respectively. The average water use for each well in the study area was approximately 5362 m^3 in 2013 and 3842 m^3 in 2014. On average, 4495 m^3 of water was used for each well in 2013 and 2014. This average amount of water was considered for each well in every month in 2021–2036 due to limited available information during that time period. In total, 9,052,930 m^3 of water would be used in 2021–2036. Assuming truck capacity of 25 m^3, 362,117 large water tank truckloads would be needed in 2021–2036 for delivering water to drill these hydraulically fractured wells. Publicly available data (www.fracfocus.ca) only reports the date and quantity of water used in HF. It does not include the time of water withdrawal, how the water was transported to the site, the location of the source of water, the type of water source, and whether any water was recycled. Similar to other studies related to water withdrawal for HF on water resources, this constitutes a limitation in our study. In order to reduce these uncertainties, the following assumptions were made:

- Monthly water use data in 2021–2036 were distributed equally among all days of the particular month for numerical simulations. Best and Lowry [19] distributed all water withdrawals uniformly over the entire year for numerical modeling.
- Only surface water (i.e., river) was selected as a potential water source.
- It was assumed that all water was extracted from one location near the time of the fracturing operations, and the location of water extraction was selected close to the water level and flow monitoring station (SW3 station) so that the maximum impacts on river water level could be estimated. The water extraction location was assumed 1 km upstream of the SW3 station in the Smoky River (Figure 1a). This location was selected because the SW3 station is located in the Montney and Duvernay formations, and water extraction from this location would have the maximum impacts on river water level fluctuations at the SW3 station. No recycling of water was considered.

One potential scenario of water use for HF was generated based on the above assumptions. This scenario does not provide exact prediction, however shows the trend and nature of prediction in order of magnitude by using the available data. This scenario was used in the developed model to compare the outputs (i.e., stream flow and groundwater discharge) in 2021–2036 under future climate change scenarios (i.e., RCP4.5 and RCP8.5) with those under sole future climate change scenarios in 2021–2036, and base modeling period (2000–2012), where no HF was used. In this study, the period of 2000–2012 was used as base modeling period because the calibration and validation of the model was done during that period. Albek et al. [50] used a similar approach to compare streamflow variation due to climate change with respect to the base model results (i.e., during 4 years model calibration and validation periods) in the Middle Seydi Suyu Watershed, Turkey. The results illustrate the maximum probable impacts on water resources under future climate change.

2.5. Generation of Future HF Associated Activities Scenarios

Future scenarios of HF associated activities (i.e., construction of roads, well pads, pipelines, seismic lines, and power transmission lines) were generated by using ALCES [27]. ALCES is a fast, user-friendly and powerful landscape simulator that creates a "what-if" modeling environment that allows stakeholders to explore the economic, ecological, land and social consequences of different land use changes on defined landscapes [51]. ALCES generates future scenarios of HF associated activities under a business as usual (BAU) management scenario. ALCES provides future outputs for every decade. However, this study assessed climate change impact for the period of 2021–2036, so that it would be consistent with the outputs of Johnson et al. [45]. Therefore, ALCES simulation was performed from 2010 to 2030. Future scenarios of ALCES outputs were generated for the decades of 2020 and 2030. In addition, we assumed 2020 ALCES BAU scenario for the hydrological simulation for the period of 2021–2029, and 2030 ALCES BAU scenario for the period of 2030–2036. These scenarios were included in year 2000 land use map to get 2021–2029 and 2030–2036 land use maps, which were not shown here because of small land use changes. Wijesekara et al. [52] also used one-year land use map for a 5-year hydrological simulation. In this way, we attained average impacts of HF associated activities on water resources.

Typically, shale gas multi-well pads require 2 acres (8093.71 m^2) to 5 acres (20,234.3 m^2) of land [53]. However, in this study one grid cell size of the developed model was 284 m by 284 m, which is equivalent to an area of 80,656 m^2. Therefore, in this study, 284 m by 284 m was assumed for each well pad size. Six wells were considered for each well pad as per other studies in HF [53,54]. The density of well pad was considered as one well pad per 2.56 km^2 [54].

2.6. Limitations and Uncertainties of the Results

There are a number of limitations and uncertainties in the results generated from HF and its associated activities under changing climate. First, uncertainties always exist in future climate change scenarios [55]. Therefore, uncertainty analysis of climate change should be considered in further studies to assess the average impact of climate change scenarios on water resources in the study area. Second, internal variability, which was not considered in this study because climate data was directly downloaded from the PCIC data portal, could affect the patterns of climate change scenarios. Third, different climate models will provide different patterns of future precipitation and temperature trends under the RCP4.5 and the RCP8.5 scenarios. Therefore, other climate models predicted precipitation and temperature should be used to compare the obtained results of this study. Fourth, because of the lack of proper information no-flow boundary condition was used for the MIKE-SHE model domain, which would affect the outputs of this study. Fifth, future HF and its associated activities scenarios were generated based on certain current assumptions and are likely to fluctuate with global energy demand, prices, extraction techniques, etc. Thus, the outputs of this study will not characterize exact prediction, but show the trend and nature of prediction in order of magnitude.

3. Results and Discussion

3.1. Results of Model Calibration and Validation

Based on the sensitivity analysis, it was found that horizontal hydraulic conductivity (loam) was the most sensitive parameter in the model. Horizontal hydraulic conductivity (clay loam), horizontal hydraulic conductivity (silty clay), specific storage (bedrock), specific yield (loam), evapotranspiration surface depth, water content at saturation (loam), degree-day melting coefficient, leakage coefficient of the bed material, channel roughness and overland surface roughness (forest) were ranked as the second, third, fourth, fifth, sixth, seventh, eighth, ninth, tenth and eleventh sensitive parameter, respectively.

Since precipitation plays a negligible role (nearly 1.6%) in the rainfall-runoff processes in the study area, most of the sensitive parameters governing the channel routing and saturated groundwater flow play the major roles. Therefore, the model set up in this study favors mostly channel routing and saturated groundwater flow parameters. These parameters were changed during the model calibration stage. The monthly model calibration (Figure 3a) resulted in $R^2 = 0.88$ and NSE= 0.76 at the outlet (SW1 station) of the study area. Santhi et al. [56] suggested an acceptable model evaluation when a $R^2 \geq 0.6$ and a NSE ≥ 0.5 are obtained. These evaluation statistics criteria showed that the developed model calibration was deemed satisfactory. The model validation resulted (Figure 3b) in $R^2 = 0.92$, and NSE= 0.89 at the outlet of the study area using monthly data. Therefore, satisfactory model validation was also achieved.

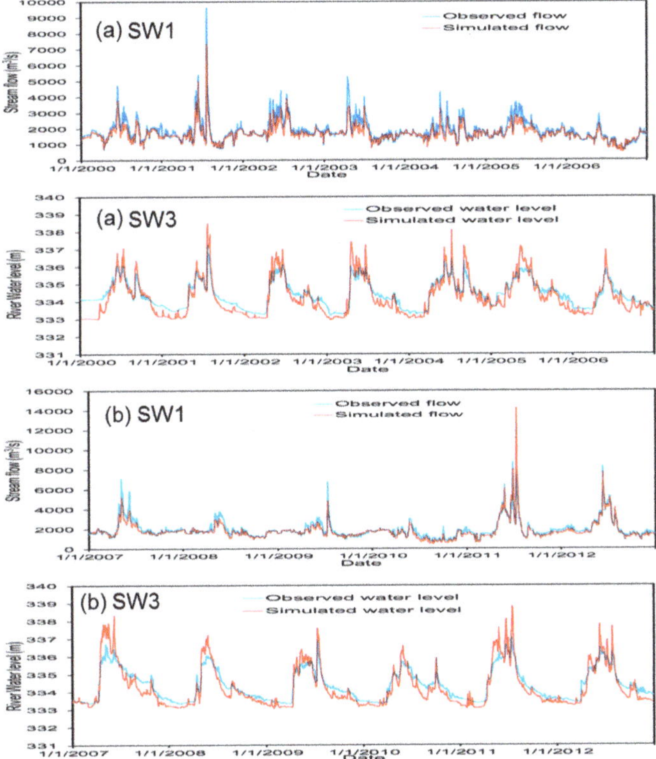

Figure 3. Observed and simulated stream flows at the SW1 (outlet of the study area) station, and river water levels at the SW3 station by the developed model during (**a**) calibration and (**b**) validation periods.

The model calibration and validation considering groundwater levels (here showing the results for the GW1 well) showed satisfactory results (Figure 4). Total water balance during the simulation period was also used as an indicator of the model performance. During both calibration and validation periods, the total water balance error was less than 1%, which indicates an adequate model performance. Table 3 presents the calculated R^2 and NSE values at various monitoring stations and wells based on monthly stream flow, stream (i.e., river) water level, and groundwater level data.

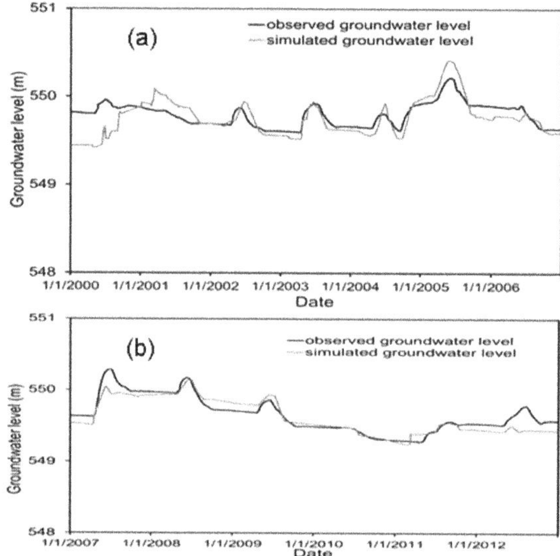

Figure 4. Observed and simulated groundwater levels by the developed model at the GW1 well during (**a**) calibration and (**b**) validation periods.

Table 3. R^2 (coefficient of determination) and NSE (Nash-Sutcliffe efficiency) values using observed and simulated stream flows, stream water levels, and groundwater levels data at various monitoring stations and wells during calibration and validation periods.

Monitoring Station/Well (Measuring Parameter)	Calibration		Validation	
	R^2	NSE	R^2	NSE
SW1 (stream flow)	0.88	0.76	0.92	0.89
SW1 (stream water level)	0.77	0.69	0.84	0.82
SW2 (stream water level)	0.48	0.56	0.52	0.59
SW3 (stream flow)	0.92	0.63	0.9	0.75
SW3 (stream water level)	0.91	0.71	0.86	0.65
GW1 (groundwater level)	0.71	0.69	0.75	0.81
GW2 (groundwater level)	0.69	0.70	0.68	0.65
GW3 (groundwater level)	0.66	0.69	0.61	0.67
GW4 (groundwater level)	0.73	0.77	0.87	0.85

3.2. Impact of Climate Change on Precipitation and Temperature

The future monthly precipitation in the study area under the RCP4.5 and RCP8.5 scenarios show variable patterns in 2021–2036 (Figure 5a) due to the anthropogenic increases in the atmospheric concentrations of GHG [57]. In this figure, the error bars of the standard deviation of monthly precipitation for the period of 2021–2036 are shown. Both the trend and peak of the mean projected monthly precipitation under the RCP4.5 and RCP8.5 scenarios in 2021–2036 follow the pattern of the base modeling period. This similar pattern also justifies why the CanESM2 projections were used to represent future climate over this region. It was also found that the mean monthly precipitation is higher under the RCP4.5 scenario than under the base modeling period for all months, except August and September. On the other hand, the mean monthly precipitation under the RCP8.5 scenario is higher for all months, except July and August, compared to the base modeling period. The mean monthly precipitation under the RCP8.5 scenario is higher for 4 months (April, May, June and September) than those under the RCP4.5 scenario. From the seasonal (winter: December–February, spring: March–May, summer: June–August, and fall: September–November) point of view, the highest and lowest seasonal precipitation under both RCP4.5 and RCP8.5 scenarios from 2021 to 2036 are expected in summer and winter, respectively (Table 4). The mean seasonal precipitation under the RCP4.5 and the RCP8.5 scenarios is also expected to increase in 2021–2036, with respect to the mean seasonal precipitation under the base modeling period. A greater increase in mean seasonal precipitation is expected during spring (33 mm and 43 mm under the RCP4.5 and RCP8.5 scenarios, respectively) than other seasons in both scenarios. The mean annual precipitation under the RCP4.5 and the RCP8.5 scenarios is expected to increase in 2021–2036, as compared to that under the base modeling period. The mean annual precipitation of 2021–2036 under the RCP4.5 and the RCP8.5 scenarios is expected to be 504 mm (σ = 92 mm), and 509 mm (σ = 102 mm), respectively. These numbers are higher than the mean annual precipitation under the base modeling period by 89 mm (21.4%) and 94 mm (22.6%), respectively.

The trend of mean monthly temperature under the RCP4.5 scenario is similar in every year, with the highest and lowest mean monthly temperature occurring in July and January, respectively, which are similar to those under the base modeling period (Figure 5b). However, under the RCP8.5 scenario this trend is similar in most of the months, except the lowest mean monthly temperature that is expected to occur in December. The mean monthly temperature under both scenarios is higher than the base modeling period for all months. The mean monthly temperature under the RCP8.5 scenario is higher in 6 months (January, April, May, June, July and August) than under the RCP4.5 scenario. The mean seasonal temperature under the RCP4.5 and the RCP8.5 scenarios is expected to increase in 2021–2036 with respect to the mean seasonal temperature under the base modeling period (Table 4) because of the anthropogenic increases in the GHG concentrations [57]. A greater increase in mean seasonal temperature is expected during spring (2.3 °C) and summer (2.92 °C) under the RCP4.5 and the RCP8.5 scenarios, respectively. The mean annual temperature would also increase under both scenarios in 2021–2036. On average, the mean annual temperature of 2021–2036 under the RCP4.5 and the RCP8.5 scenarios would be 3.46 °C (σ = 0.76 °C) and 3.62 °C (σ = 1.06 °C), respectively. The mean annual temperature under the RCP4.5 and the RCP8.5 scenarios is expected to increase by 1.5 °C and 1.66 °C, respectively, as compared to that under the base modeling period.

Table 4. Mean seasonal precipitation and temperature under the base modeling period (2000–2012), the RCP (Representative Concentration Pathways) 4.5 and the RCP8.5 scenarios in 2021–2036. The values within the parentheses are standard deviation among mean seasonal precipitation and temperature, respectively, from 2021 to 2036. The values within the angle brackets are absolute changes in mean seasonal precipitation and temperature under the RCP4.5 and the RCP8.5 scenarios in 2021–2036, with respect to the mean seasonal precipitation and temperature under the base modeling period, respectively.

	Precipitation (mm)				Temperature (°C)			
Scenario	Winter	Spring	Summer	Fall	Winter	Spring	Summer	Fall
Base modeling period (2000–2012)	66	87	172	90	−11.31	2.21	14.66	2.28
RCP4.5 (2021–2036)	91	120	191	102	−9.95	4.51	16.69	2.63
	(29)	(40)	(62)	(25)	(1.65)	(1.1)	(1.06)	(1.76)
	<25>	<33>	<19>	<12>	<1.36>	<2.3>	<2.03>	<0.35>
RCP8.5 (2021–2036)	83	130	195	101	−10.53	4.98	17.58	2.46
	(15)	(49)	(70)	(28)	(2.4)	(1.48)	(1.13)	(2.04)
	<17>	<43>	<23>	<11>	<0.78>	<2.77>	<2.92>	<0.18>

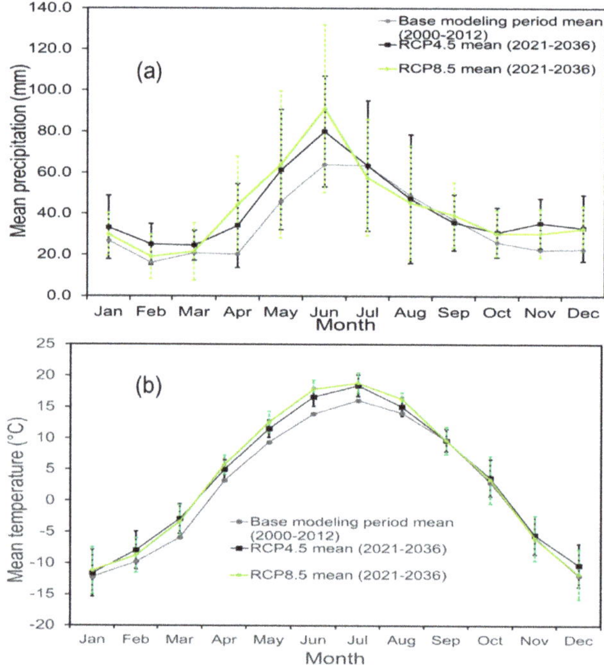

Figure 5. Comparison of (**a**) mean monthly precipitation and (**b**) mean monthly temperature under the RCP4.5 and the RCP8.5 scenarios from 2021 to 2036 and the base modeling period (2000–2012). The error bars represent the standard deviation among monthly precipitation/temperature of 2021 to 2036.

3.3. Land Use Changes due to HF Associated Activities

Based on the ALCES outputs, very little amount of land use change is predicted to arise from HF associated activities in 2030 compared to 2000 (Table 5). The change would occur in the land uses of minor roads (5.88 km^2), well pads for oil and gas exploration and extraction (32.27 km^2), and pipelines (21.22 km^2). However, the land uses of major roads, power lines and seismic lines would not change. In this study, pipelines and minor roads were not considered for future scenarios of HF associated activities because of the narrow size (diameter) of the pipelines and smaller width of minor roads, which were not possible to include in 30 m by 30 m resolution of land use map. Only the change in well pads, which is 0.13% of the total study area, was considered. The results show that forest, agriculture, and perineal crops and pasture areas would be converted into clear cut area (i.e., well pads here) in 2030. The major decrease would occur in forest (23.27 km^2). Agriculture, and perineal crops and pasture areas would decrease by 5.13 km^2 and 3.87 km^2, respectively, from 2000 to 2030 (Table 5). In Pennsylvania, Drohan et al. [58] also found similar conversion of forest and agricultural areas into gas well pads, which is clear cut area in this study.

Table 5. Land use changes due to HF (Hydraulic Fracturing) associated activities from 2000 to 2030 in the study area. Change (%) = [(Area of 2030 land use related to HF associated activities-Area of 2000 land use)/Area of 2000 land use] × 100. Negative sign indicates decrease in land area.

Land Use Type	Area (km^2) in 2000	Area (km^2) in 2030	Change (km^2)	Change (%)
Forest	8250.81	8227.54	−23.27	−0.28
Agriculture	8178.85	8170.22	−8.63	−0.11
Perineal crops and pasture	4365.25	4359.00	−6.25	−0.14
Water	1606.99	1606.99	0.00	0.00
Grassland	1175.26	1175.26	0.00	0.00
Shrub land	287.82	287.82	0.00	0.00
Road	95.94	101.82	5.88	6.13
Clear cut area	23.98	56.25	32.27	134.54
Total	23,984.90	23,984.90		

3.4. Surface Water and Groundwater Under the RCP4.5, the RCP8.5, HF and Its Associated Activities Scenarios

3.4.1. Monthly, Seasonal and Annual Stream Flows

The integrated model simulated results were analyzed on a mean monthly basis. The results show that the mean monthly stream flows in 2021–2036 under the RCP4.5 and the RCP8.5 scenarios are higher than those under the base modeling period (2000–2012) during the whole year, due to the increased precipitation and temperature predicted under the RCP4.5 and the RCP8.5 scenarios (Figure 6). At the SW1 station (outlet of the study area), the highest mean monthly stream flow occurs in June in both scenarios and the base modeling period (Figure 6a). On the other hand, at the SW3 station, the highest mean monthly stream flow occurs in May in both scenarios and the base modeling period (Figure 6b). This variation occurs because of the spatial and temporal precipitation variability in and outside of the study area. The upstream parts of the Peace River, the Smoky River and the Little Smoky River (which are outside of the study area) contribute 88%, 8.1% and 2.3% of the stream flow at the outlet of the study area, respectively, which also supports the significance of precipitation variability outside of the study area on these results. At the SW1 station, the lowest mean monthly stream flow under both scenarios and the base modeling period occurs in September and October, respectively. At the SW3 station, the lowest mean monthly stream flow under both scenarios and the base modeling period occurs in January and February, respectively. Therefore, climate change significantly affects the pattern of mean monthly stream flows in the study area.

From the seasonal point of view, the mean seasonal stream flows at the SW1 station under the RCP4.5 and the RCP8.5 scenarios are also expected to increase in 2021–2036 with respect to those under the base modeling period (Table 6). This occurs due to the increased precipitation and temperature predicted under the RCP4.5 and the RCP8.5 scenarios with respect to the base modeling period. The highest and lowest water extraction from the Peace River reach, where the SW1 station is located, under both scenarios could be possible during summer and fall, respectively. It would occur due to the highest (i.e., on average 2043.12 m^3/s and 2048.85 m^3/s under the RCP4.5 and the RCP8.5 scenarios, respectively) and lowest (i.e., on average 1424.93 m^3/s and 1424.32 m^3/s under the RCP4.5 and the RCP8.5 scenarios, respectively) mean stream flows at the SW1 station during summer and fall, respectively. However, a greater increase in mean seasonal stream flow is expected during spring (2.96% and 3.41% under the RCP4.5 and the RCP8.5 scenarios, respectively) compared to other seasons due to a greater increase in mean precipitation during spring. At the SW3 station, almost similar trends to those at the SW1 station would occur under both scenarios. However, the highest and lowest water extraction from the Smoky River reach, where the SW3 station is located, under both scenarios could be possible during summer and winter, respectively (Table 7). It would occur because the highest (i.e., on average 299.19 m^3/s and 302.09 m^3/s under the RCP4.5 and the RCP8.5 scenarios, respectively) and lowest (i.e., on average 39.42 m^3/s and 38.41 m^3/s under the RCP4.5 and the RCP8.5

scenarios, respectively) mean stream flows at the SW3 station would occur during summer and winter, respectively. The mean seasonal stream flows at the SW1 and SW3 stations under the RCP8.5 scenario are higher in spring and summer than those under the RCP4.5 scenario due to the higher precipitation under the RCP8.5 scenario during these seasons. Therefore, more water extraction from both the Peace River and the Smoky River would be possible in spring and summer under the RCP8.5 scenario than under the RCP4.5 scenario. However, the mean seasonal stream flows at the SW1 and SW3 stations under the RCP8.5 scenario are lower in winter and fall than those under the RCP4.5 scenario due to the lower precipitation under the RCP8.5 scenario during these seasons. Therefore, less water extraction from both the Peace River and the Smoky River would be possible in winter and fall under the RCP8.5 scenario than under the RCP4.5 scenario.

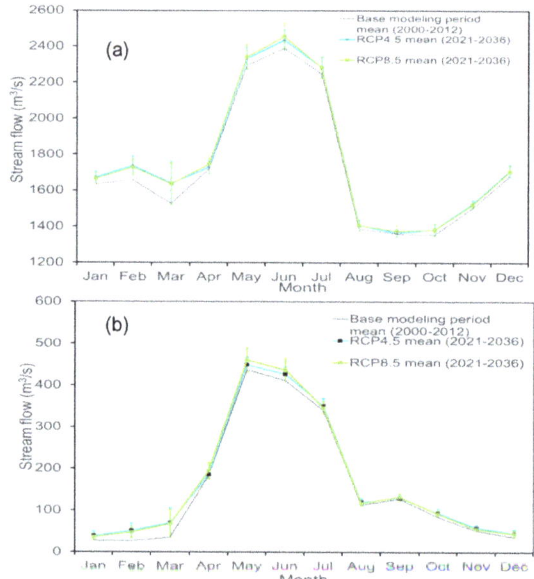

Figure 6. Comparison of projected mean monthly stream flows at the (**a**) SW1 (outlet of the study area) and (**b**) SW3 stations under the RCP4.5 and the RCP8.5 scenarios from 2021 to 2036 and the base modeling period (2000–2012). The error bars represent the standard deviation among mean monthly stream flows of 2021 to 2036.

On average, the mean annual stream flow at the SW1 and SW3 stations under the RCP4.5 scenario is expected to be 1768.72 m^3/s (σ = 23.22 m^3/s), and 168.60 m^3/s (σ = 7 m^3/s), respectively. On the other hand, the mean annual stream flow at the SW1 and SW3 stations under the RCP8.5 scenario is expected to be 1770.66 m^3/s (σ = 27.57 m^3/s), and 169.39 m^3/s (σ = 10.07 m^3/s), respectively. With respect to the mean annual stream flow at the SW1 (i.e., 1729.87 m^3/s) station under the base modeling period, the mean annual stream flow at the SW1 station under the RCP4.5 and the RCP8.5 scenarios would increase by 2.24% (i.e., 38.85 m^3/s), and 2.36% (i.e., 40.79 m^3/s), respectively. In contrast, with respect to the mean annual stream flow at the SW3 (i.e., 156.22 m^3/s) station under the base modeling period, the mean annual stream flow under the RCP4.5 and the RCP8.5 scenarios would increase by 7.92% (i.e., 12.38 m^3/s), and 8.43% (i.e., 13.17 m^3/s), respectively. It is to be noted that the upstream parts of three large rivers (which are outside of the study area) contributed 98.4% of the flow at the outlet of the study area based on the observed stream flow data in 2000–2012. The mean annual stream flow generated in the study area in the base modeling period (2000–2012) was 27.47 m^3/s, and under the RCP4.5 and the RCP8.5 scenarios the stream flow generated in the study area would be 66.32 m^3/s, and 68.26 m^3/s, respectively. The increment of stream flow generated in the study area under the

RCP4.5 and the RCP8.5 scenarios is 141.43% and 148.51%, respectively. A similar high-increase in stream flow due to the increased precipitation was found by Guo et al. [59] in the Xinjiang River basin, China, and Muhammad et al. [60] in the upper Assiniboine River Basin, Canada. Therefore, more annual water extraction from the river, and allocation to the stakeholders for future water supply could be possible under both scenarios in 2021–2036 than under the base modeling period.

Table 6. Mean seasonal stream flows at the outlet of the study area (SW1 station) under the (a) base modeling period (2000–2012), (b) RCP4.5 scenario, (c) HF and RCP4.5 scenario, (d) HF, its associated activities and RCP4.5 scenario, (e) RCP8.5 scenario, (f) HF and RCP8.5 scenario, and (g) HF, its associated activities and RCP8.5 scenario in 2021–2036. The values within the parentheses are standard deviation among mean seasonal stream flows from 2021 to 2036. The values within the angle brackets are relative changes in mean seasonal stream flows under the (i) RCP4.5 scenario, (ii) RCP8.5 scenario, (iii) HF and RCP4.5 scenario, iv) HF and RCP8.5 scenario, (v) HF, its associated activities and RCP4.5 scenario, and (vi) HF, its associated activities and RCP8.5 scenario in 2021–2036 with respect to the mean seasonal stream flows under the base modeling period (2000–2012).

Stream Flow (m^3/s) at the SW1 Station (Outlet of The Study Area)						
Season	Base Modeling Period (2000–2012)	RCP4.5 Scenario (2021–2036)	HF and RCP4.5 Scenario	HF, Its Associated Activities and RCP4.5 Scenario	RCP8.5 Scenario (2021–2036)	HF and RCP8.5 Scenario	HF, Its Associated Activities and RCP8.5 Scenario
Winter	1657.58	1706.37 (40.29) <2.94%>	1706.15 <2.93%>	1706.60 <2.95%>	1700.71 (34.62) <2.60%>	1700.49 <2.59%>	1700.87 <2.61%>
Spring	1845.88	1900.47 (72.27) <2.96%>	1900.35 <2.95%>	1901.14 <2.99%>	1908.81 (78.93) <3.41%>	1908.69 <3.40%>	1909.68 <3.46%>
Summer	2009.66	2043.12 (48.30) <1.66%>	2042.96 <1.65%>	2043.76 <1.69%>	2048.85 (53.37) <1.95%>	2048.69 <1.94%>	2049.62 <1.99%>
Fall	1406.36	1424.93 (25.49) <1.32%>	1424.72 <1.31%>	1425.18 <1.34%>	1424.32 (27.21) <1.27%>	1424.11 <1.26%>	1424.55 <1.29%>

Table 7. Mean seasonal stream flows at the SW3 station (near the water withdrawal location) under the (a) base modeling period (2000–2012), (b) RCP4.5 scenario, (c) HF and RCP4.5 scenario, (d) HF, its associated activities and RCP4.5 scenario, (e) RCP8.5 scenario, (f) HF and RCP8.5 scenario, and (g) HF, its associated activities and RCP8.5 scenario in 2021–2036. The values within the parentheses are standard deviation among mean seasonal stream flows from 2021 to 2036. The values within the angle brackets are relative changes in mean seasonal stream flows under the (i) RCP4.5 scenario, ii) RCP8.5 scenario, (iii) HF and RCP4.5 scenario, (iv) HF and RCP8.5 scenario, (v) HF, its associated activities and RCP4.5 scenario, and (vi) HF, its associated activities and RCP8.5 scenario in 2021–2036 with respect to the mean seasonal stream flows under the base modeling period (2000–2012).

Stream Flow (m^3/s) at the SW3 Station (Near the Water Withdrawal Location)						
Season	Base Modeling Period (2000–2012)	RCP4.5 Scenario (2021–2036)	HF and RCP4.5 Scenario	HF, its Associated Activities and RCP4.5 Scenario	RCP8.5 Scenario (2021–2036)	HF and RCP8.5 Scenario	HF, its Associated Activities and RCP8.5 Scenario
Winter	36.30	39.42 (12.28) <8.6%>	39.20 <8.0%>	39.48 <8.76%>	38.41 (9.78) <5.81%>	38.19 <5.21%>	38.45 <5.92%>
Spring	212.22	241.42 (21.60) <13.76%>	241.30 <13.70%>	241.62 <13.86%>	243.68 (27.26) <14.83%>	243.56 <14.77%>	243.95 <14.95%>
Summer	288.30	299.19 (15.16) <3.78%>	299.03 <3.72%>	299.38 <3.84%>	302.09 (16.15) <4.78%>	301.93 <4.73%>	302.32 <4.86%>
Fall	88.04	94.37 (7.86) <7.19%>	94.16 <6.95%>	94.45 <7.28%>	93.37 (6.86) <6.05%>	93.16 <5.81%>	93.44 <6.14%>

When the HF scenario is added to the RCP4.5 scenario in 2021–2036, the mean monthly, seasonal and annual stream flows at the SW1 and SW3 stations would decrease with respect to those under the only RCP4.5 scenario due to water withdrawals for HF from the Smoky River. Those decrements are very small (less than 1%) compared to the stream flow generated in a large study area. Therefore, these results were not possible to show in Figure 6, but the results follow similar trends to those under the sole RCP4.5 scenario. Similar results are obtained when HF scenario is added to the RCP8.5 scenario. Although the impacts of HF on stream flow are very small in the large area, the impacts could be significant in a small catchment area where large amount of water is extracted from the nearby river for HF. Cothren et al. [18] did not find any noticeable change in stream flow at a large basin scale (127,300 km^2), but found significant changes in sub-basin scale and in monthly time steps. From seasonal point of view, the highest and lowest decreases in mean seasonal stream flow at both stations under both scenarios would occur during winter (i.e., 0.22 m^3/s), and spring (i.e., 0.12 m^3/s) seasons, respectively. It would occur because the highest and lowest number of wells would be completed by HF collecting water from the Smoky River in 2021–2036 during winter and spring, respectively. However, those reductions would not decrease the mean monthly, seasonal and annual stream flows at the SW1 and SW3 stations under the effects of (i) HF and RCP4.5 scenario, and (ii) HF and RCP8.5 scenario than those under the base modeling period (Tables 6 and 7).

When HF associated activities (construction of well pads) are combined with the HF and RCP4.5 scenario in 2021–2036, the mean monthly, seasonal and annual stream flows at the SW1 and SW3 stations are expected to increase with respect to those under the sole RCP4.5 scenario. This occurs because HF associated activities results in increasing surface runoff and stream flow due to increasing area of low hydraulic conductivity soil. However, the increment is very small (less than 1%) because of small amount of land use changes (i.e., 0.13% of the study area). Similar outcomes are expected when HF associated activities are combined with the HF and RCP8.5 scenario. Cothren et al. [18] found significant increase (10%) in stream flow in sub-basin scale (area 387 km^2) due to the increase in well pad and shale gas infrastructure in South Fork of the Little Red River watershed, USA. Saha [61] also found similar increase in stream flow in the Mainstem sub-watershed (213.82 km^2) of the Kiskatinaw River watershed, Canada. Therefore, significant additive impacts on stream flow could be possible in a small catchment area where large amount of HF associated activities would occur. Since the increments in this study are very small (less than 1%), those results were not possible to show in Figure 6. The highest and lowest increases in the mean seasonal stream flow at both stations under both (i) HF, its associated activities and RCP4.5 scenario, and (ii) HF, its associated activities and RCP8.5 scenario would occur during spring and winter (Tables 6 and 7), respectively, because of the highest and lowest increases in surface runoff would occur during spring and winter, respectively. However, a slight greater increase (by 0.20 m^3/s at the SW1, and 0.07 m^3/s at the SW3) would occur during spring under the HF, its associated activities and RCP8.5 scenario than under the HF, its associated activities and RCP4.5 scenario because of higher precipitation during spring under the RCP8.5 scenario. On the other hand, a slight greater increase (by 0.07 m^3/s at the SW1, and 0.02 m^3/s at the SW3) would occur during winter under the HF, its associated activities and RCP4.5 scenario than under the HF, its associated activities and RCP8.5 scenario because of higher precipitation during fall and winter under the RCP4.5 scenario.

3.4.2. Monthly, Seasonal and Annual Groundwater Discharges

Similar to stream flow, the mean monthly groundwater discharges generated at the outlet of the study area would increase under the RCP4.5 and the RCP8.5 scenarios compared to those under the base modeling period due to the increased groundwater levels under both scenarios resulted from increased precipitation (Figure 7). The highest mean monthly groundwater discharge occurs in June in both scenarios and the base modeling period. The lowest mean monthly groundwater discharge under both scenarios and the base modeling period occurs in January and October, respectively. Because of climate change, the pattern of mean monthly groundwater discharge changes in the study area.

From the seasonal point of view, the highest and lowest mean groundwater discharges at the outlet of the study area under both scenarios would occur during summer and winter, respectively (Table 8). However, a greater increase in mean seasonal groundwater discharge is expected during summer (i.e., 131.25% and 147.12% under the RCP4.5 and the RCP8.5 scenarios, respectively) than other seasons. Among all seasons, the increase in seasonal groundwater discharge of less than 100% would occur in winter due to the lower infiltration rate resulted from snow covered land area. The mean seasonal groundwater discharge under the RCP8.5 scenario is higher in spring and summer than those under the RCP4.5 scenario due to the higher precipitation under the RCP8.5 scenario during these seasons. On the other hand, the mean seasonal groundwater discharge under the RCP8.5 scenario is lower in winter and fall than those under the RCP4.5 scenario due to the lower precipitation under the RCP8.5 scenario during these seasons. The mean annual groundwater discharge generated at the outlet of the study area under the RCP4.5 and the RCP8.5 scenarios would be 35.96 m^3/s (σ = 0.98 m^3/s) and 36.55 m^3/s (σ = 1.06 m^3/s), respectively. The mean annual groundwater discharge under the RCP4.5 and the RCP8.5 scenarios would increase by 117.56% (i.e., 19.43 m^3/s) and 121.12% (i.e., 20.02 m^3/s), respectively, with respect to the base modeling period (i.e., 16.53 m^3/s). Consequently, more annual water extraction from the Smoky River and the Peace River for the oil and gas industries in the study area would be possible under both scenarios than under the base modeling period without causing any substantial negative impact on regional groundwater levels and groundwater discharge.

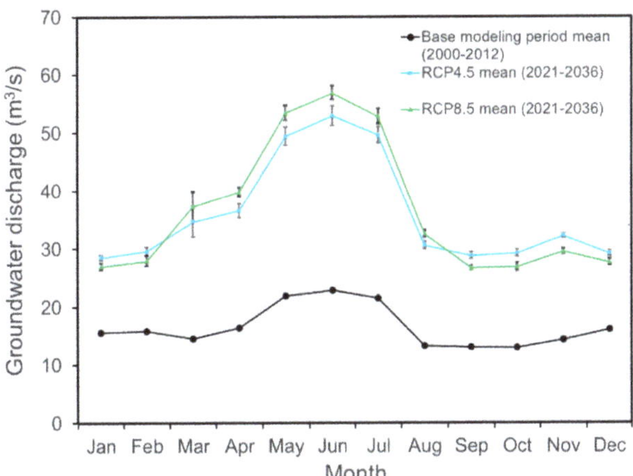

Figure 7. Comparison of projected mean monthly groundwater discharges generated at the outlet of the study area under the RCP4.5 and the RCP8.5 scenarios from 2021 to 2036 and the base modeling period (2000–2012). The error bars represent the standard deviation among mean monthly groundwater discharges of 2021 to 2036.

Although adjacent groundwater levels of the Smoky River reach would decrease under the effects of HF and RCP4.5 scenario than those under the sole RCP4.5 scenario, the mean monthly groundwater discharge of the study area would increase under the effects of HF and RCP4.5 scenario than those under the sole RCP4.5 scenario. This would occur because the minor decrease would happen in groundwater level compared to surface water level due to low groundwater velocity and low hydraulic conductivity of soils. Therefore, the gradient between the groundwater and surface water systems would increase, and would result in increased groundwater discharge under the effects of HF and RCP4.5 scenario. However, these increments are very small (less than 1%) compared to the groundwater discharge generated in a large study area. Therefore, these results were not possible to show in Figure 7, but the results follow similar trends to those under the sole RCP4.5 scenario. From the seasonal point of view, the highest and lowest increases in mean seasonal groundwater discharge in the study area under the effects of HF and RCP4.5 scenario would occur during winter (0.06 m^3/s) and spring (0.02 m^3/s), respectively (Table 8). This would occur because the highest and lowest number of wells would be completed by using water-intensive HF in 2021–2036 during winter and spring, respectively. Similar results would happen when HF scenario is added to the RCP8.5 scenario. However, the mean seasonal groundwater discharge increases more (i.e., 0.01 m^3/s) under the HF and RCP8.5 scenario during winter than that under the HF and RCP4.5 scenario. This happens because lower soil moisture condition would occur due to the lower precipitation in the study area during winter months under the RCP8.5 scenario than under the RCP4.5 scenario, which resulted in higher river water level declines in winter months under the HF and RCP8.5 scenario. Therefore, a higher gradient between groundwater and surface water would occur under the HF and RCP8.5 scenario, and result in higher groundwater discharge during winter. Similar trends occur across all seasons. The mean annual groundwater discharge at the outlet of the study area under the HF and RCP4.5 scenario, and the HF and RCP8.5 scenario would increase by 0.11% (i.e., 0.04 m^3/s), with respect to the sole RCP4.5 and RCP8.5 scenarios, respectively. Although the impacts of HF on groundwater discharge are very small in this large study area, the impacts could be significant in a small catchment area where large amount of water is extracted from the nearby river for HF. The mean annual groundwater discharge of the study area under the HF and RCP4.5 scenario, and the HF and RCP8.5 scenario would increase by 117.80% (i.e., 19.47 m^3/s) and 121.37% (i.e., 20.06 m^3/s), respectively, with respect to the base modeling period. This increased groundwater discharge under both scenarios may result in some positive effects on stream water quality, such as cooler stream temperature during warm months (i.e., months in late spring, summer and fall), and warmer stream temperature during cold months (i.e., months in winter and early spring) in the river [62,63].

Table 8. Mean seasonal groundwater discharges at the outlet of the study area under the (a) base modeling period (2000–2012), (b) RCP4.5 scenario, (c) HF and RCP4.5 scenario, (d) HF, its associated activities and RCP4.5 scenario, (e) RCP8.5 scenario, (f) HF and RCP8.5 scenario, (g) HF, its associated activities and RCP8.5 scenario in 2021–2036. The values within the round and square brackets are absolute and relatives changes in mean seasonal groundwater discharges under the (i) RCP4.5 scenario, (ii) RCP8.5 scenario, (iii) HF and RCP4.5 scenario, (iv) HF and RCP8.5 scenario, (v) HF, its associated activities and RCP4.5 scenario, and (vi) HF, its associated activities and RCP8.5 scenario in 2021–2036 with respect to the mean seasonal groundwater discharges under the base modeling period (2000–2012), respectively.

		Groundwater Discharge (m^3/s) at the Outlet of the Study Area					
Season	Base Modeling Period (2000–2012)	RCP4.5 Scenario (2021–2036)	HF and RCP4.5 Scenario	HF, Its Associated Activities and RCP4.5 Scenario	RCP8.5 Scenario (2021–2036)	HF and RCP8.5 Scenario	HF, Its Associated Activities and RCP8.5 Scenario
Winter	15.84	29.10 (13.26) [83.71%]	29.16 (13.32) [84.09%]	29.22 (13.38) [84.47%]	27.53 (11.69) [73.82%]	27.60 (11.76) [74.21%]	27.62 (11.78) [74.37%]
Spring	17.64	40.25 (22.61) [128.17%]	40.27 (22.63) [128.29%]	40.43 (22.79) [129.19%]	43.57 (25.93) [147.00%]	43.58 (25.94) [147.05%]	43.78 (26.14) [148.20%]
Summer	19.20	44.40 (25.20) [131.25%]	44.43 (25.23) [131.42%]	44.59 (25.39) [132.24%]	47.44 (28.24) [147.12%]	47.46 (28.26) [147.19%]	47.69 (28.49) [148.40%]
Fall	13.44	30.10 (16.66) [123.96%]	30.15 (16.71) [124.34%]	30.26 (16.82) [125.15%]	27.65 (14.21) [105.77%]	27.71 (14.27) [106.16%]	27.80 (14.36) [106.89%]

The mean monthly groundwater discharges in the study area under the effects of HF, its associated activities, and RCP4.5 scenario would increase with respect to those under the sole RCP4.5 scenario. However, the increment is less than 1% because of small land use changes. Therefore, those results were not possible to show in Figure 7. Although increasing well pads of low hydraulic conductivity soil generally results in increasing surface runoff and decreasing groundwater discharge, the outputs in this study found increasing groundwater discharge at the outlet of the study area. This occurs because most of the land use changes would occur in the forest area, which would provide additional precipitation from canopy interception in the study area. When well pads will be built in forested areas such as surrounding the GW1 and GW2 monitoring wells, canopy rain and snow interception in those areas will decrease and provide additional precipitation amount from canopy interception on those areas. This additional precipitation would result in increased soil moisture, which in turn would increase surface runoff, infiltration, and groundwater levels at those wells. The mean annual groundwater level at the GW1 well under the effects of HF, its associated activities, and RCP4.5 scenario would increase by 0.33 m compared to that under the sole RCP4.5 scenario. The mean annual groundwater level at the GW1 well under the effects of HF, its associated activities, and RCP8.5 scenario would increase by 0.36 m compared to that under the sole RCP8.5 scenario. Because of higher precipitation under the RCP8.5 scenario, the mean annual groundwater level at the GW1 well under the effects of HF, its associated activities and RCP8.5 scenario would increase more than under the effects of HF, its associated activities and RCP4.5 scenario. Therefore, HF associated activities would provide significant additive impacts on groundwater resources mainly in forest clear-cut area (i.e., forest area converted into well pads). Evans et al. [64] found groundwater level increase in forest-harvested area in northeast Alberta, Canada. In contrast, when well pads will be built in agricultural and pasturelands such as surrounding the GW3 and GW4 monitoring wells, additional precipitation from canopy interception will not generate in these areas similar to forested area. Therefore, groundwater level would decrease in those wells due to the increasing areas of low hydraulic conductivity soil, and surface runoff would increase. The mean annual groundwater level at the GW4 well under the effects of HF, its associated activities and RCP4.5 scenario would decrease by 0.04 m compared to that under the sole RCP4.5 scenario. The mean annual groundwater level at the GW4 well under the effects of HF, its associated activities and RCP8.5 scenario would decrease by 0.05 m compared to that under the sole RCP8.5 scenario. Because of higher precipitation and temperature under the RCP8.5 scenario, more surface runoff would occur, and consequently, the mean annual groundwater level at the GW4 well under the effects of HF, its associated activities and RCP8.5 scenario would decrease more than under the effects of HF, its associated activities and RCP4.5 scenario. Therefore, associated activities related to HF affect temporal groundwater levels locally due to various relationships of land use types with precipitation.

The highest and lowest increases in the mean seasonal groundwater discharge in the study area under the effects of HF, its associated activities and RCP4.5 scenario would occur during summer and winter, respectively (Table 8). Similar results would happen under the HF, its associated activities and RCP8.5 scenario. The mean annual groundwater discharge at the outlet of the study area under the effects of (i) HF, its associated activities and RCP4.5 scenario, and (ii) HF, its associated activities and RCP8.5 scenario would increase by 0.44% (i.e., 0.16 m^3/s) and 0.45% (i.e., 0.17 m^3/s), respectively, with respect to the sole RCP4.5 and RCP8.5 scenarios, respectively. The mean annual groundwater discharge at the outlet of the study area under the effects of (i) HF, its associated activities and RCP4.5 scenario and (ii) HF, its associated activities and RCP8.5 scenario would increase by 118.54% (i.e., 19.59 m^3/s) and 122.15% (i.e., 20.19 m^3/s), respectively, with respect to the base modeling period. This increased groundwater discharge under both scenarios of the CanESM2 may result in some positive effects on water temperature of the river (i.e., cooler stream temperature during warm months and vice versa) [62,63]. Therefore, HF and its associated activities in the RCP4.5 and the RCP8.5 scenarios would provide more groundwater discharge in the study area for future water supply for oil and gas exploration, and better water quality (stream temperature) compared to those under both base modeling period, and sole RCP4.5 and RCP8.5 scenarios.

3.5. Potential Regarding the Results

The results of this study provide a good prospect for future HF in the study area under the RCP4.5 and the RCP8.5 scenarios of the CanESM2 in 2021–2036 without causing any substantial negative impacts on stream flow and groundwater discharge compared to the base modeling period (2000–2012). These results provide valuable preliminary information to the watershed manager for developing future water allocation plans in the study area for HF and other development activities, irrigation, and forestry. Although the impacts of HF and its associated activities (construction of well pads) on stream flow and groundwater discharge are very small, significant impacts on stream flow and groundwater discharge could be possible in a small catchment area where large amount of HF and its associated activities would occur. In addition, this integrated modeling approach can be used for assessing future water resources in changing climate under the effects of water extraction from river for other water uses, such as irrigation, mining, manufacturing industries and municipal water supply, where water is more used than in HF.

4. Conclusions

This study evaluated how HF and its associated activities would affect surface water and groundwater in 2021–2036 under changing climate (i.e., RCP4.5 and RCP8.5 scenarios of the CanESM2) in a shale gas and oil play area of northwestern Alberta, Canada as a case study by using an integrated hydrologic model (i.e., MIKE-SHE and MIKE-11 models), and a cumulative effects landscape simulator (i.e., ALCES). The simulation results show climate change (i.e., RCP4.5 and RCP8.5 scenarios) during 2021–2036 would significantly increase precipitation and temperature in the study area, and therefore would result in increases in stream flow and groundwater discharge, with respect to those under the base modeling period (2000–2012). Stream flow and groundwater discharge under the RCP8.5 scenario would be higher during spring and summer (due to the higher precipitation), and lower during winter and fall (due to the lower precipitation) as compared to those under the RCP4.5 scenario. The simulation results show very small (less than 1%) reduction on stream flow due to water withdrawals for HF under both RCP4.5 and RCP8.5 scenarios because of the large size of the study area. However, groundwater discharge would increase negligibly (less than 1%) because of the increase in the gradient between groundwater and surface water systems. The offsetting impacts of HF would not decrease stream flow under the effects of both HF and RCP4.5 scenario, and HF and RCP8.5 scenario than those under the base modeling period. The results also demonstrate a very little (less than 1%) positive impact of HF related associated activities on stream flow and groundwater discharge because of insignificant changes in land use, although the impacts on groundwater levels are locally controlled and closely connected to land use type change. Therefore, associated activities would provide additive impacts on stream flow and groundwater discharge under the effects of both (i) HF, its associated activities and RCP4.5 scenario, and (ii) HF, its associated activities and RCP8.5 scenario. The results obtained from this study provide useful information to the oil and gas industries to expand their shale oil and shale gas exploration in the study area in 2021–2036, without facing public pressure on water extraction for HF. The results also provide useful information for developing future water resources management plan at regional level for HF and its associated activities to meet future energy demand by considering future climate change.

Author Contributions: G.C.S.: conceptualization, methodology, investigation, validation, formal analysis, writing—original draft preparation; M.Q.: conceptualization, supervision, writing—review and editing. All authors have read and agreed to the published version of the manuscript.

Funding: This research received no external funding.

Acknowledgments: The authors would like to thank Connie Van der Byl, Brad Stelfox, Greg Chernoff, Linda Smith, Sonia Portillo, Qiao Ying and Kevin Beneteau for their help.

Conflicts of Interest: The authors have declared no conflict of interest exists.

References

1. McCarthy, J.J.; Canziani, O.F.; Leary, N.A.; Dokken, D.J.; White, K.S. *Climate Change 2001: Impacts, Adaptation and Vulnerability*; Intergovernmental Panel on Climate Change: Geneva, Switzerland, 2001.
2. Solomon, S.; Qin, D.; Manning, M.; Chen, Z.; Marquis, M.; Averyt, K.B.; Tignor, M.; Miller, H.L. *Climate Change 2007: The Physical Science Basis, Contribution of Working Group I to the Fourth Assessment Report of the IPCC*; Cambridge University Press: Cambridge, UK, 2007.
3. Oikonomou, P.D.; Kallenberger, J.A.; Waskom, R.M.; Boone, K.K.; Plombon, E.N.; Ryan, J.N. Water acquisition and use during unconventional oil and gas development and the existing data challenges: Weld and Garfield counties, CO. *J. Environ. Manag.* **2016**, *181*, 36–47. [CrossRef] [PubMed]
4. Kennedy, M. *BC Oil & Gas Commission—Experiences in Hydraulic Fracturing*; Ministry of Economy: Warsaw, Poland, 2011.
5. Jinno, K.; Tsutsumi, A.; Alkaeed, O.; Saita, S.; Berndtsson, R. Effects of land use change on groundwater recharge model parameters. *Hydrol. Sci. J.* **2009**, *54*, 300–315. [CrossRef]
6. Saha, G.C. Groundwater-Surface Water Interaction under the Effects of Climate and Land Use Changes. Ph.D. Thesis, University of Northern British Columbia, Prince George, BC, Canada, 2014. [CrossRef]
7. Brisk Insights. Hydraulic Fracturing Market Analysis by Shale Type, by Fracturing (Sliding Sleeve), Industry Size, Growth, Share and Forecast to 2022. 2016. Available online: http://www.briskinsights.com/report/hydraulic-fracturing-market-forecast-2015-2022 (accessed on 12 March 2019).
8. Massachusetts Institute of Technology. The Future of Natural Gas. 2015. Available online: http://web.mit.edu/mitei/research/studies/natural-gas-2011.shtml (accessed on 12 March 2019).
9. U.S. Energy Information Administration. *Annual Energy Outlook 2020 with Projections to 2050*; U.S. Department of Energy: Washington, DC, Canada, 2020. Available online: https://www.eia.gov/outlooks/aeo/pdf/aeo2020.pdf (accessed on 17 August 2020).
10. Gallegos, T.J.; Varela, B.A. *Trends in Hydraulic Fracturing Distributions and Treatment Fluids, Additives, Proppants, and Water Volumes Applied to Wells Drilled in the United States from 1947 through 2010—Data Analysis and Comparison to the Literature*; U.S. Geological Survey Scientific Investigations, Report 2014–5131; U.S. Geological Survey: Reston, VA, USA, 2015; p. 15. [CrossRef]
11. Entrekin, S.; Evans-White, M.; Johnson, B.; Hagenbuch, E. Rapid expansion of natural gas development poses a threat to surface waters. *Front. Ecol. Environ.* **2011**, *9*, 503–511. [CrossRef]
12. Patterson, L.A.; Konschnik, K.E.; Wiseman, H.; Fargione, J.; Maloney, K.O.; Kiesecker, J.; Nicot, J.; Baruch-Mordo, S.; Entrekin, S.; Trainor, A.; et al. Unconventional oil and gas spills: Risks, mitigation priorities, and state reporting requirements. *Environ. Sci. Technol.* **2017**, *51*, 2563–2573. [CrossRef] [PubMed]
13. Mayer, A.; Malin, S.; McKenzie, L.; Peel, J.; Adgate, J. Understanding self-rated health and unconventional oil and gas development in three Colorado communities. *Soc. Nat. Resour.* **2020**. [CrossRef]
14. Shank, M.K.; Stauffer, J.R., Jr. Land use and surface water withdrawal effects on fish and macroinvertebrate assemblages in the Susquehanna River basin, USA. *J. Freshw. Ecol.* **2015**, *30*, 229–248. [CrossRef]
15. Barth-Naftilan, E.; Aloysius, N.; Saiers, J.E. Spatial and temporal trends in freshwater appropriation for natural gas development in Pennsylvania's Marcellus Shale Play. *Geophys. Res. Lett.* **2015**, *42*, 6348–6356. [CrossRef]
16. MacQuarrie, A. Case Study Analysis on the Impacts of Surface Water Allocations for Hydraulic Fracturing on Surface Water Availability of the Upper Athabasca River. Master's Thesis, Royal Roads University, Victoria, BC, Canada, 2018. [CrossRef]
17. Entrekin, S.; Trainor, A.; Saiers, J.; Patterson, L.; Maloney, K.; Fargione, J.; Kiesecker, J.; Baruch-Mordo, S.; Konschnik, K.; Wiseman, H.; et al. Water stress from high-volume hydraulic fracturing potentially threatens aquatic biodiversity and ecosystem services in Arkansas, United States. *Environ. Sci. Technol.* **2018**, *52*, 2349–2358. [CrossRef]
18. Cothren, J.; Thoma, G.; Diluzio, M.; Limp, F. *Integration of Water Resource Models with Fayetteville Shale Decision Support and Information System*; University of Arkansas and Blackland Texas A&M Agrilife, Final Technical Report; University of Arkansas System: Fayetteville, AR, USA, 2013; p. 161. [CrossRef]
19. Best, L.C.; Lowry, C.S. Quantifying the potential effects of high-volume water extractions on water resources during natural gas development: Marcellus Shale, NY. *J. Hydrol. Reg. Stud.* **2014**, *1*, 1–16. [CrossRef]

20. Sharma, S.; Shrestha, A.; McLean, C.E.; Martin, S.C. Hydrologic Modelling to Evaluate the Impact of Hydraulic Fracturing on Stream Low Flows: Challenges and Opportunities for a Simulation Study. *Am. J. Environ. Sci.* **2015**, *11*, 199–215. [CrossRef]

21. Shrestha, A.; Sharma, S.; McLean, C.E.; Kelly, B.A.; Martin, S.C. Scenario analysis for assessing the impact of hydraulic fracturing on stream low flows using the SWAT model. *Hydrol. Sci. J.* **2016**, *62*, 849–861. [CrossRef]

22. Lin, Z.; Lin, T.; Lim, S.H.; Hove, M.H.; Schuh, W.M. Impacts of Bakken shale oil development on regional water uses and supply. *J. Am. Water Resour. Assoc.* **2018**, *54*, 225–239. [CrossRef]

23. Buchanan, B.P.; Auerbach, D.A.; McManamay, R.A.; Taylor, J.M.; Flecker, A.S.; Archibald, J.A.; Fuka, D.R.; Walter, M.T. Environmental flows in the context of unconventional natural gas development in the Marcellus Shale. *Ecol. Appl.* **2017**, *27*, 37–55. [CrossRef] [PubMed]

24. Vandecasteele, I.; Marí Rivero, I.; Sala, S.; Baranzelli, C.; Barranco, R.; Batelaan, O.; Lavalle, C. Impact of Shale Gas Development on Water Resources: A Case Study in Northern Poland. *Environ. Manag.* **2015**, *55*, 1285–1299. [CrossRef]

25. IPCC. Summary for Policymakers. In *Climate Change 2014: Mitigation of Climate Change: Contribution of Working Group III to the Fifth Assessment Report of the Intergovernmental Panel on Climate Change*; Cambridge University Press: Cambridge, UK; New York, NY, USA, 2014.

26. Abbott, M.B.; Bathurst, J.C.; Cunge, J.A.; O'Connell, P.E.; Rasmussen, J. An introduction to the European Hydrological System Systeme Hydrologique Europeen, "SHE", 1: History and philosophy of a physically-based, distributed modelling system. *J. Hydrol.* **1986**, *87*, 45–59. [CrossRef]

27. ALCES Group. *ALCES 5 Technical Manual*; ALCES Group: Fort McMurray, AB, Canada, 2013.

28. Alberta Environment. Current and Future Water Use in Alberta. 2007. Available online: http://www.assembly.ab.ca/lao/library/egovdocs/2007/alen/164708.pdf (accessed on 18 August 2020).

29. Gray, D.M.; Male, D.H. *Handbook of Snow: Principles, Processes, Management and Use*; Pergamon Press: Toronto, ON, Canada, 1981; p. 776.

30. DHI. *Mike She User Manual*; Reference Guide; DHI: Horsholm, Denmark, 2009; Volume 2.

31. Yan, J.J.; Smith, K.R. Simulation of integrated surface water and ground water systems—Model formulation. *Water Resour. Bull.* **1994**, *30*, 1–12. [CrossRef]

32. Abbott, M.B.; Ionescu, F. On the numerical computation of nearly-horizontal flows. *J. Hyd. Res.* **1967**, *5*, 97–117. [CrossRef]

33. DHI. *Mike 11 A Modelling System for Rivers and Channels*; Reference Manual; DHI: Horsholm, Denmark, 2017.

34. Schut, G.H. Review of interpolation methods for digital terrain modelling. *Can. Surv.* **1976**, *30*, 389–412. [CrossRef]

35. Task Committee on Hydrology Handbook of Management Group D of the American Society of Civil Engineers. *Hydrology Handbook*, 2nd ed.; ASCE: New York, NY, USA, 1996.

36. Zeng, X. Global vegetation root distribution for land modeling. *J. Hydrometeorol.* **2001**, *2*, 525–530. [CrossRef]

37. Myneni, R.; Knyazikhin, Y.; Glassy, J.; Votava, P.; Shabanov, N. *User's Guide FPAR, LAI (ESDT: MOD15A2) 8-Day Composite NASA MODIS Land Algorithm*; FPAR, LAI User's Guide, Terra MODIS Land Team (Report); Boston University: Boston, MA, USA, 2003; p. 17.

38. Duan, Q.; Sorooshian, S.; Gupta, V. Effective and efficient global optimization for conceptual rainfall-runoff models. *Water Resour. Res.* **1992**, *28*, 1015–1031. [CrossRef]

39. Pacific Climate Impacts Consortium. Statistically Downscaled Climate Scenarios. 2014. Available online: https://data.pacificclimate.org/portal/downscaled_gcms_archive/map/ (accessed on 15 April 2016).

40. Meinhausen, M.; Smith, S.J.; Calvin, K.; Daniel, J.S.; Kainuma, M.L.T.; Lamarque, J.; Matsumoto, K.; Montzka, S.A.; Raper, S.C.B.; Riahi, K.; et al. The RCP greenhouse gas concentrations and their extensions from 1765 to 2300. *Clim. Chang.* **2011**, *109*, 213–241. [CrossRef]

41. Riahi, K.; Rao, S.; Krey, V.; Cho, C.; Chirkov, V.; Fischer, G.; Kindermann, G.; Nakicenovic, N.; Rafaj, P. RCP 8.5—A scenario of comparatively high greenhouse gas emissions. *Clim. Chang.* **2011**, *109*, 33. [CrossRef]

42. Roy, L.; Leconte, R.; Brissette, F.P.; Marche, C. The impact of climate change on seasonal floods of a southern Quebec River Basin. *Hydrol. Process.* **2001**, *15*, 3167–3179. [CrossRef]

43. Saha, G.C. Climate change induced precipitation effects on water resources in the Peace Region of British Columbia, Canada. *Climate* **2015**, *3*, 264–282. [CrossRef]

44. Saha, G.C.; Li, J.; Thring, R.W.; Hirshfield, F.; Paul, S.S. Temporal dynamics of groundwater-surface water interaction under the effects of climate change: A case study in the Kiskatinaw River Watershed, Canada. *J. Hydrol.* **2017**, *551*, 440–452. [CrossRef]

45. Johnson, L.; Kralovic, P.; Romaniuk, A. *Canadian Crude Oil and Natural Gas Production and Supply Costs Outlook (2016–2036)*; Study No. 159; Canadian Energy Research Institute: Calgary, AB, Canada, 2016; p. 66.

46. Rivard, C.; Lavoie, D.; Lefebvre, R.; Sejourne, S.; Lamontagne, C.; Duchesne, M. An overview of Canadian Shale gas production and environmental concerns. *Int. J. Coal Geol.* **2014**, *126*, 64–76. [CrossRef]

47. Ycharts Inc. Average Crude Oil Spot Price. 2015. Available online: https://ycharts.com/indicators/average_crude_oil_spot_price (accessed on 22 August 2015).

48. Alberta Energy Regulator. Alberta Drilling Activity Monthly Statistics. December 2013. Available online: https://www.aer.ca/documents/sts/st59/ST59-2013.pdf (accessed on 15 August 2016).

49. Alberta Energy Regulator. Alberta Drilling Activity Monthly Statistics. December 2014. Available online: https://www.aer.ca/documents/sts/st59/ST59-2014.pdf (accessed on 15 August 2016).

50. Albek, M.; Ogutveren, U.B.; Albek, E. Hydrological modeling of Seydi Suyu watershed (Turkey) with HSPF. *J. Hydrol.* **2004**, *285*, 260–271. [CrossRef]

51. Carlson, M.; Stelfox, B.; Purves-Smith, N.; Straker, J.; Berryman, S.; Braker, T.; Wilson, B. ALCES online: Web-delivered scenario analysis to inform sustainable land-use decisions. In Proceedings of the 7th International Environmental Modelling and Software Society, San Diego, CA, USA, 15–19 June 2014.

52. Wijesekara, G.N.; Gupta, A.; Valeo, C.; Hasbani, J.G.; Qiao, Y.; Delaney, P.; Marceau, D.J. Assessing the impact of future land-use changes on hydrological processes in the Elbow River watershed in southern Alberta, Canada. *J. Hydrol.* **2012**, *412–413*, 220–232. [CrossRef]

53. ALL Consulting. *Horizontal Drilling and Hydraulic Fracturing Considerations for Shale Gas Wells*; Bureau D'audiences Publiques Sur L'environement (BAPE): Saint-Hyacinthe, QC, Canada, 2010.

54. NYSDEC. Statewide Spacing Unit Sizes and Setbacks. 2013. Available online: https://www.dec.ny.gov/energy/1583.html (accessed on 25 November 2016).

55. Christensen, J.H.; Christensen, O.B. A summary of the PRUDENCE model projections of changes in European climate by the end of this century. *Clim. Chang.* **2007**, *81*, 7–30. [CrossRef]

56. Santhi, C.; Arnold, J.G.; Williams, J.R.; Dugas, W.A.; Srinivasan, R.; Hauck, L.M. Validation of the SWAT model on a large river basin with point and nonpoint sources. *J. Am. Water Resour. Assoc.* **2001**, *37*, 1169–1188. [CrossRef]

57. IPCC. *IPCC Special Report*; Emission Scenarios; Cambridge University Press: Cambridge, UK, 2000; p. 570.

58. Drohan, P.J.; Brittingham, M.; Bishop, J.; Yoder, K. Early trends in land cover change and forest fragmentation due to shale-gas development in Pennsylvania: A potential outcome for the Northcentral appalachains. *Environ. Manag.* **2012**, *49*, 1061–1075. [CrossRef] [PubMed]

59. Guo, H.; Hu, Q.; Jiang, T. Annual and seasonal streamflow responses to climate and land-cover changes in the Poyang Lake basin, China. *J. Hydrol.* **2008**, *355*, 106–112. [CrossRef]

60. Muhammad, A.; Evenson, G.R.; Unduche, F.; Stadnyk, T.A. Climate change impacts on reservoir inflow in the Prairie Pothole Region: A watershed model analysis. *Water* **2020**, *12*, 271. [CrossRef]

61. Saha, G.C. Investigation of Groundwater Contribution to Stream Flow under Climate and Land Use Changes: A Case Study in British Columbia, Canada. *Int. J. Environ. Clim. Chang.* **2015**, *5*, 1–22. [CrossRef]

62. Price, K.; Leigh, D.S. Morphological and sedimentological responses of streams to human impact in the southern Blue Ridge Mountains, USA. *Geomorphology* **2006**, *78*, 142–160. [CrossRef]

63. The Freshwater Blog. How Groundwater Influences Europe's Surface Waters. 2017. Available online: https://freshwaterblog.net/2017/01/13/how-groundwater-influences-europes-surface-waters/ (accessed on 16 April 2018).

64. Evans, J.E.; Prepas, E.E.; Devito, K.J.; Kotak, B.G. Phosphorus dynamics in shallow subsurface waters in an uncut and cut subcatchment of a lake on the Boreal Plain. *Can. J. Fish. Aquat. Sci.* **2000**, *57* (Suppl. S2), 60–72. [CrossRef]

 *hydrology*

Article

Analysis of Groundwater Level Variations Caused by the Changes in Groundwater Withdrawals Using Long Short-Term Memory Network

Mun-Ju Shin *, Soo-Hyoung Moon, Kyung Goo Kang, Duk-Chul Moon and Hyuk-Joon Koh

Water Resources Research Team, Jeju Province Development Corporation, 1717-35, Namjo-ro, Jocheon-eup, Jeju-si, Jeju-do 63345, Korea; justdoit74@jpdc.co.kr (S.-H.M.); kgkang11@jpdc.co.kr (K.G.K.); waterfeel@jpdc.co.kr (D.-C.M.); hjkoh@jpdc.co.kr (H.-J.K.)
* Correspondence: munjushin@jpdc.co.kr; Tel.: +82-(0)-64-7803-751

Received: 26 August 2020; Accepted: 4 September 2020; Published: 7 September 2020

Abstract: To properly manage the groundwater resources, it is necessary to analyze the impact of groundwater withdrawal on the groundwater level. In this study, a Long Short-Term Memory (LSTM) network was used to evaluate the groundwater level prediction performance and analyze the impact of the change in the amount of groundwater withdrawal from the pumping wells on the change in the groundwater level in the nearby monitoring wells located in Jeju Island, Korea. The Nash–Sutcliffe efficiency between the observed and simulated groundwater level was over 0.97. Therefore, the groundwater prediction performance of LSTM was remarkably high. If the groundwater level is simulated on the assumption that the future withdrawal amount is reduced by 1/3 of the current groundwater withdrawal, the range of the maximum rise of the groundwater level would be 0.06–0.13 m compared to the current condition. In addition, assuming that no groundwater is taken, the range of the maximum increase in the groundwater level would be 0.11–0.38 m more than the current condition. Therefore, the effect of groundwater withdrawal on the groundwater level in this area was exceedingly small. The method and results can be used to develop new groundwater withdrawal sources for the redistribution of groundwater withdrawals.

Keywords: Long Short-Term Memory; groundwater level prediction; groundwater withdrawal impact; groundwater level variation; machine learning

---

1. Introduction

Groundwater is an important water resource that can be used in tandem with surface water, and research on groundwater is especially important in island areas because it comprises most of the water for used living and agriculture. On Jeju Island in Korea, groundwater is an exceedingly important water resource that accounts for about 80% of the total water resource use. Therefore, research on the effects of withdrawals on groundwater levels is especially important for its continuous and stable use. The variability in the groundwater level is influenced by seasonal variations in precipitation, groundwater withdrawals, and river water level changes [1]. Jeju Island is a region with a high amount of precipitation, but most of its rivers are dry streams because of the soil's high water permeability, so the water level of the rivers has a negligible impact on the groundwater level. Therefore, to provide information for the proper management of Jeju Island's groundwater resources, it is necessary to analyze the effects of precipitation and withdrawals on the groundwater level.

Excessive withdrawals from a specific pumping well in an island region creates problems for sustainable groundwater yields, such as rapid decreases in groundwater levels and seawater intrusions [2]. To solve these problems, Oki [2,3] conducted a study to reduce groundwater level decreases by redistributing withdrawals using a numerical model and various withdrawal scenarios. As a result, groundwater level increased as the withdrawal amount decreased. Therefore, a direct relationship between the thickness of the aquifer and the groundwater withdrawal reduction amount as well as the effect of the withdrawal redistribution were confirmed. To analyze the impact of groundwater withdrawals using a numerical model, spatial data with various physical properties such as detailed land use maps and geological maps for the study area are required; however, in reality, these data are often insufficient. If an explicit model-driven approach cannot be used because of the difficulty of collecting these heterogeneous data, the explicit model-driven approach can be supplemented through an implicit data-driven approach, i.e., machine learning [4,5].

Various machine learning models can be used for groundwater analysis, from the traditional Artificial Neural Network (ANN) and Support Vector Machine (SVM) to novel deep learning methods such as Long Short-Term Memory (LSTM). ANN has been widely used in various research fields as a representative machine learning method [6]. However, ANN cannot learn long-term data (long-term dependency problem) [7,8], so the appropriate prediction period is not as long as 1 to 3 time steps (e.g., one- to three-day prediction) [9]. The SVM proposed by Vapnik [10] is more predictive than ANN [11] and is a data-driven model widely used in the field of machine learning [12]. A previous study analyzed the effect of changes in groundwater withdrawals on the extent of seawater intrusion into coastal aquifers using SVM [13], but SVM reportedly demonstrated a lower predictive capacity than LSTM [14].

The purpose of this study was to evaluate the groundwater prediction performance of the LSTM network [15], which comprises a novel recurrent neural network (RNN) that can remember input data over a long period [16] by solving the long-term dependency problem [17] of existing RNNs. Moreover, we analyzed the effects of changes in withdrawals from pumping wells on changes in the groundwater level of monitoring wells using LSTM to reduce decreases in groundwater supplies. LSTM is a deep learning model suitable for learning continuous data, such as time series data, and has recently been used in various groundwater analysis studies [18–25]. The details of the LSTM are described in Section 2.2.

2. Materials and Methods

2.1. Study Area and Data

The study area comprised of two groundwater monitoring well points located in the central mountainous region of Pyoseon watershed in the southeast portion of Jeju Island (Figure 1). This study used daily precipitation data from the Seongpanak and Gyorae rainfall stations in the vicinity of groundwater monitoring wells, daily groundwater withdrawal data from two groundwater pumping wells (pumping well 1 (PW1) and pumping well 2 (PW2)), and daily groundwater level data from two groundwater monitoring wells (monitoring well 1 (MW1) and monitoring well 2 (MW2)) (Table 1). Data from the Seongpanak rainfall station (automatic weather station), which is operated by the Korea Meteorological Administration (http://www.weather.go.kr/), and Gyorae rainfall station, which is operated by the Jeju Island Disaster and Safety Countermeasures Headquarters (http://bangjae.jeju119. go.kr/), are provided online. The withdrawal data of the pumping wells and the groundwater level data of the monitoring wells were observed and operated by the Jeju Province Development Corporation. The locations of rainfall stations, groundwater pumping wells, and groundwater monitoring wells in the study area are shown in a diagram (Figure 1). Precipitation at the rainfall stations and groundwater level in the monitoring wells are shown in Figures 2 and 3, respectively. In Figure 2, the precipitation at the Seongpanak rainfall station is greater than that at the Gyorae rainfall station. Accordingly, the Seongpanak rainfall station (El. 763 m, Figure 1) is located at a higher elevation than the Gyorae rainfall station (El. 400 m, Figure 1), and the former is more affected by the orographic lift effect than the latter, resulting in greater precipitation. In Figure 3, the degrees in groundwater level variation of MW1 and MW2 are similar; however, the groundwater level of MW1 is higher than that of MW2.

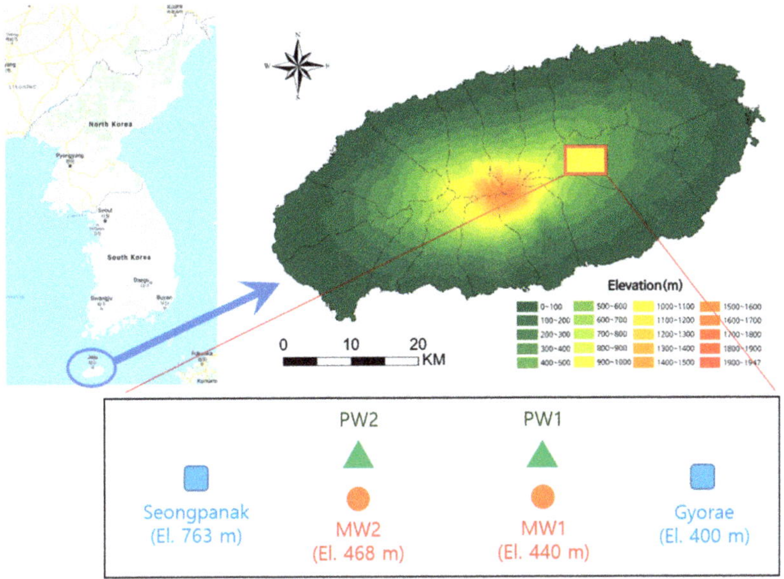

Figure 1. Schematic of the positions of rainfall stations, groundwater pumping wells, and groundwater monitoring wells.

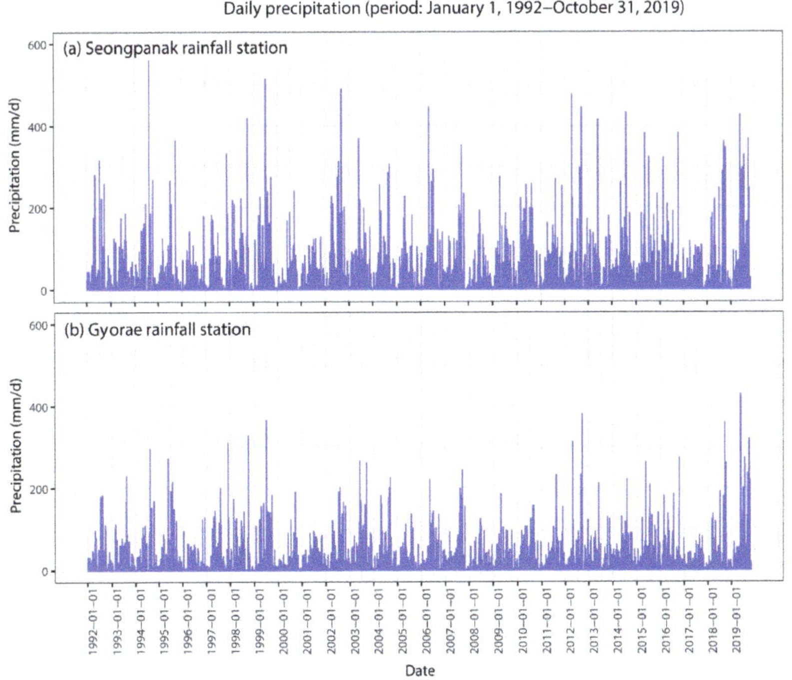

Figure 2. Comparison of daily precipitation in Seongpanak and Gyorae rainfall stations.

Table 1. Data period of rainfall stations, groundwater pumping wells, and groundwater monitoring wells.

Classification	Station Name	Data Period	Remarks
Rainfall Station	Seongpanak	1 January 1992–31 October 2019	Precipitation (mm/day)
	Gyorae	1 January 1992–31 October 2019	
Groundwater Pumping Well	PW1	1 January 2001–31 October 2019	Groundwater pumping rate (m³/day)
	PW2	31 July 2013–31 October 2019	
Groundwater Monitoring Well	MW1	11 February 2001–31 October 2019	Groundwater level (m)
	MW2	19 March 2012–31 October 2019	

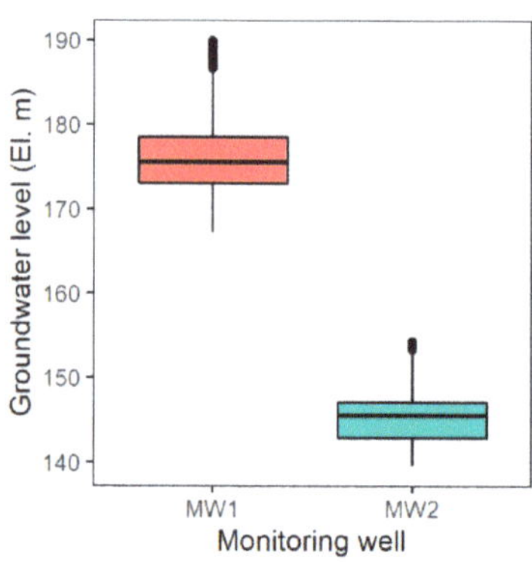

Figure 3. Boxplot for the variability of groundwater level in monitoring wells.

2.2. Long Short-Term Memory (LSTM)

Long Short-Term Memory [15] is a modification of the RNN developed to solve the vanishing gradient problem [17], which makes it impossible for RNN to learn long-term dependencies. LSTM uses a conveyor belt (carry track, Figure 4) placed parallel to the sequence being processed so that the information from the sequence can be transported to the carry track at any point in time and then transported to a later time to be reused when needed. The information is stored in a carry track for later use, and therefore, there is no problem if the old signal gradually disappears during the calculation process [26]. LSTM was developed to remember long-term information and has a recurrent structure similar to the existing standard RNN, but the former differs from the latter in that it learns long-term information using four unique transformations. First, the output of the cell for time t ($output_t$) is calculated as follows:

$$output_t = activation(Wo{\cdot}input_t + Uo{\cdot}state_t + Vo{\cdot}c_t + bo) \tag{1}$$

where $input_t$ is the input data for time t; $state_t$ is the state for time t, which is the state of output for time $t-1$; c_t is the carry value for time t; Wo, Uo, and Vo are the weight matrices of $input_t$, $state_t$, and c_t, respectively, for output calculation; and $\cdot$ is the dot product. The weights are the parameters of the LSTM and play the most important role in calculating the output. bo is the bias followed by the weight matrices, and *activation* is the activation function using the sigmoid (σ) function and *tanh* function.

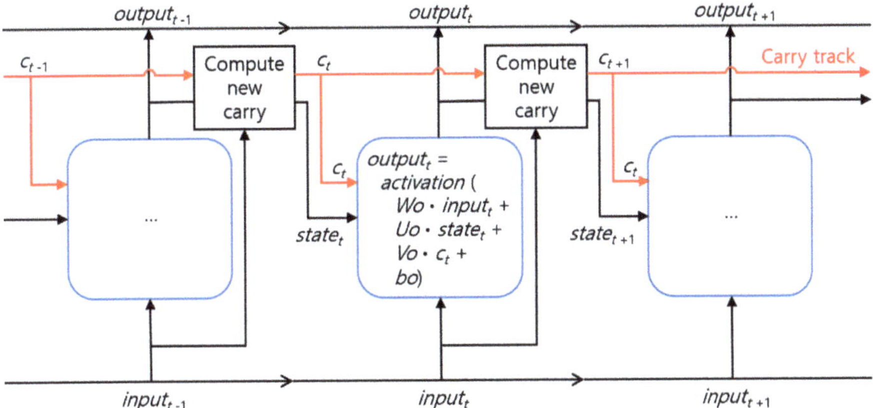

Figure 4. LSTM architecture. Reproduced from Chollet and Allaire [26].

The carry value is updated using three separate transformations, and all three individual transformations take the form of a simple RNN cell. The three individual transformations are as follows, all of which have unique weight (parameter) matrices.

$$i_t = \sigma(Wi{\cdot}input_t + Ui{\cdot}state_t + bi) \tag{2}$$

$$f_t = \sigma(Wf{\cdot}input_t + Uf{\cdot}state_t + bf) \tag{3}$$

$$k_t = tanh(Wk{\cdot}input_t + Uk{\cdot}state_t + bk) \tag{4}$$

$$c_{t+1} = i_t k_t + c_t f_t \tag{5}$$

where i_t is the newly added information (ranging from 0 to 1) through a sigmoid function (σ), f_t is the deleted information (ranging from 0 to 1) through a sigmoid function, and k_t is the importance of information (ranging from −1 to 1) through a *tanh* function. LSTM multiplies i_t and k_t to obtain new information, c_t and f_t are multiplied to remove irrelevant carry information, and finally $i_t k_t$ and $c_t f_t$ are added to obtain a new carry value. Therefore, LSTM uses the carry track to modulate the next output and next state [26].

Figure 5 shows LSTM's supervised learning procedure. The data input to the layer of the LSTM is converted by weights that are parameters of the layer. The groundwater level simulated by LSTM is compared with the observed groundwater level using the objective function (here loss function), and the weights are adjusted (updated) through the optimizer until the simulated groundwater level is closest to the observed groundwater level (i.e., until loss is minimized). The process of gradually adjusting the weights is called training, and LSTM is learned through training. Therefore, the weights contain the learned information, and the optimizer uses the backpropagation algorithm. In this study, LSTM included in the Keras package [27], which is an R language-based deep learning framework, was used.

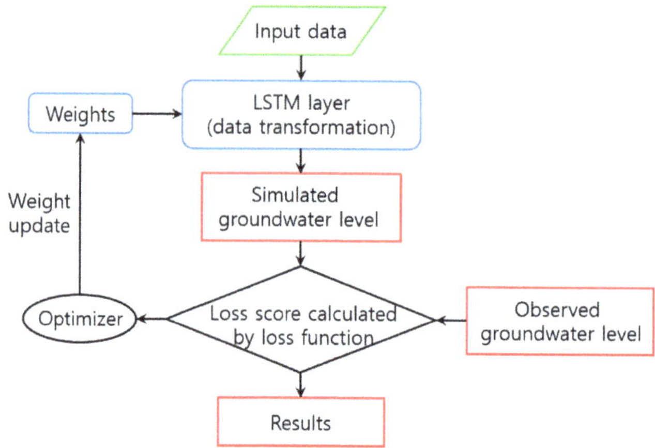

Figure 5. LSTM modeling process. Reproduced from Chollet and Allaire [26].

2.3. Method

First, this study predicted the groundwater level for one day ahead via the LSTM for groundwater monitoring wells and evaluated the model's predictive performance. The reason for selecting the one day ahead prediction was that the LSTM had the highest predictive performance with this prediction type, so the calibrated LSTM parameters best reflected the variation characteristics of the observed groundwater level. In addition, the reason that periods of more than one day ahead were not predicted is that the purpose of this study is not to predict long term future groundwater levels via LSTM, but to analyze the effect of changes in withdrawals on variations in groundwater levels. To prevent overfitting of LSTM parameters with respect to training period data among the entire period of observation data, the estimated parameters were validated using validation period data through a callback method. In addition, test periods were used to evaluate the predictive performance of the LSTM. Training, validation, and test periods were deemed independent, and each period is shown in Table 2.

Table 2. Training, validation, and test periods for LSTM modeling.

Station Name	Training Period	Validation Period	Test Period
MW1 [a]	11 February 2001–31 December 2013	1 January 2014–31 December 2016	1 January 2017–31 October 2019
MW2 [b]	31 July 2013–31 December 2015	1 January 2016–31 December 2017	1 January 2018–31 October 2019

[a] The training, validation, and test periods of precipitation and groundwater withdrawal data for simulating the groundwater level in MW1 were the same as the respective periods in MW1. [b] The training, validation, and test periods of precipitation and groundwater withdrawal data for simulating the groundwater level in MW2 were the same as the respective periods in MW2.

For the training of LSTM, the values of hyper-parameters required for model building must be set, and there are no clear instructions on how to set them [26]. In this study, hyper-parameters were set as shown in Table 3. The n_timesteps denote the number of prediction days, and as described above, the number of prediction days was set to 1. The n_units represent the number of hidden units in the LSTM layer, i.e., the dimension of the layer, which denotes the degree of freedom a neural network can have for learning [26]. The larger the value of n_units, the more complicated the data that can be learned; thus, the convergence of the objective function values is faster, but the calculation time is longer and overfitting problems for training data can occur. In this study, dropout and callback methods were used to prevent overfitting, so sufficient n_units were used. As a result of trial and error, the n_units amount was set to 100. LSTM does not process the entire dataset during the training period at one time, thus dividing it into a small sample group, or a mini batch, to process the dataset during the training period. The batch_size refers to the number of data points in the mini batch to be processed once in the training dataset for training of LSTM. In this study, this amount was set to 50 considering the characteristics of daily training data and the duration of training data. A dropout was used to prevent overfitting during the training process by randomly removing multiple output features of the layer [28]; generally, dropouts use a value between 0.2 and 0.5. In this study, both the dropout and recurrent_dropout were set to 0.5 as a result of trial and error. For the optimization algorithm, Adam [29] was used, which is widely used for optimization in recent deep learning fields [30], and the mean absolute error was used for the objective function. Adam optimizer uses the stochastic gradient decent procedure. For the learning_rate, which denotes the learning rate of the Adam optimizer, the default value of 0.001 was used. For the optimization process, one iteration for the entire training dataset is called an epoch, and the n_epochs, which defines the maximum number of iterations, was set to 50 as a result of trial and error. The callback is a method of early termination when simulation results are no longer improved by repeating training as much as an arbitrarily selected threshold value. The patience, which is a hyper-parameter for the callback method, denotes this threshold, and the patience was set to 10 as a result of trial and error.

Table 3. Setting values of hyper-parameters in LSTM.

Hyper-Parameter	Range	Setting Value	Description
n_timesteps	-	1	Number of prediction step
n_units	-	100	Number of hidden units in LSTM layer
batch_size	-	50	Number of samples fed to LSTM in one sub-simulation
dropout	0–1	0.5	Fraction of the units to drop for the linear transformation of the inputs
recurrent_dropout	0–1	0.5	Fraction of the units to drop for the linear transformation of the recurrent state
learning_rate	float ≥ 1	0.001	Learning rate of Adam optimizer
n_epochs	-	50	Number of iterations
patience	-	10	Number of epochs for early termination of training when simulation values do not improve

The Nash–Sutcliffe efficiency (NSE) [31] and root mean square error (RMSE), which are widely used in the field of hydrology, were used to evaluate the simulation performance of LSTM by comparing the simulated groundwater level with the observed groundwater level. The NSE provides overall information on simulation results [32], and RMSE indicates how closely simulated values match observed values [30]. The NSE and RMSE are defined as follows:

$$NSE = 1 - \frac{\sum_{i=1}^{n}\left(Q_{obs,i} - Q_{sim,i}\right)^2}{\sum_{i=1}^{n}\left(Q_{obs,i} - \overline{Q_{obs}}\right)^2} \tag{6}$$

$$RMSE = \sqrt{\frac{1}{n}\sum_{i=1}^{n}\left(Q_{obs,i} - Q_{sim,i}\right)^2} \tag{7}$$

where n is the number of time steps, $Q_{obs,i}$ and $Q_{sim,i}$ are the observed and simulated groundwater levels in time step i (daily here), respectively, and $\overline{Q_{obs}}$ is the average value of the observed groundwater level. The range of NSE is $-\infty$ to 1; 1 means that the simulated values exactly match the observed values, and 0 means that the simulated values are equal to the mean of the observed values. An RMSE of 0 means that the simulated values exactly match the observed values.

Second, the estimated parameter values and groundwater withdrawal scenarios were used to calculate the groundwater level of the monitoring wells and to analyze the effect of changes in withdrawals on the groundwater level variability. Withdrawal scenarios were set based on the assumption that a new groundwater pumping well will be constructed in the future to reduce the decrease in the groundwater level in the monitoring wells caused by the withdrawals from pumping wells. We assumed that the current withdrawal rate for both PW1 and PW2 was 2300 m^3/d, and for the future scenario, we assumed that PW1 and PW2 would take as much as 2/3 of the current withdrawal rate considering the development of a new pumping well. Therefore, withdrawal Scenario1 is a case where no groundwater is taken, Scenario2 is a case where the groundwater of 1533 m^3/d (2/3 of 2300 m^3/d) is taken daily, and Scenario3 is a case where the groundwater of 2300 m^3/d is taken daily. This study used these three withdrawal scenarios to analyze the effect of the amount of withdrawal reduction on the degree of groundwater level increase and the linearity or nonlinearity between the amount of withdrawal reduction and the groundwater level increase. In addition, this study analyzed the limitations of LSTM, a data-driven approach, for modeling groundwater levels.

3. Results

3.1. Analysis of the Prediction Performance of LSTM

Figures 6 and 7 show the one-day prediction performance of the groundwater level by LSTM for the training, validation, and test periods of MW1 and MW2, respectively. In the figures, the horizontal axis represents the observed groundwater level, the vertical axis represents the simulated groundwater level, and the red dashed line represents the one-to-one line.

In the case of MW1 with a relatively long data period (approximately 19 years), the one-day prediction performances of groundwater level by LSTM for the training, validation, and test periods were exceedingly high, with NSE values of 0.99 or higher (Figure 6, Table 4). This was because the training period required for model calibration was sufficiently long (approximately 13 years), and the training period contained enough information to predict the groundwater level during the validation and test periods.

In the case of MW2 with a relatively short data period (approximately 6 years), the one-day groundwater level prediction performances by LSTM for training and validation periods were also high, with NSE values of 0.99 or higher (Figure 7, Table 4). However, the prediction performance for the test period was relatively low, with an NSE of 0.976, and the one-day prediction performance for relatively high groundwater levels above 150 m was relatively low (Figure 7, Table 4). This was because the training period required for model calibration was relatively short (approximately 2 years and 6 months), so the information designated for predicting the relatively high observed groundwater level included in the test period was not sufficient for the training period.

Although the groundwater level prediction performance during the test period of MW2 was relatively lower than that of MW1, the NSE was 0.97 or higher and the RMSE was 0.5 or lower, showing a sufficiently high prediction performance. In addition, the results of MW1 also showed a high prediction performance. Therefore, LSTM's ability to predict groundwater levels was sufficiently high. This study analyzed the effect of changes in groundwater withdrawal on variations in groundwater level using estimated parameter values, as shown in Section 3.2.

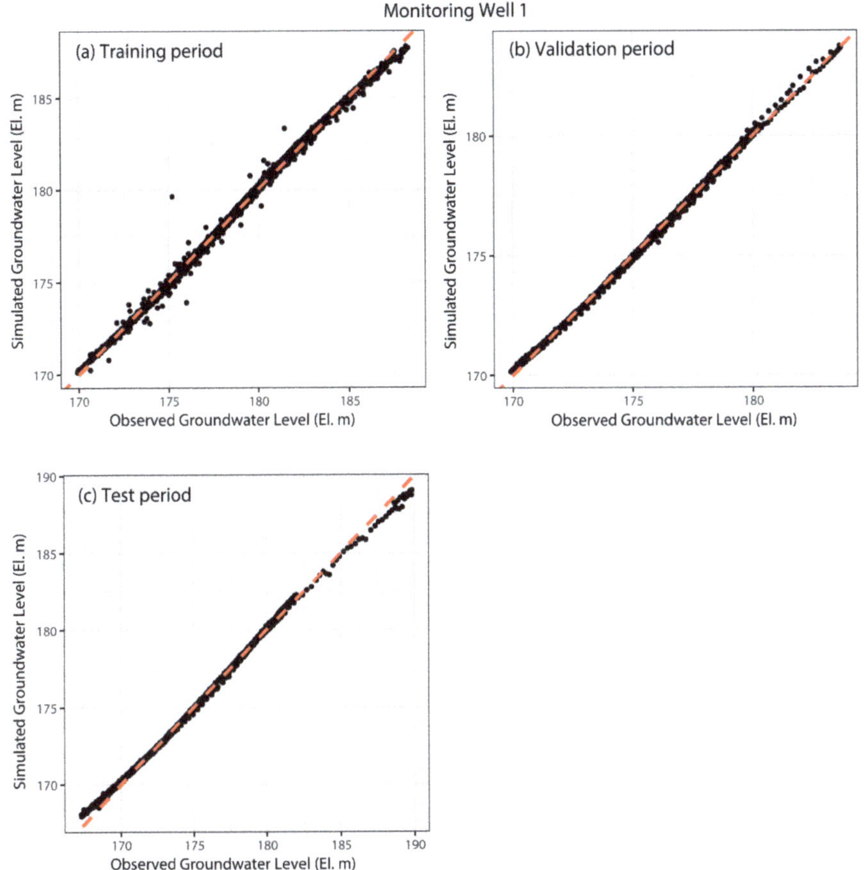

Figure 6. Simulated and observed groundwater levels for monitoring well 1.

Table 4. Modeling performance of LSTM for the one-day groundwater level prediction.

Monitoring Well	Statistics	Training Period	Validation Period	Test Period
MW1	NSE	0.998	0.999	0.995
	RMSE	0.166	0.120	0.327
MW2	NSE	0.999	0.998	0.976
	RMSE	0.084	0.103	0.494

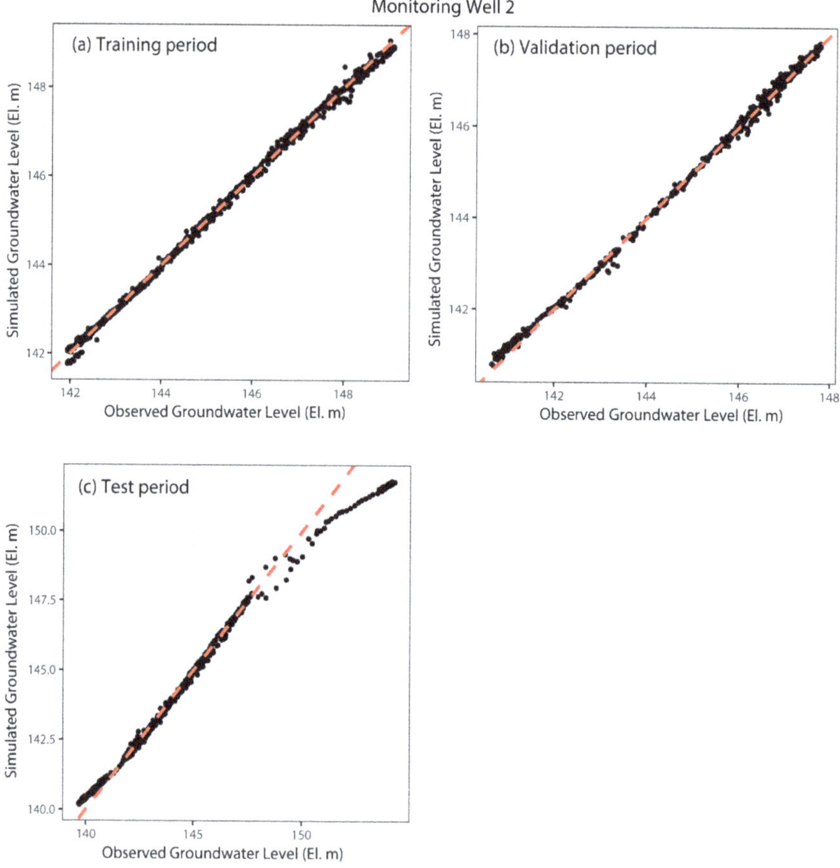

Figure 7. Simulated and observed groundwater levels for monitoring well 2.

3.2. Analysis of the Effect of Changes in Groundwater Withdrawals on the Variations in Groundwater Levels

Figure 8 shows the representative section (1–30 September 2018) of the simulated groundwater level variations according to the change in the withdrawal amounts for MW1 and MW2. As expected, the groundwater level was highest for Scenario1 with no withdrawal, followed by Scenario2 (withdrawal of 1533 m^3/d) and Scenario3 (withdrawal of 2300 m^3/d).

Table 5 shows the differences between the simulated groundwater levels by withdrawal scenarios for the entire data period for each monitoring well. In the case of MW1, when the withdrawal amount was reduced by 1/3 through taking 1533 m^3/d (Scenario2) from the withdrawal of 2300 m^3/d (Scenario3), the increase in groundwater level was estimated to be up to 0.13 m and an average of 0.03 m. In addition, when the groundwater was not taken (Scenario1), the groundwater level rose to 0.38 m and an average of 0.09 m compared to the case where 2300 m^3/d was withdrawn (Scenario3). Therefore, the increase in groundwater level between Scenario1 and Scenario3 was estimated to be approximately three times larger than that for Scenario2 and Scenario3. This linear relationship between the change in the withdrawal amount and the variation in the groundwater level indicates that the permeability of the subsurface geology and the groundwater yield around MW1 are high, so the groundwater level will increase as the withdrawal amount decreases. In addition, the maximum difference in groundwater level between Scenario1 and Scenario3 was 0.38 m, which means that in this area, the effect of the withdrawal on the groundwater level was small.

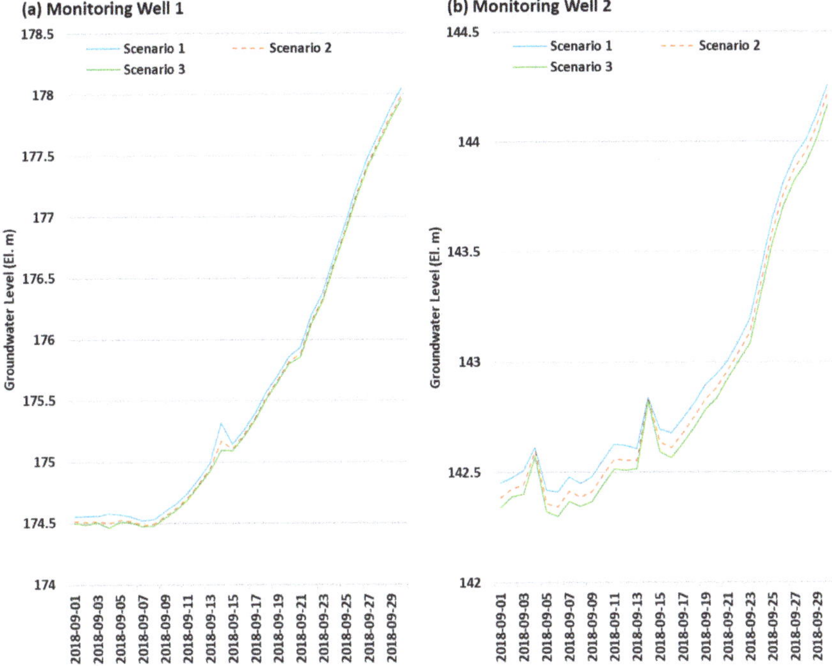

Figure 8. Groundwater levels predicted using LSTM and groundwater withdrawal scenarios for MW1 and MW2.

Table 5. Differences between the simulated groundwater levels in groundwater withdrawal scenarios for the entire data period.

Monitoring Well	Statistics	Groundwater Level Difference (M) (Scenario1–Scenario3)	Groundwater Level Difference (M) (Scenario2–Scenario3)
MW1	Mean	0.09	0.03
	Max	0.38	0.13
MW2	Mean	0.09 [a]	0.05
	Max	0.11 [a]	0.06

[a] The statistical values for the difference in groundwater level between Scenario1 and Scenario3 of MW2 represent the results of using only the simulated groundwater levels, except for periods when the groundwater level simulated by Scenario1 was lower than the groundwater levels simulated by Scenario2 and Scenario3 (approximately 2.5% of the entire data period).

In the case of MW2, when the withdrawal amount was reduced by 1/3 from 2300 m³/d (Scenario2–Scenario3), the increase in groundwater level was estimated to be 0.06 m at the maximum and 0.05 m on average. In addition, when the groundwater was not taken (Scenario1), the groundwater level increased to 0.11 m and an average of 0.09 m compared to the case where the groundwater was taken at 2300 m³/d (Scenario3). Therefore, the increase in groundwater level between Scenario1 and Scenario3 was estimated to be twice that for Scenario2 and Scenario3. Therefore, given the uneven differences in reduction rates and groundwater level increases between scenarios, the change in groundwater withdrawal had a nonlinear relationship with the variation in groundwater level. This means that the geological properties around MW2 are different from those of MW1, and therefore, there is a difference in permeability and groundwater yield. In addition, in the case of MW2, the maximum difference in groundwater level between Scenario1 and Scenario3 was only 0.11 m, and hence, the effect of the withdrawal rate on the groundwater level in this area was exceedingly small.

However, in the case of MW2, the periods when the groundwater level simulated by Scenario1 was lower than the groundwater levels simulated by Scenario2 and Scenario3 (inversion phenomenon of simulated groundwater level) accounted for approximately 2.5% of the entire data period (Figure 9). Therefore, LSTM had limitations in its ability to simulate the groundwater level for the groundwater withdrawal scenarios. However, the period in which the inversion of the simulated groundwater level occurred was short, at approximately 2.5% of the entire simulation period. Hence, the effect of this inversion on calculating the difference in groundwater level for Scenario1–Scenario3 of MW2 was minor. Therefore, the statistical values for Scenario1–Scenario3 of MW2 in Table 5 represent the results of calculating the difference in groundwater level using simulated groundwater level data, excluding the inversion period.

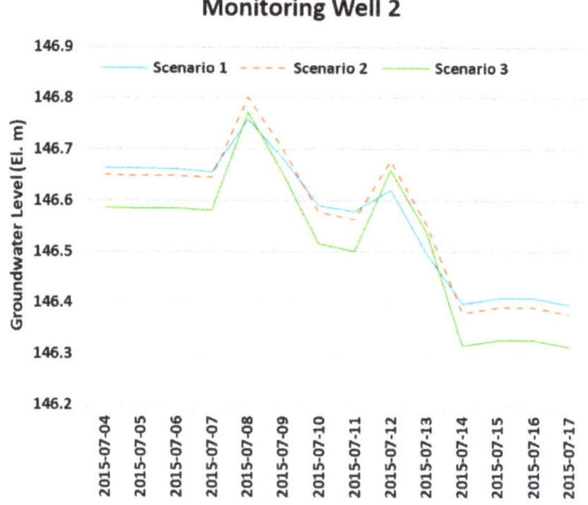

Figure 9. Inversion of simulated groundwater levels using the groundwater withdrawal scenarios for MW2.

4. Discussion

The training period of LSTM affected the prediction performance of the groundwater level during the test period for two monitoring wells. The implicit data-driven approach, LSTM, is not a hydrological model with an explicit model-driven approach. However, as per the case of the split sample test for hydrological models by Klemeš [33], and as suggested in the study by Yapo et al. [34], when data of a sufficiently long period of 8 years or more are used for the training of the model, this is a sufficient condition for model validation and testing. Research on estimating a sufficient training period for LSTM is necessary to obtain a proper test result of the model, but this is outside the scope of this study and will be performed in the future.

The LSTM, a data-driven approach, does not use a hydrological process in the modeling process. Therefore, for some periods, a groundwater level inversion occurred, in which the groundwater level without withdrawal was lower than that with withdrawal. Because the analysis of the simulated groundwater level inversion problem by LSTM was outside the scope of this study, the cause of this phenomenon will be analyzed through an uncertainty analysis of machine learning [35] in the future.

In this study, the NSE of one-day prediction by LSTM for the validation period ranged from 0.999 to 0.998, and was therefore higher than the NSE of 0.977, which was the result of ANN [9]. Of course, the study of Hidayat et al. [9] predicted river runoff in Indonesia, and the target areas and types of data used were different from those of this study. In addition, the study of Gangwar et al. [14], which claimed that LSTM had higher predictive performance than SVM, was a study on wind speed prediction, so the prediction target is different from that of this study. Therefore, comparative analysis between LSTM and other machine learning methods using the same data will be performed in the future.

5. Conclusions

In this study, the results of one-day groundwater level predictions via LSTM were evaluated for two groundwater monitoring wells located in the central mountainous area of the Pyoseon watershed on Jeju Island, Korea. Moreover, the effects of changes in groundwater withdrawal rates through withdrawal scenarios on the variability of groundwater levels were analyzed.

As a result of the groundwater level predictions for training, validation, and test periods via LSTM, NSE values were 0.99 or higher for MW1, with a relatively long data period (approximately 19 years); NSE values were 0.97 or higher for MW2, with a relatively short data period (approximately 6 years). Therefore, the overall prediction performance of LSTM was sufficiently high.

As a result of analyzing the effect of changes in the amount of withdrawal rates on the variations in the groundwater level, the groundwater level was highest in the order of not withdrawing groundwater (Scenario1), withdrawing at 1533 m^3/d (Scenario2), and withdrawing at 2300 m^3/d (Scenario3). Therefore, LSTM properly simulated the variations in groundwater levels according to the change in the withdrawal amount.

When the withdrawal amount was reduced by 1/3 from 2300 m^3/d to 1533 m^3/d, the groundwater level increase for MW1 was 0.13 m maximum and 0.03 m on average, and that for MW2 was 0.06 m maximum and 0.05 m on average. In addition, when the withdrawal amount changed from 2300 m^3/d to 0 m^3/d, the groundwater level increase for MW1 was 0.38 m maximum and 0.09 m on average, and those for MW2 were 0.11 m and 0.09 m, respectively. Therefore, the variation in the groundwater level was within a maximum of 0.38 m, and therefore, the effect of the withdrawal rate in this area was exceedingly small.

In the case of MW1, there was a linear relationship between the change in the withdrawal rate and the variation in the groundwater level; however, in the case of MW2, there was a nonlinear relationship between them. Therefore, there was a difference in the permeability of the subsurface geology and the groundwater yield for the surrounding areas of MW1 and MW2.

However, in the case of MW2, the periods when the groundwater level simulated by Scenario1 was lower than the groundwater levels simulated by Scenario2 and Scenario3 accounted for approximately 2.5% of the entire simulation period. This means that LSTM, which is a data-driven approach, does not use a hydrological process when simulating groundwater levels, and thus it has limitations in simulating groundwater levels for withdrawal scenarios. The reversal of the simulated groundwater level via LSTM will be analyzed through an uncertainty analysis of machine learning in the future. The method and results of analyzing the impact of groundwater withdrawal by LSTM in this study can be useful in the development of new groundwater pumping wells for redistributing groundwater withdrawals in the future.

Author Contributions: Conceptualization, M.-J.S. and H.-J.K.; Methodology, M.-J.S.; Software, M.-J.S.; Validation, M.-J.S.; Formal Analysis, M.-J.S.; Investigation, M.-J.S.; Resources, M.-J.S. and D.-C.M.; Data Curation, M.-J.S. and D.-C.M.; Writing—Original Draft Preparation, M.-J.S.; Writing—Review and Editing, M.-J.S., S.-H.M., K.G.K., D.-C.M. and H.-J.K.; Visualization, M.-J.S.; Supervision, M.-J.S. and S.-H.M.; Project Administration, S.-H.M. and K.G.K. All authors have read and agreed to the published version of the manuscript.

Funding: This research received no external funding.

Acknowledgments: We would like to thank the researcher Hee-Joo Han of Jeju Province Development Corporation for helping us create a map of the location of the study area.

Conflicts of Interest: The authors declare no conflict of interest.

References

1. Todd, D.K.; Larry, W.M. *Groundwater Hydrology*, 3rd ed.; John Wiley & Sons Inc.: Hoboken, NJ, USA, 2004; pp. 1–656.
2. Oki, D.S. *Numerical Simulation of the Effects of Low-Permeability Valley-Fill Barriers and the Redistribution of Ground-Water Withdrawals in the Pearl Harbor Area, Oahu, Hawaii: U.S. Geological Survey Scientific Investigations Report*; 2005-5253; U.S. Department of the Interior: Washington, DC, USA, 2005; p. 111.
3. Oki, D.S. *Numerical Simulation of the Hydrologic Effects of Redistributed and Additional Ground-Water Withdrawal, Island of Molokai, Hawaii: U.S. Geological Survey Scientific Investigations Report*; 2006–5177; U.S. Department of the Interior: Washington, DC, USA, 2006; p. 49.
4. Marçais, J.; De Dreuzy, J.R. Prospective Interest of Deep Learning for Hydrological Inference. *Ground Water* **2017**, *55*, 688–692. [CrossRef]
5. Siade, A.J.; Cui, T.; Karelse, R.N.; Hampton, C. Reduced-Dimensional Gaussian Process Machine Learning for Groundwater Allocation Planning Using Swarm Theory. *Water Resour. Res.* **2020**, *56*, e2019WR026061. [CrossRef]
6. Nayak, P.C.; Satyaji Rao, Y.R.; Sudheer, K.P. Groundwater Level Forecasting in a Shallow Aquifer Using Artificial Neural Network Approach. *Water Resour. Manag.* **2006**, *20*, 77–90. [CrossRef]
7. Fang, K.; Shen, C.; Kifer, D.; Yang, X. Prolongation of SMAP to Spatiotemporally Seamless Coverage of Continental U.S. Using a Deep Learning Neural Network. *Geophys. Res. Lett.* **2017**, *44*, 11–30. [CrossRef]
8. Fan, H.; Jiang, M.; Xu, L.; Zhu, H.; Cheng, J.; Jiang, J. Comparison of Long Short Term Memory Networks and the Hydrological Model in Runoff Simulation. *Water* **2020**, *12*, 175. [CrossRef]
9. Hidayat, H.; Hoitink, A.J.F.; Sassi, M.G.; Torfs, P.J.J.F. Prediction of Discharge in a Tidal River Using Artificial Neural Networks. *J. Hydrol. Eng.* **2014**, *19*, 04014006. [CrossRef]
10. Vapnik, V.N. *The Nature of Statistical Learning Theory*; Springer: New York, NY, USA, 1995; p. 314.
11. Liong, S.-Y.; Sivapragasam, C. Flood stage forecasting with support vector machines. *JAWRA J. Am. Water Resour. Assoc.* **2002**, *38*, 173–186. [CrossRef]
12. Asefa, T.; Kemblowski, M.; McKee, M.; Khalil, A. Multi-Time scale stream flow predictions: The support vector machines approach. *J. Hydrol.* **2006**, *318*, 7–16. [CrossRef]
13. Lal, A.; Datta, B. Development and Implementation of Support Vector Machine Regression Surrogate Models for Predicting Groundwater Pumping-Induced Saltwater Intrusion into Coastal Aquifers. *Water Resour. Manag.* **2018**, *32*, 2405–2419. [CrossRef]
14. Gangwar, S.; Bali, V.; Kumar, A. Comparative Analysis of Wind Speed Forecasting Using LSTM and SVM. *EAI Endorsed Trans. Scalable Inf. Syst.* **2018**, *7*, 1–9. [CrossRef]
15. Hochreiter, S.; Schmidhuber, J. Long Short-Term Memory. *Neural Comput.* **1997**, *9*, 1735–1780. [CrossRef] [PubMed]
16. LeCun, Y.; Bengio, Y.; Hinton, G. Deep learning. *Nature* **2015**, *521*, 436. [CrossRef] [PubMed]
17. Bengio, Y.; Simard, P.; Frasconi, P. Learning long-term dependencies with gradient descent is difficult. *IEEE Trans. Neural Netw.* **1994**, *5*, 157–166. [CrossRef] [PubMed]
18. Zhang, J.; Zhu, Y.; Zhang, X.; Ye, M.; Yang, J. Developing a Long Short-Term Memory (LSTM) based model for predicting water table depth in agricultural areas. *J. Hydrol.* **2018**, *561*, 918–929. [CrossRef]
19. Bowes, B.D.; Sadler, J.M.; Morsy, M.M.; Behl, M.; Goodall, J.L. Forecasting Groundwater Table in a Flood Prone Coastal City with Long Short-term Memory and Recurrent Neural Networks. *Water* **2019**, *11*, 1098. [CrossRef]
20. Huang, X.; Gao, L.; Crosbie, R.S.; Zhang, N.; Fu, G.; Doble, R. Groundwater Recharge Prediction Using Linear Regression, Multi-Layer Perception Network, and Deep Learning. *Water* **2019**, *11*, 1879. [CrossRef]
21. Jeong, J.; Park, E. Comparative applications of data-driven models representing water table fluctuations. *J. Hydrol.* **2019**, *572*, 261–273. [CrossRef]
22. Afzaal, H.; Farooque, A.A.; Abbas, F.; Acharya, B.; Esau, A.T. Groundwater Estimation from Major Physical Hydrology Components Using Artificial Neural Networks and Deep Learning. *Water* **2020**, *12*, 5. [CrossRef]
23. Jeong, J.; Park, E.; Chen, H.; Kim, K.-Y.; Han, W.S.; Suk, H. Estimation of groundwater level based on the robust training of recurrent neural networks using corrupted data. *J. Hydrol.* **2020**, *582*, 124512. [CrossRef]
24. Manandhar, A.; Greeff, H.; Thomson, P.; Hope, R.; Clifton, D.A. Shallow aquifer monitoring using handpump vibration data. *J. Hydrol. X* **2020**, *8*, 100057. [CrossRef]

25. Müller, J.; Park, J.; Sahu, R.; Varadharajan, C.; Arora, B.; Faybishenko, B.; Agarwal, D. Surrogate optimization of deep neural networks for groundwater predictions. *J. Glob. Optim.* **2020**, 1–29. [CrossRef]

26. Chollet, F.; Allaire, J.J. *Deep Learning with R*; Manning Publications: Shelter Island, NY, USA, 2018; p. 360.

27. Falbel, D.; Allaire, J.J.; Chollet, F.; Tang, Y.; Van Der Bijl, W.; Studer, M.; Keydana, S. R Interface to 'Keras' R. Package Version 2.2.4.1. 2019. Available online: https://CRAN.R-project.org/package=keras (accessed on 5 April 2019).

28. Srivastava, N.; Hinton, G.; Krizhevsky, A.; Sutskever, I.; Salakhutdinov, R. Dropout: A simple way to prevent neural networks from overfitting. *J. Mach. Learn. Res.* **2014**, *15*, 1929–1958.

29. Kingma, D.P.; Ba, J. Adam: A Method for Stochastic Optimization. *arXiv preprint* **2014**, arXiv:1412.6980.

30. Le, X.H.; Ho, H.V.; Lee, G.; Jung, S. Application of Long Short-Term Memory (LSTM) Neural Network for Flood Forecasting. *Water* **2019**, *11*, 1387. [CrossRef]

31. Nash, J.E.; Sutcliffe, J.V. River Flow forecasting through conceptual models-Part I: A discussion of principles. *J. Hydrol.* **1970**, *10*, 282–290. [CrossRef]

32. Moriasi, D.N.; Arnold, J.G.; Van Liew, M.W.; Bingner, R.L.; Harmel, R.D.; Veith, T.L. Model Evaluation Guidelines for Systematic Quantification of Accuracy in Watershed Simulations. *Trans. ASABE* **2007**, *50*, 885–900. [CrossRef]

33. Klemeš, V. Operational testing of hydrological simulation models. *Hydrol. Sci. J.* **1986**, *31*, 13–24. [CrossRef]

34. Yapo, P.O.; Gupta, H.V.; Sorooshian, S. Automatic calibration of conceptual rainfall-runoff models: Sensitivity to calibration data. *J. Hydrol.* **1996**, *181*, 23–48. [CrossRef]

35. Sahoo, S.; Russo, T.A.; Elliott, J.; Foster, I.T. Machine learning algorithms for modeling groundwater level changes in agricultural regions of the U.S. *Water Resour. Res.* **2017**, *53*, 3878–3895. [CrossRef]

Article

Analytical and Numerical Groundwater Flow Solutions for the FEMME-Modeling Environment

Mustafa El-Rawy [1,2,*], **Okke Batelaan** [3], **Kerst Buis** [4], **Christian Anibas** [5,6], **Getachew Mohammed** [7], **Wouter Zijl** [8] and **Ali Salem** [1,9]

[1] Civil Engineering Department, Faculty of Engineering, Minia University, Minia 61111, Egypt; alisalem@gamma.ttk.pte.hu

[2] Civil Engineering Department, College of Engineering, Shaqra University, Dawadmi 11911, Ar Riyadh, Saudi Arabia

[3] National Centre for Groundwater Research and Training, College of Science and Engineering, Flinders University, GPO Box 2100, Adelaide, SA 5001, Australia; okke.batelaan@flinders.edu.au

[4] Department of Biology, Ecosystem Management Research Group (ECOBE), University of Antwerp, Universiteitsplein 1, B-2610 Antwerpen, Belgium; kerst.buis@uantwerpen.be

[5] Water Research Laboratory, School of Civil & Environmental Engineering, UNSW Sydney, Sydney, NSW 2052, Australia; c.anibas@unsw.edu.au

[6] Connected Waters Initiative Research Centre, UNSW Sydney, Sydney, NSW 2052, Australia

[7] Golder Associates Ltd. 700 2 St. Southwest, Calgary, AB T2P 2W2, Canada; getachew_mohammed@golder.com

[8] Department of Hydrology and Hydraulic Engineering, Vrije Universiteit Brussel, Pleinlaan 2, 1050 Brussels, Belgium; VUB@zijl.be

[9] Doctoral School of Earth Sciences, University of Pécs, Ifjúság útja 6, H-7624 Pécs, Hungary

* Correspondence: mustafa.elrawy@mu.edu.eg

Received: 3 April 2020; Accepted: 8 May 2020; Published: 12 May 2020

Abstract: Simple analytical and numerical solutions for confined and unconfined groundwater-surface water interaction in one and two dimensions were developed in the STRIVE package (stream river ecosystem) as part of FEMME (flexible environment for mathematically modelling the environment). Analytical and numerical solutions for interaction between one-dimensional confined and unconfined aquifers and rivers were used to study the effects of a 0.5 m sudden rise in the river water level for 24 h. Furthermore, a two-dimensional groundwater model for an unconfined aquifer was developed and coupled with a one-dimensional hydrodynamic model. This model was applied on a 1 km long reach of the Aa River, Belgium. Two different types of river water level conditions were tested. A MODFLOW model was set up for these different types of water level condition in order to compare the results with the models implemented in STRIVE. The results of the analytical solutions for confined and unconfined aquifers were in good agreement with the numerical results. The results of the two-dimensional groundwater model developed in STRIVE also showed that there is a good agreement with the MODFLOW solutions. It is concluded that the facilities of STRIVE can be used to improve the understanding of groundwater-surface water interaction and to couple the groundwater module with other modules developed for STRIVE. With these new models STRIVE proves to be a powerful example as a development and testing environment for integrated water modeling.

Keywords: groundwater-surface water interaction; analytical; numerical; FEMME; STRIVE; MODFLOW

---

1. Introduction

There is a need to evaluate groundwater-surface water (GW-SW) interaction for water and ecosystem management. This is essential because linkages and feedback between groundwater and surface water systems affect both the quantity and quality of available water required by humans and ecosystems [1–4]. Therefore, the research topic of GW-SW interaction has gained importance in the last two decades, because of its role in conjunctive use, riparian zone management, and ecohydrology. In addition, understanding the interaction between groundwater and surface water can be important in the determination of migration pathways for contaminants [5]. The hydraulics of groundwater interaction with adjoining streams, canals, and drains is an important aspect of many hydrogeologic systems. Examples are the support of groundwater discharge to stream flow, bank storage attenuation of flood waves, and how groundwater discharge to streams lowers water tables maintains favorable root-zone salinity levels and prevents water logging of soil [6].

A variety of investigation methods have been used to study the hydraulic interaction of stream-aquifer systems including analytical, numerical, chemical, and field methods [7–11]. Recent examples of improved capabilities of MODFLOW [12–14] for stream-aquifer interaction are GSFLOW [15], HYDRUS [16–18] and unsaturated-zone flow (UZF1) packages [19], MIN3P [20], PARFLOW [21], the integrated water flow model (IWFM) [22], and SWAT-MODFLOW [23]. Numerical, chemical, and field methods have been widely used in different regions [24–30]. To study the interaction of groundwater and surface water flow in a river and adjacent areas, analytical solutions are often advantageous because of their simplicity. They are more general than site-specific field experiments but yet easier to implement for a particular site than numerical models [6]. In fact, several analytical solutions have been published for evaluation of the interaction of groundwater systems and hydraulically connected surface water bodies such as streams, lakes, reservoirs, drains, and canals. These solutions can be useful for understanding base flow processes, determining aquifer hydraulic properties, and predicting the response of aquifers to changing stream stage. Most of the solutions have been developed for confined and unconfined aquifers, such as [6,31–37].

The goal of this paper is to develop and test modules for groundwater-surface water interaction as part of the STRIVE (stream river ecosystem) package within the flexible environment for mathematically modelling the environment (FEMME) software [38]. Both numerical and analytical solutions have been developed to evaluate hydraulic interaction of river-aquifer systems. The analytical solutions from [31,32] for an unconfined aquifer and from [33] for a confined aquifer to calculate groundwater heads and discharges of the aquifer are described. The numerical solutions are based on the explicit finite difference approximation of the transient flow equation in a saturated, homogeneous, and isotropic aquifer [39]. A two-dimensional groundwater module for an unconfined aquifer is coupled to a one-dimensional hydrodynamic model [40]. This model was tested for a part of the Aa River, Belgium. Inter-model comparison is performed with MODFLOW. We present the modeling methodologies (FEMME, STRIVE package, hydrodynamic model, analytical and numerical solutions) as well as applications and comparison between analytical and numerical solutions, coupling with a hydrodynamic model, and comparison between STRIVE and MODFLOW.

2. Materials and Methods

2.1. FEMME Modeling Environment and STRIVE Package

FEMME was developed by the Netherlands Institute of Ecology (NIOO) [38]. FEMME is a modeling environment for the development and application of ecological time-dependent processes by using numerical integration of time-dependent differential equations. FEMME is constructed for ecosystem modeling and enables the simulation of different physical, biogeochemical, and transport processes of river ecosystems, like retention or exchange of matter [38,41]. The program is written in FORTRAN; it is designed to implement 0- to multi-dimensional, time-dependent models. FEMME is open source and facilitates the use of pre-defined integration tools in a modular FORTRAN environment [38]. FEMME consists of a wide range of numerical functions as integration, forcing, and calibration functions, as well as data manipulation functions. These technical possibilities allow the user to focus on the scientific part of model development rather than having to deal with complex programming issues. Hence, the environment allows rapid and efficient code development for environmental applications.

The STRIVE-package is a set of modules incorporated in the FEMME environment. It consists of different modules for macrophyte growth, water quality, and hyporheic processes, which can be coupled to form numerical models for specific research questions regarding river ecosystems. Hence, the module for hydrodynamic flow can, e.g., interact with a macrophyte growth routine, which influences the Manning coefficient and therefore the flow simulation. A heat transport module was implemented [42,43] in the STRIVE package for studying the vertical interaction in the hyporheic zone between rivers and aquifers. In this article, the STRIVE package is extended with groundwater modules, which are necessary to understand the interaction between a river and its riparian margin.

2.2. Hydrodynamic Module in STRIVE Package

A hydrodynamic surface water flow module [40] was developed for the STRIVE package and applied for a part of the Aa River, Belgium. The module is based on the Saint-Venant equations for one dimensional unsteady open channel flow [44]. Based on the capabilities of STRIVE, we hypothesize that the implementation of simple analytical and numerical groundwater flow solutions coupled with the hydrodynamic surface water module will allow the investigation of groundwater-river exchange processes in more detail.

2.3. Analytical Solutions for Groundwater-Surface Water Interaction

In this section, analytical solutions [31–33] for the interaction between surface water and unconfined and confined aquifers are presented.

2.3.1. Edelman Analytical Solution

We describe here the interaction between a river and a one-dimensional homogeneous semi-infinite and unconfined aquifer, which is bounded at one side by a fully-penetrating stream and below by an impermeable stratum (Figure 1). Under steady-state conditions, the water table in the aquifer and the water level in the river coincide, and there is no flow in or out of the aquifer. A sudden rise in the water level of the river induces a flow from the river towards the aquifer. As a result, the water table in the aquifer starts rising until it reaches the same level as that in the river. The flow from the river to the aquifer is unsteady and one-dimensional.

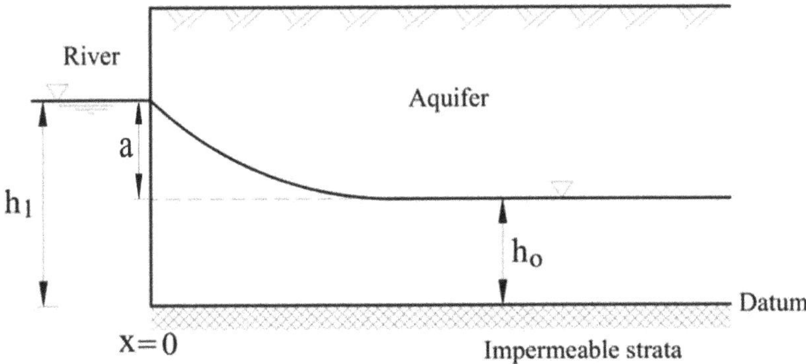

Figure 1. Unsteady, one-dimensional flow in a semi-infinite unconfined aquifer.

If the Dupuit approximation, i.e., (a) the equipotential lines are vertical and its consequence; (b) the slope of the groundwater table is equivalent to the hydraulic gradient and is invariant with depth, holds [45], then the flow problem can be described by the partial differential equation:

$$\frac{\partial h}{\partial t} = \frac{kb}{S_y}\frac{\partial^2 h}{\partial x^2}$$

(1)

where $kb = T$ is the transmissivity of the homogeneous aquifer [L^2T^{-1}], h is the hydraulic head in the aquifer [L], t is the time [T], x is the distance from the river bank [L], and S_y is the specific yield (−). A general solution to Equation (1) does not exist and integration is possible only for specific boundary conditions [46].

Edelman [31] presented a solution for the case of a sudden change of the water level in the river and a constant water level thereafter, $h(x, t)$ = head.

$$h(x,t) = a(1 - \frac{2}{\sqrt{\pi}}\int_0^u e^{-\mu^2}d\mu)$$

(2)

where $h(x, t)$ is the head in the aquifer [L] at the horizontal coordinate x [L] and time t [T], a is the sudden change in the water level of the river [L], and u is a dimensionless auxiliary variable.

$$u = \sqrt{x^2 S_y / 4Kbt}$$

(3)

$$erf(u) = \frac{2}{\sqrt{\pi}}\int_0^u e^{-\mu^2}d\mu$$

(4)

where $erf(u)$ is the error function, and $1\text{-}erf(u) = erfc(u)$ is the complementary error function. Hence, the head in the aquifer can be formulated:

$$h(x,t) = a \cdot erfc(u)$$

(5)

The flux in the aquifer per unit length of river at distance x is:

$$q(x,t) = \frac{a}{\sqrt{\pi}}\frac{\sqrt{KbS_y}}{\sqrt{t}}e^{-u^2}$$

(6)

Equation (6) gives the discharge from one side of the river. This equation can also be used if the water level in the river suddenly drops, inducing a flow from the aquifer to the river, resulting in a fall of the water table in the aquifer.

2.3.2. Lockington Analytical Solution

Lockington [32] derived simple analytical solutions for the one-dimensional Boussinesq equation using a weighted residual method. His approach can be applied to both a recharging and dewatering semi-infinite unconfined, homogeneous aquifer with a fully penetrating trench. Only the analytical solution for a recharging aquifer is discussed here (Figure 1) [33].

$$h = h_0 + (h_1 - h_0)\left(1 - \frac{x}{\lambda}\sqrt{\frac{S_y}{Kt}}\right)^{\frac{1}{\mu}}$$

(7)

where h is piezometric head [L], h_1 is the water level in the river [L], h_0 is the initial water level in the river and aquifer [L], K is the hydraulic conductivity of the aquifer [LT^{-1}], Sy is the specific yield of the aquifer [–], and λ and μ are parameters, which are defined as:

$$\mu = -\frac{3}{4}(1+N) + \frac{N}{(2-A)} + \frac{\left[(2-A)^2(1+2N) + N^2(2+A)^2\right]^{\frac{1}{2}}}{4(2-A)} \tag{8}$$

$$\lambda^2 = \frac{(1+\mu)(1+2\mu)}{2\mu^2}(h_0 + h_1) \tag{9}$$

where A is a constant defined as:

$$A = \frac{4[h_0 + (1+N)h_1]}{(1+N)(2+N)(h_1 + h_0)} \tag{10}$$

In which the exponent N is estimated as in Parlange et al. [47]:

$$N = 2.27932 - \frac{3h_0}{(h_1 + 2h_0)} \tag{11}$$

The flux from one side of the river is given by:

$$q = \frac{C_r(h_1 - h_0)\sqrt{KS_y}}{2\sqrt{t}} \tag{12}$$

where C_r is a recharge coefficient, given by:

$$C_r^2 = \frac{(1+2\mu)(h_1 + h_0)}{2(1+\mu)} \tag{13}$$

2.3.3. Bruggeman Analytical Solution

Bruggeman [33] derived the following general analytical solution:

$$h(x,t) = a \cdot 2^n \Gamma(1 + \frac{n}{2}) t^{\frac{n}{2}} erfc(u) \tag{14}$$

where:

$$u = \sqrt{x^2 S/4Kbt} \tag{15}$$

S is the storage coefficient of the aquifer (–), and $n = 0, 1, 2, 3\ldots$ depends on the type of the water level change in the river. For $n = 0$, the change in the water level is assumed to be a sudden change. Similarly, an n value of 1 and 2 indicate a linear and parabolic water level change, respectively. Using the following relationship:

$$i^n erfc(0) = \frac{1}{2^n \Gamma(1 + \frac{n}{2})}, \quad n = -1, 0, 1, 2, \ldots \tag{16}$$

Equation (14) can be simplified:

$$h(x,t) = a \cdot t^{\frac{n}{2}} \frac{i^n erfc(u)}{i^n erfc(0)} \tag{17}$$

The flux is calculated based on Darcy's law:

$$q(x,t) = \frac{a}{2} \cdot t^{\frac{n-1}{2}} \sqrt{KbS} \frac{i^{n-1} erfc(u)}{i^n erfc(0)} \tag{18}$$

where:

$$i^n erfc(u) = -\frac{u}{n}i^{n-1}erfc(u) + \frac{1}{2n}i^{n-2}erfc(u), n = 1, 2, 3, \ldots \tag{19}$$

with:

$$i^0 erfc(u) = erfc(u), \tag{20}$$

$$i^{-1}erfc(u) = \frac{2}{\sqrt{\pi}}e^{-u^2} \tag{21}$$

For $n = 0$, a sudden change in the surface water level, Equation (17) simplifies to:

$$h(x,t) = a \cdot erfc(u) \tag{22}$$

and Equation (18) for the flux from one side of the aquifer to the river reduces to:

$$q(x,t) = \frac{a}{\sqrt{\pi}}\frac{\sqrt{KbS}}{\sqrt{t}}e^{-u^2} \tag{23}$$

2.4. Numerical Solutions for Groundwater-Surface Water Interaction

Numerical solutions for transient groundwater flow (Equation (24)) are implemented in STRIVE. The solutions are based on the explicit finite difference approximation of transient flow in a saturated, homogeneous, incompressible, and isotropic aquifer [39].

$$\frac{\partial^2 h}{\partial x^2} + \frac{\partial^2 h}{\partial y^2} = \frac{S}{T}\frac{\partial h}{\partial t} - \frac{R(x,y,t)}{T} \tag{24}$$

where x and y are the spatial coordinates [L], and R is recharge [L].

In order to solve this diffusion equation (Equation (24)), it is necessary to prescribe boundary and initial conditions [39]. Finite difference approximations for different unsteady-state groundwater flow problems are developed in the following sections.

2.4.1. One-Dimensional Flow in a Confined Aquifer

The explicit finite difference approximation for the head for one-dimensional flow in a confined aquifer can be calculated from the heads at time moment n.

$$h_i^{n+1} = h_i^n + F(h_{i+1}^n - 2h_i^n + h_{i-1}^n) \tag{25}$$

with:

$$F = T\Delta t / S(\Delta x)^2 \tag{26}$$

In order to implement Equation (25) in STRIVE, the equation should be written as the rate of change in head:

$$\frac{dh_i}{dt} = G(h_{i+1} - 2h_i + h_{i-1}) \tag{27}$$

where G is defined as:

$$G = T/S(\Delta x)^2 \tag{28}$$

The criterion for stability of the numerical solution of Equation (25) is $F < 0.25$. Δt is the time step [T], Δx is the size of the grid cell [L], h_i^n and h_i^{n+1} are the heads in the center of grid cell i at respectively time step n and $n + 1$, h_{i+1}^n, h_{i-1}^n are the heads at times step n in the centers of respectively grid cell $i+1$ and $i-1$. The discharge from the river to the aquifer or vice versa is calculated by Darcy's law:

$$Q = -KA\frac{dh}{dl} \tag{29}$$

where Q is the discharge from or into the aquifer [L^3T^{-1}], K is the hydraulic conductivity [LT^{-1}], and A is the cross-sectional area normal to the flow direction [L^2].

2.4.2. One and Two-Dimensional Flow in an Unconfined Aquifer

The explicit finite difference approximation for the head at time step $n+1$ in terms of v for one-dimensional flow in unconfined aquifers is [39]

$$v_i^{n+1} \;=\; v_i^n + F\sqrt{v_i^n}(v_{i+1}^n - 2v_i^n + v_{i-1}^n) \tag{30}$$

with:

$$v \;=\; h^2 \tag{31}$$

where F is defined as in Equation (26) but with Sy instead of S.

In order to implement Equation (30) in STRIVE, the equation is written as a time derivative:

$$\frac{dv_i}{dt} \;=\; G\sqrt{v_i}(v_{i+1} - 2v_i + v_{i-1}) \tag{32}$$

where G is defined as in Equation (28) but with Sy instead of S. Considering a finite set of points on a regularly spaced grid (Figure 2), the explicit finite difference approximation for unconfined two-dimensional flow is:

$$v_{ij}^{n+1} \;=\; v_{ij}^n + F\sqrt{v_{ij}^n}(v_{i+1,j}^n + v_{i,j+1}^n - 4v_{ij}^n + v_{i-1,j}^n + v_{i,j-1}^n) \tag{33}$$

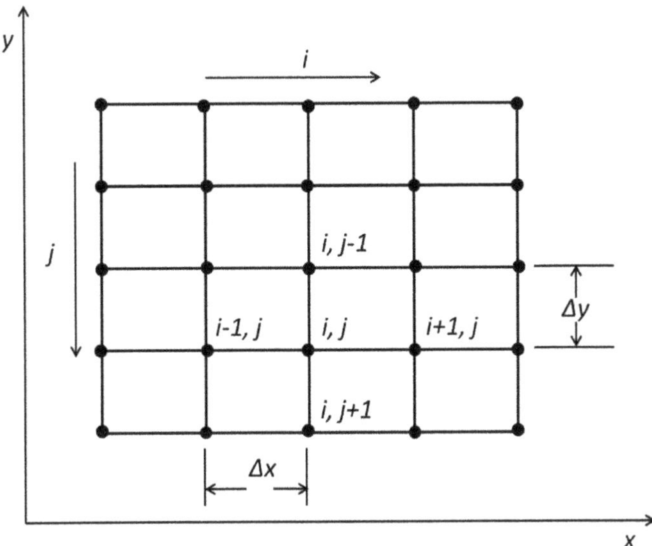

Figure 2. A diagram of the finite difference grid cells.

Here, F is equal to:

$$F \;=\; K\Delta t / S_y a^2 \tag{34}$$

where $a = \Delta x = \Delta y$.

The formulation in STRIVE is:

$$\frac{dv_{ij}}{dt} = \frac{K\sqrt{v_{ij}}}{S_y a^2}\left(v_{i+1,j} + v_{i,j+1} - 4v_{ij} + v_{i-1,j} + v_{i,j-1}\right) \tag{35}$$

3. Application and Discussion

3.1. One-Dimensional Analytical and Numerical Solutions for Confined and Unconfined Aquifers

The implemented one-dimensional analytical and numerical solutions for groundwater heads and discharges in unconfined or confined aquifers are compared as a function of time and distance. The solutions were set up in the STRIVE package for a domain perpendicular to the river to calculate the rise in the head and flow in the unconfined or confined aquifer at a distance x from the river after a sudden water level rise of 0.5 m. Table 1 specifies the details for the unconfined and confined aquifer.

Table 1. Parameters and dimensions of the unconfined and confined aquifer.

Parameter	Value	Dimension
Hydraulic conductivity (K)	10	m d^{-1}
Storage coefficient (S)	0.2	(−)
Thickness of the aquifer (b)	10	m
Specific yield (S_y)	0.2	(−)
Initial head everywhere in the aquifer	10.4	m
Head in the river	10.9	m

The heads are calculated at the center of cells, and the discharges are calculated using Darcy's equation at the interface of the cells based on the head calculations for a one-day simulation with a time step of 0.001 day. The results of this application are presented below as a comparison between the analytical and numerical solutions.

Figure 3 shows the numerical solution for one-dimensional transient flow through an unconfined aquifer and the analytical solutions of [31,32] for the heads at different lateral distances from the river at 1.5, 12, and 24 h The differences between the solutions of [31,32] and the numerical one are negligible; the maximum difference is 0.021 m between the Edelman [31] and the numerical solutions, and the root mean squared error (RMSE) is 0.0075 m after 1.5 h. The maximum difference between the Lockington [32] and the numerical solutions was 0.015 m, and the RMSE was 0.0048 m.

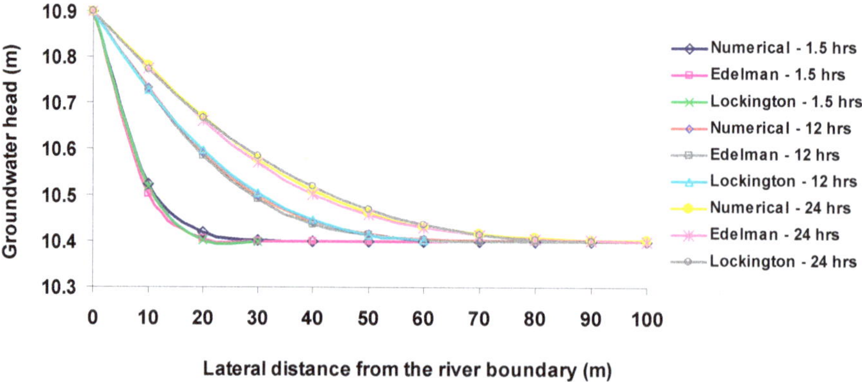

Figure 3. Groundwater head versus lateral distance from river at specific times for numerical and analytical solutions (Edelman [31] and Lockington [32]) for a semi-infinite unconfined aquifer.

Figure 4 shows the match between the three solutions for the discharges at the river-aquifer boundary for a simulation period of 1 day. The maximum difference between the analytical (Edelman [31] and Lockington [32]) and the numerical solutions for the groundwater flux per meter of river length is 0.17 m^2 d^{-1}, while the RMSE is 0.1 m^2 d^{-1} at 1.5 h. The difference decreases with time.

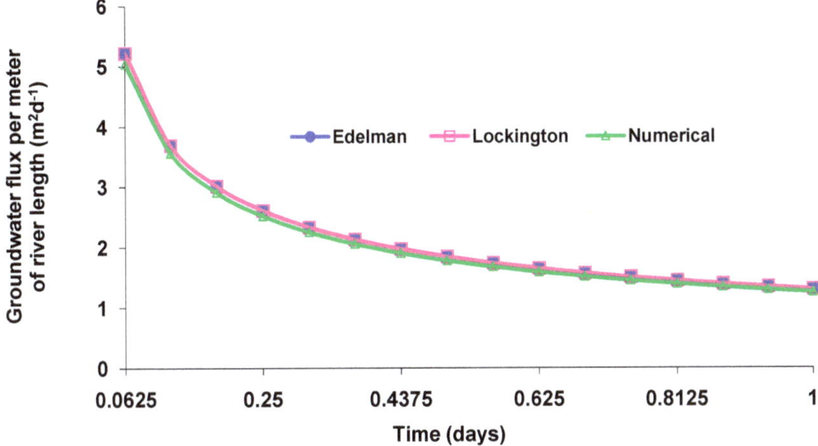

Figure 4. Groundwater discharge versus time at the river-aquifer boundary for numerical and analytical solutions (Edelman [31] and Lockington [32]) for a semi-infinite unconfined aquifer.

The groundwater heads and fluxes simulated with the numerical solution for one-dimensional transient flow in a confined aquifer and the analytical solution of Bruggeman [35] are presented in Figures 5 and 6. Figure 5 shows the relationship between groundwater heads as a function of distance from the river at different times (1.5, 12, 24 h). It is observed that the effect of a 0.5 m sudden rise in the river stage is negligible beyond a distance of 80 m from the river-aquifer boundary. The distance increases with time; at 1.5 h the effect disappears after 50 m and at 24 h it is negligible after 80 m from the river-aquifer boundary. The maximum change in head in the aquifer is attained after one day for locations further than 10 m away from the river. The maximum difference between the numerical and Bruggeman [33] solutions was 0.01 m at 1.5 h.

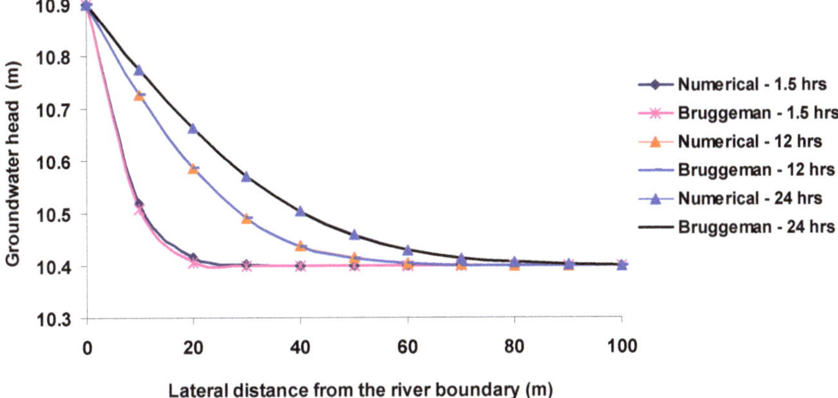

Figure 5. Groundwater head versus lateral distance from river at specific times for numerical and Bruggeman [33] solutions for a confined aquifer.

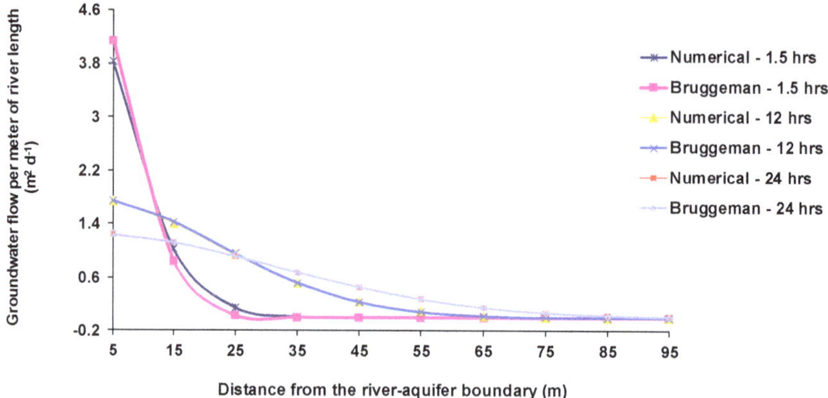

Figure 6. Groundwater fluxes versus lateral distance from river at specific times (1.5, 12, and 24 h) for numerical and Bruggeman [33] solutions for a confined aquifer.

Figure 6 shows the groundwater flux as a function of distance from the river at different times (1.5, 12, 24 h). It is noticed that the maximum discharges are attained within the first 1.5 h and range from 4.13 to 1.25 m^2 d^{-1} per meter of river length for a location 5 m away from the river-aquifer boundary. The maximum difference between the numerical and Bruggeman solutions was 0.3 m^2 d^{-1} and the RMSE was 0.0145 m at 1.5 h. The accuracy of the numerical solution is mainly determined by the space and time discretization. It is concluded that the results of the analytical solutions of Edelman [31], Lockington [32], and Bruggeman [33] are in a good agreement with the numerical solution.

3.2. Two-Dimensional Numerical Solution in an Unconfined Aquifer

The two-dimensional groundwater module for an unconfined aquifer is applied to the Aa River aquifer, Belgium (based on data from [42,43]) (Figure 7).

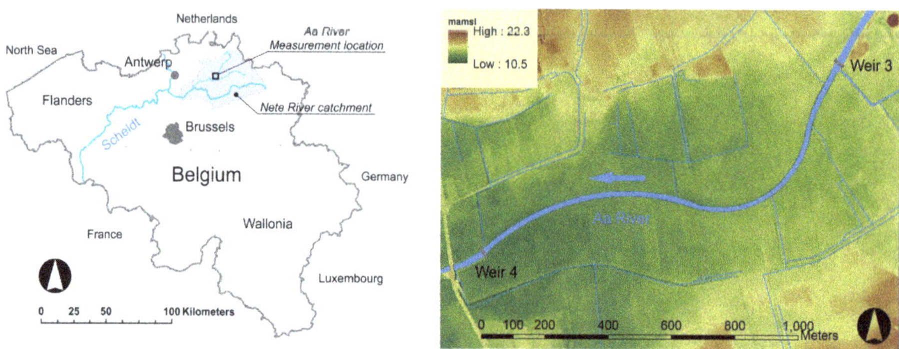

Figure 7. Location and topography map of the Aa River in Belgium.

The simulated area is simplified to be as Figure 8; it has a length of 1000 m (between weirs 3 and 4 in Figure 7), a width of 200 m, a thickness of 20 m, a hydraulic conductivity K_x and K_y of 10 m d^{-1}, a specific yield S_y of 0.2, and a grid cell size of 10 m. To simplify implementation of the problem in STRIVE, we simulated the reach of 1000 m of the Aa River as a straight line, as shown in Figure 8. The groundwater module developed in STRIVE was applied for two cases of the river boundary conditions. In case 1, the river boundary is estimated by interpolation based on the two head values assigned to the upstream and downstream points of the river. The model was run in steady state by using boundary conditions based on interpolation of heads assigned at the corner cells (a = 10.5 m, b = 10.2 m, c = 10.5 m, d = 10.8 m) (Figure 8). The boundary condition heads are interpolated linearly based on the heads assigned at the corner cells. The initial head everywhere in the aquifer is specified as 10.4 m. A time step of 10 seconds was used for a duration of 1000 days.

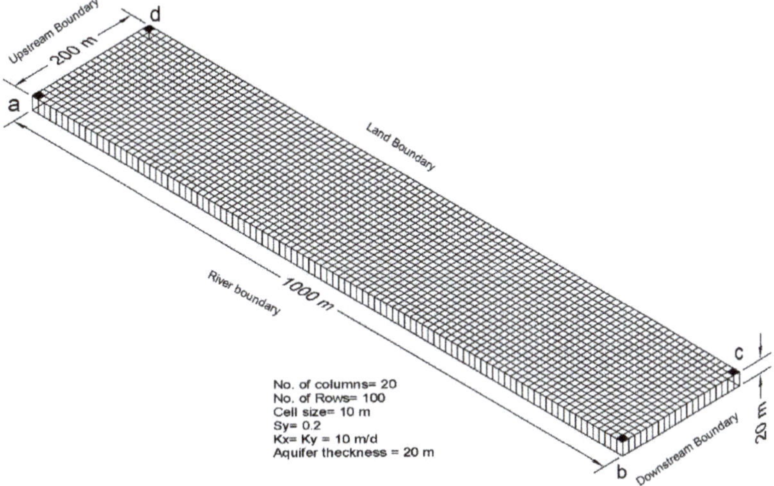

Figure 8. Concept of the groundwater model as simulated with STRIVE and MODFLOW. The heads along the upstream (ad), land side (dc) and downstream (bc) boundaries are specified and interpolated based on four corner heads derived from the original groundwater model (a = 10.5 m, b =10.2 m, c = 10.5 m, d = 10.8 m).

The simulated groundwater head and the lateral flows are presented in Figures 9 and 10 respectively. From Figure 9, it is observed that the hydraulic gradient is directed towards the river boundary, due to the selected boundary conditions.

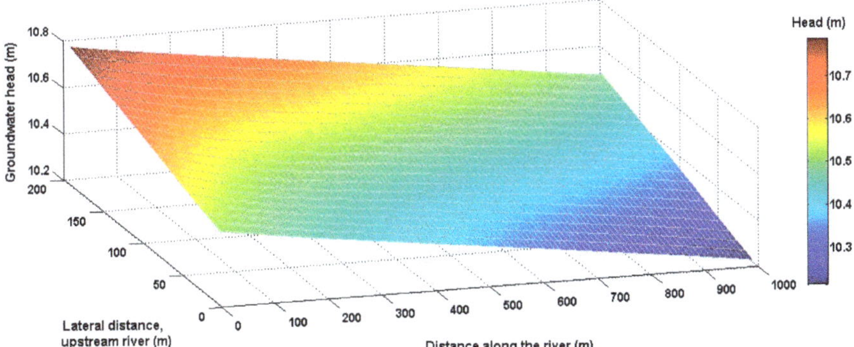

Figure 9. Groundwater head obtained from STRIVE for the Aa River based on interpolated river boundary with a steady-state simulation.

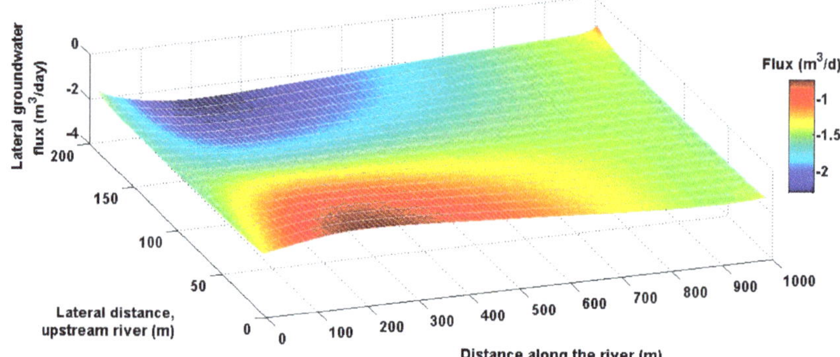

Figure 10. Lateral groundwater flux obtained from STRIVE implementation for the Aa River based on interpolated river boundary with a steady-state simulation.

Figure 10 shows in two dimensions the lateral groundwater fluxes obtained from the STRIVE model. The fluxes are directed towards the river (values are negative); the maximum and minimum discharges from the aquifer to the river along the river are respectively 1.49 and 0.74 m^3 d^{-1} at 900 m and 200 m viewed from the upstream boundary respectively.

In the second case, the river boundary is based on water levels from a hydrodynamic surface water model, which is also integrated in STRIVE [40]. The water level in the hydrodynamic surface water model was calculated based on an upstream discharge condition, formulated as a half sine wave (Figure 11). Other boundary conditions and initial conditions are the same as those applied in case 1. The model was run in transient for a period of 30 days with a time step of 1 min and output interval of 1 h. The river boundary values in this case were obtained from the output of the hydrodynamic model.

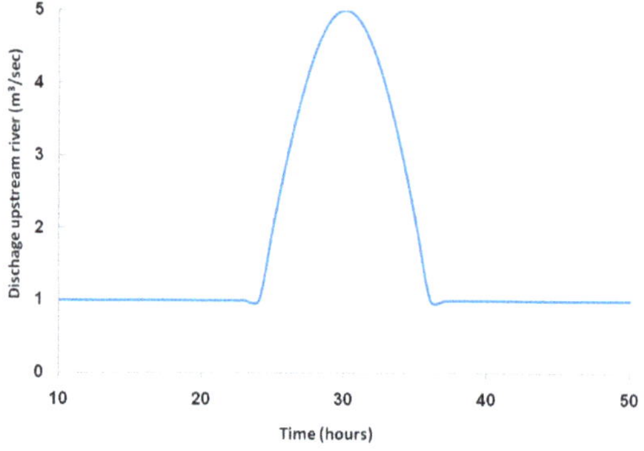

Figure 11. Analytical hydrograph formulated as a half sine wave introduced in hydrodynamic surface water model.

Figure 12 presents the two-dimensional groundwater heads as simulated with STRIVE for this second case. The groundwater heads are presented at two times, in order to study the effect of the river stages on the groundwater heads in the aquifer and to test the response of the interaction between the river and the aquifer. Figure 12a shows the groundwater heads in the aquifer at 24 h (the beginning of the pulse of the upstream discharge in the river). It is observed that the hydraulic gradient is directed towards the river boundary. Figure 12b presents the groundwater heads in the aquifer at 32 h (the maximum effect of the pulse of the upstream discharge in the river). It is observed that at the river-aquifer boundary, the hydraulic gradient is directed from the river boundary to the aquifer.

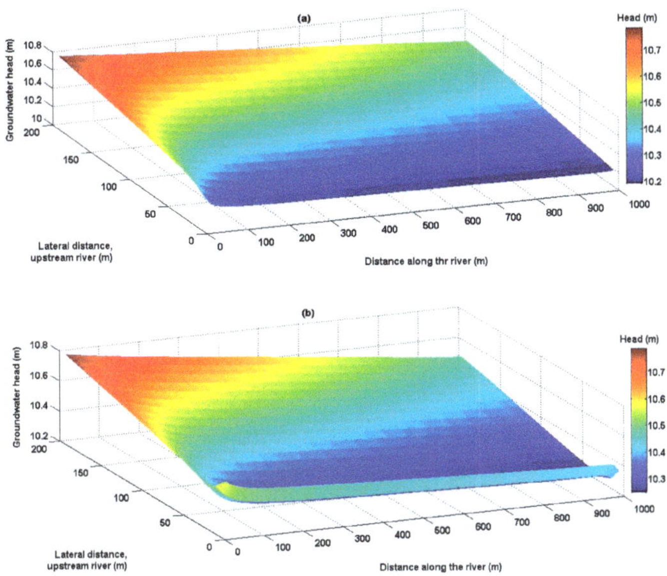

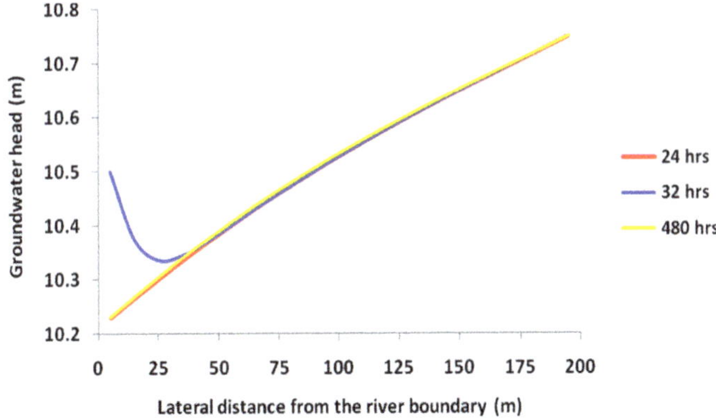

Figure 12. Groundwater head obtained from STRIVE for the Aa River based on the hydrodynamic river boundary with a transient simulation: (**a**) at the beginning of the pulse (24 h), and (**b**) at the maximum effect of the pulse (32 h).

In Figure 13, the effect of the river pulse on groundwater heads at different times (24, 32, and 480 h) and at different distances along the river at 100 m (near to the upstream model boundary) are shown. At 24 and 480 h (before the beginning and after the ending of the pulse), there is no effect of the river pulse on the groundwater heads, and the flow is directed towards the river boundary. However, at 32 h, the effect of the river pulse on the groundwater heads in the model appears clearly at the edge of the river-aquifer boundary (30 m) and disappears after a distance of 30 m from the river-aquifer boundary.

Figure 13. STRIVE simulated groundwater head versus lateral distance from the river-aquifer boundary at 100 m (near to the upstream model boundary in longitudinal direction of the river), at times 24, 32, and 48 h, using a hydrodynamic river boundary with a transient state simulation.

The river-aquifer fluxes along the river at 24, 32, and 480 h are shown in Figure 14. At the beginning of the pulse, the Aa River gains water from the aquifer, while at 32 h, the aquifer gains water from the Aa River. The Aa River comes back again to gain water from the aquifer after 480 h. The direction and the rate of flow depend on the difference between the river stage and the groundwater heads.

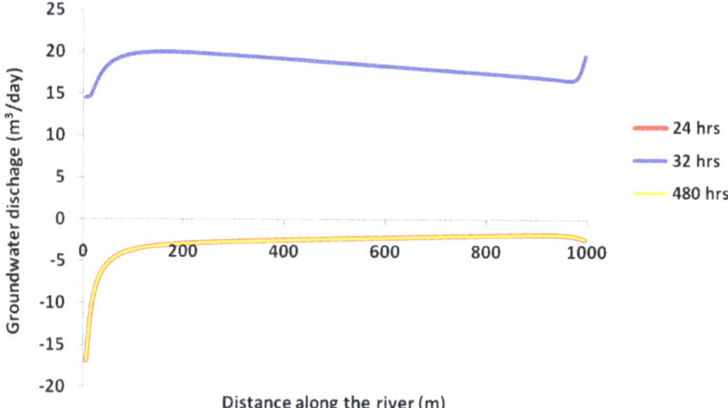

Figure 14. Groundwater discharge at a lateral distance of 10 m of the river-aquifer boundary obtained from STRIVE along the Aa River based on the hydrodynamic river boundary with a transient simulation at times of 24, 32, and 480 h.

3.3. Comparison between STRIVE and MODFLOW Results

The groundwater flow model MODFLOW-2000 [12] was used for model comparison of the STRIVE two-dimensional groundwater flow implementation. A simple MODFLOW model was set up using the same data as was used for the second case example in STRIVE. The PCG2 solver was used and both steady-state and transient state simulations were performed in order to compare these results with STRIVE results.

In order to see the agreement between the STRIVE and MODFLOW results, the groundwater heads in lateral direction from the river-aquifer boundary at different distances along the river at 5 m (upstream river), 505 m (the middle of the river), and 995 m (downstream river) are shown in Figure 15. Figure 16 shows the groundwater heads in lateral direction from the river-aquifer boundary at distance of 505 m along the river at different simulation times (1, 6, 12, 24 h). The maximum head difference between STRIVE and MODFLOW was 0.004 m, and the RMSE is 7.1×10^{-6} m and 1.8×10^{-5} m for steady-state and transient state simulations, respectively.

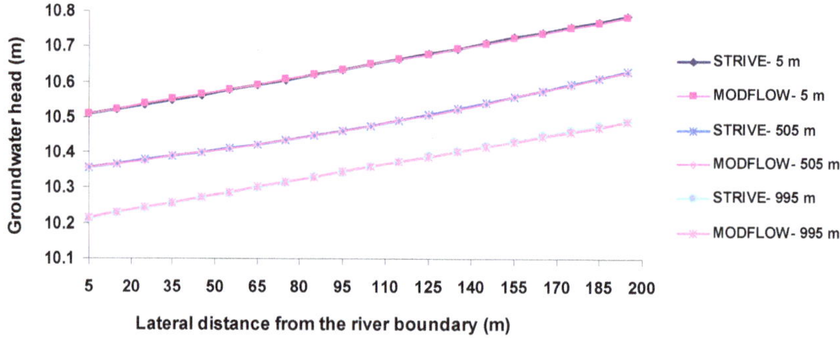

Figure 15. Groundwater head versus lateral distance from the river-aquifer boundary at fixed distances in the longitudinal direction of the river at 5 m (near the upstream model boundary), 505 m (at the middle of the river length), and 995 m (near the downstream model boundary), using STRIVE and MODFLOW, based on interpolated river boundary with a steady-state simulation.

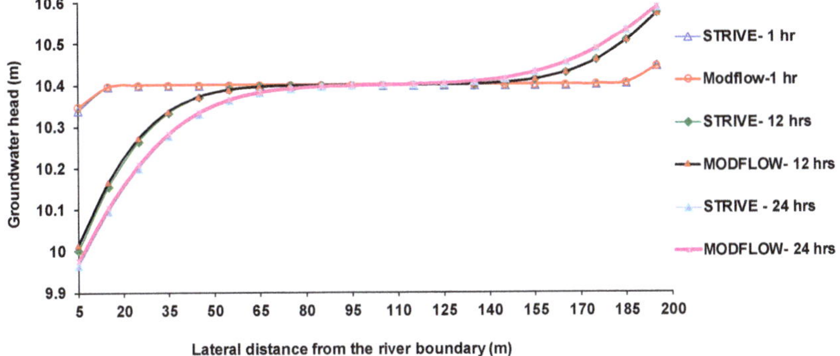

Figure 16. Groundwater head versus lateral distance from the river-aquifer boundary at 505 m from the upstream river for FEMME and MODFLOW, based on interpolated river boundary with a transient simulation.

4. Conclusions

The facilities of STRIVE were tested for the interaction between groundwater flow in confined and unconfined aquifers with streams in one and two dimensions. The analytical solutions implemented in STRIVE include Edelman [31] and Lockington [32] for unconfined aquifers and Bruggeman [33] for confined aquifers. Groundwater heads and discharges in the aquifers were calculated based on a sudden change in the river stage. The results of the analytical solutions were compared with one-dimensional numerical solutions which were implemented in STRIVE for confined and unconfined aquifers. The results of the analytical solutions for confined and unconfined aquifers were in good agreement with the numerical results.

In addition, a two-dimensional groundwater model for an unconfined aquifer was developed in STRIVE and coupled with a one-dimensional hydrodynamic model in STRIVE. This model was applied on a 1 km long reach of the Aa River. The model was tested for two cases with different boundary conditions. In the first case, the river boundary was interpolated, and the model was simulated in steady-state. In the second case, the river boundary was linked with the water levels from a hydrodynamic surface water model. MODFLOW models were set up for these cases as well, in order to check the implementation in STRIVE. The results of the two-dimensional groundwater model developed in STRIVE showed that there is a very good agreement with MODFLOW.

It is concluded that analytical and numerical solutions for groundwater-surface water interaction for unconfined and confined aquifers have been successfully implemented in STRIVE. Hence, STRIVE is extended in terms of modeling facilities for groundwater-surface water interaction, but also due to these implemented sub-models, new integration possibilities with existing modules such as the hydrodynamic, hyporheic zone and sediment appear. The flexibility of STRIVE has proven to be a major advantage in developing these new sub-modules. With these new models, STRIVE increases its capabilities without becoming a dedicated type of model with a graphical user interface. Rather, it is a powerful example of a development and testing environment for integrated water modeling.

Author Contributions: Conceptualization, M.E.-R. and O.B.; methodology M.E.-R., O.B. and K.B.; software, M.E.-R.; K.B.; and C.A.; validation, M.E.-R., O.B. and G.M.; formal analysis, M.E.-R.; O.B.; and K.B.; investigation, M.E.-R.; C.A.; G.M. and W.Z.; data curation, C.A.; G.M.; and W.Z.; writing—original draft preparation, M.E.-R.; writing—review and editing, M.E.-R.; O.B.; K.B; C.A.; W.Z.; and A.S.; visualization, M.E.-R.; and O.B.; supervision, O.B.; K.B.; C.A. and G.M; project administration, O.B. All authors have read and agreed to the published version of the manuscript.

Funding: This research received no external funding.

Conflicts of Interest: The authors declare no conflict of interest.

References

1. Winter, T.C.; Harvey, J.W.; Franke, O.L.; Alley, W.M. *Ground Water and Surface Water: A Single Resource, Circular 1139*; U.S. Geological Survey: Reston, VA, USA, 1998; p. 87.
2. European Commission. *E.U. Water Framework Directive, European Parliament and Commission, Official Journal, Directive 2000/60/Ec on 22 December (Directive 2000/60/EC of the European Parliament and of the Council of 23 October 2000 Establishing a Framework for Community Action in the Field of Water Policy)*; European Commission: Brussels, Belgium, 2000.
3. Sophocleous, M. Interactions between groundwater and surface water: The state of the science. *J. Hydrol.* **2002**, *10*, 52–67.
4. Makovníková, J.; Kanianska, R.; Kizeková, M. The ecosystem services supplied by soil in relation to land use. *Hung. Geogr. Bull.* **2017**, *66*, 37–42. [CrossRef]
5. Oxtobee, J.P.A.; Novakowski, K. A field study of groundwater/surface water interaction in a fractured bedrock environment. *J. Hydrol.* **2002**, *269*, 169–193. [CrossRef]
6. Barlow, P.M.; Moench, A.F. *Analytical Solutions and Computer Programs for Hydraulic Interaction of Stream-Aquifer Systems*; U.S. Geological Survey: Open-File Report 98-415A; United States Geological Survey: Reston, VA, USA, 1998; p. 85.
7. Woessner, W.W. Stream and fluvial plain ground water interactions: Rescaling hydrogeologic thought. *Ground Water* **2000**, *38*, 423–429. [CrossRef]
8. Moutsopoulos, K.N.; Tsihrintzis, V.A. Approximate analytical solutions of the Forchheimer equation. *J. Hydrol.* **2005**, *309*, 93–103. [CrossRef]
9. Kalbus, E.; Reinstorf, F.; Schirmer, M. Measuring methods for groundwater-surface water interactions: A review. *Hydrol. Earth Syst. Sci.* **2006**, *10*, 873–887. [CrossRef]
10. Intaraprasong, T.; Zhan, H. A general framework of stream-aquifer interaction caused by variable stream stages. *J. Hydrol.* **2009**, *373*, 112–121. [CrossRef]
11. Moutsopoulos, K.N. Solutions of the Boussinesq equation subject to a nonlinear robin boundary condition. *Water Resour. Res.* **2013**, *49*, 7–18. [CrossRef]
12. Harbaugh, A.W.; Banta, E.R.; Hill, M.C.; McDonald, M.G. *Modflow-2000, the U.S. Geological Survey Modular Ground-Water Model—User Guide to Modularization Concepts and the Ground-Water Flow Process*; Open-File Report 00-92; U.S. Geological Survey: Reston, VA, USA, 2000; p. 121.
13. Harbaugh, A.W. *MODFLOW-2005, the U.S. Geological Survey Modular Ground-Water Model—The Ground-Water Flow Process*; U.S Geological Survey Techniques and Methods 6-A16; U.S. Geological Survey: Reston, VA, USA, 2005; p. 253.
14. Niswonger, R.G.; Panday, S.; Ibaraki, M. *MODFLOW-NWT, a Newton Formulation for MODFLOW-2005*; U.S. Geological Survey: Reston, VA, USA, 2011.
15. Markstrom, S.; Niswonger, R.; Regan, R.; Prudic, D.; Barlow, P. *GSFLOW: Coupled Ground-Water and Surface-Water Flow Model Based on the Integration of the Precipitation-Runoff Modeling System (PRMS) and the Modular Ground-Water Flow Model (MODFLOW-430 2005)*; U.S. Geological Survey Techniques and Methods 6-d1; U.S. Geological Survey: Reston, VA, USA, 2008; p. 240.
16. Seo, H.; Šimůnek, J.; Poeter, E. Documentation of the HYDRUS package for MODFLOW-2000, the US geological survey modular ground-water model. IGWMI 2007-01. Integr. In *Ground Water Modeling Ctr.*; Colorado School of Mines: Golden, CO, USA, 2007; p. 96.
17. Twarakavi, N.K.C.; Šimůnek, J.; Seo, H.S. Evaluating Interactions between Groundwater and Vadose Zone Using the HYDRUS-Based Flow Package for MODFLOW. *Vadose Zone J.* **2008**, *7*, 757–768. [CrossRef]
18. Šimůnek, J.; Van Genuchten, M.T.; Šejna, M. Recent Developments and Applications of the HYDRUS Computer Software Packages. *Vadose Zone J.* **2016**, *15*, 15. [CrossRef]
19. Niswonger, R.G.; Prudic, D.E.; Regan, R.S. *Documentation of the Unsaturated-Zone Flow (UZF1) Package for Modeling Unsaturated Flow between the Land Surface and the Water Table with MODFLOW—2005*; US Geological Survey Techniques and Methods 6-A19, Book 6, Chapter A19; USGS: Reston, VA, USA, 2006; 62p.
20. Mayer, K.; Amos, R.T.; Molins, S.; Gerard, F. Reactive Transport Modeling in Variably Saturated Media with MIN3P: Basic Model Formulation and Model Enhancements. In *Groundwater Reactive Transport Models*; Zhang, F., Yeh, G.T., Parker, J.C., Shi, X., Eds.; Bentham Science Publishers Ltd.: Sharjah, UAE, 2012; pp. 187–212.

21. Maxwell, R.M.; Kollet, S.J.; Smith, S.G.; Woodward, C.S.; Falgout, R.D.; Ferguson, I.M.; Ferguson, N.; Condon, L.E.; Hector, B.; Lopez, S.; et al. *User's Manual. Integrated GroundWater Modeling Center Report GWMI 2016-01*; Free Software Foundation: Boston, MA, USA, 2016; p. 167.
22. Dogrul, E.C. Integrated Water Flow Model. (IWFM–2015): Theoretical Documentation. Modeling Support Branch, Bay-Delta Office, Department of Water Resources: Sacramento, CA, USA, 2016. Available online: http://baydeltaoffice.water.ca.gov/modeling/hydrology/iwfm/iwfm2015/v2015_0_475/downloadables/iwfm-2015.0.475_theoreticaldocumentation.pdf (accessed on 28 September 2016).
23. Bailey, R.T.; Wible, T.C.; Arabi, M.; Records, R.M.; Ditty, J. Assessing regional-scale spatio-temporal patterns of groundwater–surface water interactions using a coupled SWAT-MODFLOW model. *Hydrol. Process.* **2016**, *30*, 4420–4433. [CrossRef]
24. Yang, L.; Song, X.; Zhang, Y.; Han, D.; Zhang, B.; Long, D. Characterizing interactions between surface water and groundwater in the Jialu river basin using major ion chemistry and stable isotopes. *Hydrol. Earth Syst. Sci.* **2012**, *16*, 4265–4277. [CrossRef]
25. Tian, Y.; Zheng, Y.; Wu, B.; Wu, X.; Liu, J.; Zheng, C. Modeling surface water-groundwater interaction in arid and semi-arid regions with intensive agriculture, environ. *Model. Softw.* **2015**, *63*, 170–184. [CrossRef]
26. El-Rawy, M.; Zlotnik, V.A.; Al-Raggad, M.; Al-Maktoumi, A.; Kacimov, A.; Abdalla, O. Conjunctive use of groundwater and surface water resources with aquifer recharge by treated wastewater: Evaluation of management scenarios in the Zarqa River Basin, Jordan. *Environ. Earth Sci.* **2016**, *75*, 1146. [CrossRef]
27. Gannett, M.; Lite, K.J.; Risley, J.; Pischel, E.; La Marche, J. *Simulation of Groundwater and Surface-Water Flow in the Upper Deschutes Basin, Oregon, Scientific Investigations Report*; U.S. Geological Survey: Portland, OR, USA, 2017; pp. 2017–5097.
28. Salem, A.; Dezső, J.; Lóczy, D.; El-Rawy, M.; Słowik, M. Modeling surface water-groundwater interaction in an oxbow of the Drava floodplain. In Proceedings of the HIC 2018—13th International Conference on Hydroinformatics, Palermo, Italy, 20 September 2018. [CrossRef]
29. Awad, A.; Eldeeb, H.; El-Rawy, M. Assessment of surface water and groundwater interaction using field measurements: A case study of dairut city, Assuit, Egypt. *J. Eng. Sci. Technol.* **2020**, *1*, 406–425.
30. Salem, A.; Dezső, J.; El-Rawy, M.; Lóczy, D. Hydrological Modeling to Assess the Efficiency of Groundwater Replenishment through Natural Reservoirs in the Hungarian Drava River Floodplain. *Water* **2020**, *12*, 250. [CrossRef]
31. Edelman, J.H. Over de Berekening van Grondwaterstroomingen (about the Calculation of Groundwater Flow). Ph.D. Thesis, Delft University of Technology, Delft, The Netherlands, 1947.
32. Lockington, D.A. Response of unconfined aquifer to sudden change in boundary head. *J. Irrig. Drain. Eng.* **1997**, *123*, 24–27. [CrossRef]
33. Bruggeman, G.A. *Analytical Solutions of Geohydrological Problems: Developments in Water Science 46*; Elsevier: Amsterdam, The Netherlands; Oxford, UK; New York, NY, USA, 1999; 959p.
34. Cooper, H.; Rorabaugh, M. *Groundwater Movements and Bank Storage Due to Flood Stages in Surface Streams*; US Government Printing Office: Washington, DC, USA, 1963.
35. Sahuquillq, A. *Quantitative Characterization of the Interaction between Groundwater and Surface Water: Conjunctive Water Use (Proceedings of the Budapest Symposium)*; IAHS Publ: Wallingford, UK, 1986.
36. Hogarth, W.L.; Parlange, J.Y.; Parlange, M.B.; Lockington, D. Approximation analytical solution of the boussinesq equation with numerical validation. *Water Resour. Res.* **1999**, *23*, 3193–3197. [CrossRef]
37. Barlow, P.M.; Moench, A.F. Aquifer response to stream-stage and recharge variations: I. Analytical step-response functions. *J. Hydrol.* **2000**, *230*, 192–210. [CrossRef]
38. Soetaert, K.; Declippele, V.; Herman, P.M.J. FEMME, a flexible environment for mathematically modelling the environment. *Ecol. Model.* **2002**, *152*, 177–193. [CrossRef]
39. Wang, H.F.; Anderson, M.P. *Introduction to Groundwater Modeling: Finite Difference and Finite Element Methods*; Academic Press: San Diego, CA, USA, 1995.
40. De Doncker, l.; Troch, P.; Buis, K. *A Fundamental Study on Exchange Processes in River Ecosystems: FWO Project, Progress Report, Femme-Modelling*; Faculty of Engineering, Ghent University: Ghent, Belgium, 2007.

41. Buis, K.; Anibas, C.; Bal, K.; Banasiak, R.; De doncker, L.; Desmet, N.; Gerard, M.; van belleghem, S.; Batelaan, O.; Troch, P.; et al. Fundamentele studie van uitwisselingsprocessen in rivierecosystemen-geïntegreerde modelontwikkeling (a fundamental study on exchange processes in river ecosystems). Congres Watersysteemkennis, studiedag WSK8 'Modellen voor integraal waterbeheer', Universiteit Antwerpen. *Water Tijdschr. Integraal Waterbeleid* **2007**, *32*, 51–54.
42. Anibas, C.; Fleckenstein, J.H.; Volze, N.; Buis, K.; Verhoeven, R.; Meire, P.; Batelaan, O. Transient or steady-state? Using vertical temperature profiles to quantify groundwater–surface water exchange. *Hydrol. Process.* **2009**, *23*, 2165–2177. [CrossRef]
43. Anibas, C.; Tolche, A.D.; Ghysels, G.; Nossent, J.; Schneidewind, U.; Huysmans, M.; Batelaan, O. Delineation of spatial-temporal patterns of groundwater/surface-water interaction along a river reach (Aa River, Belgium) with transient thermal modeling. *Hydrogeol. J.* **2018**, *26*, 819–835. [CrossRef]
44. Chow, V.T.; Maidment, D.R.; Mays, I.W. *Applied Hydrology*; Mcgraw-Hill: New York, NY, USA, 1988.
45. Vekerdy, Z.; Meijerink, A.M.J. Statistic and analytical study of the propagation of flood-induced groundwater rise in an alluvial aquifer. *J. Hydrol.* **1998**, *205*, 112–125. [CrossRef]
46. De Ridder, N.A.; Zijlstra, G. *Seepage and Groundwater Flow*; Ritzema, H.P., Ed.; Drainage Principles and Applications; ILRI 16: Wageningen, The Netherlands, 1994; pp. 305–339.
47. Parlange, J.Y.; Barry, D.A.; Parlange, M.B.; Lockington, D.A.; Haverkamp, R. Sorptivity calculation for arbitrary diffusivity. *Transp. Porous Media* **1994**, *15*, 197–208. [CrossRef]

MDPI
St. Alban-Anlage 66
4052 Basel
Switzerland
Tel. +41 61 683 77 34
Fax +41 61 302 89 18
www.mdpi.com

Hydrology Editorial Office
E-mail: hydrology@mdpi.com
www.mdpi.com/journal/hydrology